AF615140

Temporal Dynamics and Ecological Process

In contrast with the fundamental ecological expectation that similarity induces competition and loss of species, temporal dynamics allow similar species to co-occur. In fact, the coexistence of similar species contributes significantly to species diversity and could affect ecosystem response to climate change. However, because temporal processes take place over time, they have often been a challenge to document or even to identify.

Temporal Dynamics and Ecological Process brings together studies that have met this challenge and present two specific aspects of temporal processes: reproductive scheduling and the stable coexistence of similar species. By using plants to extract general principles, these studies uncover deep ties between temporal niche dynamics and the above central ecological issues, thereby providing a better understanding of what drives temporal processes in nature.

Written by leading scientists in the field, this title will be a valuable source of reference to research ecologists and to those interested in temporal ecology.

COLLEEN K. KELLY is a Senior Research Associate in the Department of Zoology at the University of Oxford. Her research interests focus on temporal dynamics, community assembly and ecosystem function. She has published widely on these topics and has been a principal investigator or co-PI on various funded projects in the United States, the UK and México.

MICHAEL G. BOWLER has worked in the Department of Physics at the University of Oxford for over 40 years. His research interest focused on particle physics. He has taken part in several major collaborations in this field, most recently the Sudbury Neutrino Observatory. He has authored or co-authored 150 papers and has written five books.

GORDON A. FOX is Associate Professor in the Department of Integrative Biology at the University of South Florida. His main area of research is plant ecology and theoretical population biology. His recent work has focused on demographic heterogeneity and its causes and consequences, including both theoretical studies and research with populations of pines, lilies and Florida scrub-jays.

Temporal Dynamics and Ecological Process

Edited by
COLLEEN K. KELLY
University of Oxford, UK

MICHAEL G. BOWLER
University of Oxford, UK

GORDON A. FOX
University of South Florida, USA

CAMBRIDGE
UNIVERSITY PRESS

University Printing House, Cambridge CB2 8BS, United Kingdom

Published in the United States of America by Cambridge University Press, New York

Cambridge University Press is part of the University of Cambridge.

It furthers the University's mission by disseminating knowledge in the pursuit of education, learning and research at the highest international levels of excellence.

www.cambridge.org
Information on this title: www.cambridge.org/9780521198639

First published 2013

Printed in the United Kingdom by TJ International Ltd. Padstow Cornwall

A catalogue record for this publication is available from the British Library

Library of Congress Cataloguing in Publication data
Temporal dynamics and ecological process / edited by Colleen K. Kelly, University of Oxford, UK, Michael G. Bowler, University of Oxford, UK, Gordon A. Fox, University of South Florida, USA.
pages cm
Includes index.
ISBN 978-0-521-19863-9 (Hardback) – ISBN 978-0-521-12170-5 (Paperback.)
1. Population biology. 2. Biotic communities. 3. Plant populations. I. Kelly, Colleen K., editor of compilation. II. Bowler, M. G., editor of compilation. III. Fox, Gordon A., editor of compilation.
QH352.T38 2013
581.7′88–dc23
2013023446

ISBN 978-0-521-19863-9 Hardback
ISBN 978-0-521-12170-5 Paperback

Contents

Contributors

Peter B. Adler
Utah State University, College of Natural Resources,
5230 Old Main Hill, Logan, UT 84322–5230, USA

Michael G. Bowler
Department of Physics, University of Oxford, Keble Road,
Oxford, OX1 3RH, UK

Felix Breden
Department of Biological Sciences, Simon Fraser University,
Burnaby, British Columbia, V5A 1S6, Canada

Peter Chesson
Department of Ecology and Evolutionary Biology,
University of Arizona, Tucson, AZ 85721, USA

Jose M. Facelli
Terrestrial Plant Ecology, School of Earth and Environmental
Sciences, The University of Adelaide, SA 5005, Australia

Norma L. Fowler
The University of Texas at Austin, Section of Integrative
Biology, 1 University Station A6700, Austin, TX 78712, USA

Gordon A. Fox
Department of Integrative Biology, University of South
Florida, 4202 E. Fowler Ave, Tampa, FL 33620, USA

Mick E. Hanley
School of Biological Sciences, University of Plymouth,
Drake Circus, Plymouth, PL4 8AA, UK

Nancy J. Huntly
Director, USU Ecology Center, Utah State University,
5205 Old Main Hill, Logan, UT 84322–5205, USA

Yoh Iwasa
Department of Biology, Faculty of Sciences, Kyushu University,
Fukuoka, Japan

Jeffrey B. Joy
Department of Biological Sciences, Simon Fraser University,
Burnaby, British Columbia, V5A 1S6, Canada

Colleen K. Kelly
Department of Zoology, Niko Tinbergen Building, South Parks
Road, Oxford, OX1 3PS, UK

Sarah Kimball
University of California, Irvine BIO SCI – Center for
Environmental Biology, Irvine, CA 92697–1450, USA

Pilar Lopera Blair
Department of Integrative Biology, University of South Florida,
4202 E. Fowler Ave, Tampa, FL 33620, USA

Marissa Pantastico-Caldas
Science Department, Los Angeles Trade-Tech College, 400
W. Washington Blvd, Los Angeles, CA 90015, USA

Craig M. Pease
Vermont Law School, Environmental Law Center, S Royalton,
VT 05068, USA

J. Marcela Ramos-Tapia
Plan de Ayala #87, Francisco Villa, Municipio de La Huerta,
Jalisco, C.P. 48878, México

Stephen H. Roxburgh
CSIRO Sustainable Ecosystems, GPO Box 284, Canberra, ACT 2601,
Australia

Akiko Satake
Hokkaido University, Graduate School of Environmental Science,
Sapporo, Hokkaido 0600810, Japan

Susanne Schwinning
Department of Biology, Texas State University, San Marcos, TX 78666, USA

Robin E. Snyder
Department of Biology, Case Western Reserve University, 10900 Euclid Avenue, Cleveland, OH 44106–7080, USA

J. Arturo Solís-Magallanes
Laboratorio de Botánica del Departamento de Ecología y Recursos Naturales, Centro Universitario Costa Sur, Universidad de Guadalajara, Avenida Independencia # 151 Autlán, Jalisco, C.P. 48 900, México

Rebecca J. Sykes
Principal Ecological Consultant, Ecological Planning & Research Ltd, The Barn, Micheldever Station, Winchester, SO21 3AR, UK

Yuuya Tachiki
Kyushu University, Faculty of Science, Department of Biology, Fukuoka 8128581, Japan

D. Lawrence Venable
Department of Ecology and Evolutionary Biology, University of Arizona, Tucson, AZ 85721, USA

John N. Williams
Environmental Science & Policy, University of California, One Shields Avenue, Davis, CA 95616, USA

COLLEEN K. KELLY, MICHAEL G. BOWLER
AND GORDON A. FOX

1

Introduction

Temporal fluctuations in populations can have significant consequences for the stable coexistence of competing species, as well as for the evolution of life-history traits in a number of ways. In this volume, we examine two specific products of temporal processes: reproductive scheduling and the stable coexistence of similar species. The studies contained here principally use plants to extract general principles, with the added benefit that the observed temporal processes may also be applied to a better understanding of the resource base on which depends the larger community. The range of topics dealt with here present, as a whole, a foundation of the workings of temporal processes in nature.

Temporal cycling of plant dynamics and their role in ecosystem processes and services have potential impacts up and down the foodweb (Chesson and Kuang 2008). Taken together, these studies indicate deep ties between temporal niche dynamics (sometimes just temporal dynamics, or TND for short) and a number of fundamental ecological issues. In order to make the jump to those fundamental issues we need a better handle on what drives temporal processes in nature, and the studies included here point the way to doing so. The first section of this book addresses population persistence and species coexistence regulated by environmentally generated fluctuations. The second section of this book examines plant reproduction driven by internal processes leading to synchronised, oscillating behaviour at the population and community levels – masting in forest trees.

Temporal Dynamics and Ecological Process, ed. C. K. Kelly, M. G. Bowler, G. A. Fox.
Published by Cambridge University Press.

1.1 STABLE COEXISTENCE

Temporal niche dynamics can allow the stable coexistence of similar species. In temporal niche dynamics, different species can coexist through best recruiting in different types of years, resulting in condition-dependent fluctuations in recruitment to the population. This can play an important role in increasing the diversity of communities by allowing for the coexistence of species that otherwise would be in strong competition with one another. Seemingly simple as this notion appears at first glance, TND is not strongly integrated into the thinking of most ecologists. The recent proliferation of field studies – including some reported here – shows it to be an important factor in the assembly of communities.

There is no a priori reason to think that the stable coexistence of temporal dynamics is less important in structuring communities than the unstable dynamics of competitive exclusion. The signature of temporal processes is ubiquitous, written in the year-to-year variation of flowering species in local meadows and grasslands or the yearly differences in numbers of insects. This sort of year-type specialisation is well studied as a life-history problem. It has been less widely recognised that such specialisation can mediate species' coexistence; only relatively recently has a critical weight of empirical studies made evident the potentially significant role that temporal dynamics can play in species distributions and assembly at multiple scales.

Temporal dynamics has a substantial history of theoretical development. The idea of temporal dynamics as we now think of it was put forth more than 50 years ago, when Hutchinson (1961) suggested that if environmental fluctuations occurred at a frequency able to change the order of competitive superiority before one competitor drove the other extinct, competing species could coexist indefinitely. More than a decade later, in 1974, Koch published two papers on this idea, dealing first with coexistence of two competing predators (consumers) through externally driven fluctuations (Koch 1974a) and secondly, through internally driven oscillations (Koch 1974b). In the second paper, Koch noted that 'Potentially any model that will work must give one predator an advantage under some circumstances and put it at enough of a disadvantage in others.' This captures the essence of coexistence through either internally or externally induced fluctuation (that is, intrinsic fluctuations including cycles and chaos, or extrinsic fluctuations which might be summarised as environmental stochasticity). A number of such models now exist.

Soon after Koch's papers appeared, Armstrong and McGehee developed the idea of internally induced dynamics set out in Koch's second paper (Armstrong and McGehee 1976a, 1976b, 1980, McGehee and Armstrong 1977). Various studies have examined the ideas experimentally, but thus far long-term coexistence through internally induced dynamics has been observed only where the systems have been buffered from externally fluctuating factors, and the role of internally induced oscillations in coexistence under natural conditions is still unclear (Huisman and Weissing 1999, Huisman *et al.* 2006, Benincá *et al.* 2008).

It wasn't until the early '80s that Chesson and Warner (1981) combined Sale's lottery model of coexistence in reef fish (Sale 1977, 1978) with Hutchinson's fluctuating resources idea to produce a lottery-model treatment of coexistence through externally induced fluctuations. Chesson and Warner initially specified that their model applies to species with overlapping generations, but this requirement was later broadened to 'storage': persistence over hard times through long-lived reproductive capacity or resistant propagules such as dormant seeds (Warner and Chesson 1985). The term 'storage' has come to be associated with dynamics due to externally induced fluctuations, and the temporal aspect of that in particular, but storage can be an essential component of either internally or externally induced fluctuations, and in spatial as well as temporal processes (Snyder 2007).

The essential ideas important for understanding temporal niche dynamics were thus fairly well laid out by the early 1980s, but little empirical evidence for their operation had been reported until the late 1990s. Why such a lag? We suspect that part of the answer is that it can be generally quite easy to find conditions in a community that are highly suggestive of coexistence through temporal niche dynamics, but it can be rather difficult to find a smoking gun (see Chapters 4–7). A lack of clarity on some conceptual issues adds to this difficulty. For example, the term 'storage' can be – and is – used in two different ways. 'Storage' can refer merely to storage of reproductive capacity over a period of years through long-lived adults or in a seed bank. If all adults die or are killed at the end of a growing season and there is no seed or seedling bank, then there is no storage; no species could survive in these circumstances. Storage in this first sense is expected to be a property of any species, but the efficiency of storage will affect its abundance. The second sense in which 'storage' is used is as a promoter of coexistence through environmental fluctuations. The difference is nicely illustrated by an early paper on a deliberate search for coexistence through the storage effect in a community of *Daphnia* inhabiting Oneida Lake in the northeastern United States (Cáceres 1997). Fluctuations in populations and in the environment occurred, and data were put through the filter of Warner and Chesson (1985). Because one species had no observed adults in one particular year, it was concluded that storage was essential for its persistence in the community. The population did not go extinct because eggs out of an egg bank hatched and survived – i.e. storage (first sense). Although conditions existed that were very suggestive of persistence of competing species as a result of differing responses to environmental fluctuations, the storage effect in the sense relevant to coexistence of competitors, the second sense, was not documented.

Direct observation of the temporal process is difficult in most communities because the time available to observe those communities is short relative to the length of the process. In consequence, as with many areas of ecology, much of the study of temporal dynamics has made inferences on the bases of patterns and assumptions (Kelly and Bowler 2002, 2009, Adler *et al.* 2006). Additionally, even where direct observation has been possible – such as in the desert annual research (Chapter 7), the specifics that demonstrate TND may not be translatable directly to other systems or life-history types.

Nonetheless, the accumulation of the different sorts of evidence across habitats and taxonomic groupings is such that the action of temporal niche dynamics as a general process is no longer in question, and brings us to the point where we are able to start thinking productively about generalities. Better understanding of the determining factors of temporal niches, both general and specific, are required in order to build more realistic and usable models of the action of TND within and across existing communities. Taken together, the studies and literature reviewed herein can be used to construct a priority 'to do list' for this purpose, which we have put forth in our concluding, summary chapter.

1.2 MASTING

Masting provides an example of an ecological effect from internally driven oscillations. Masting refers to intermittent and synchronous production of a fruit crop, generally with regard to trees. Synchrony of fruit production may be at the level of the species, but synchrony may also occur among different, co-occurring species. Two common examples of masting species are oak acorns and beech mast, historically known for their importance to rural economies in providing forage to large herbivores, both wild and domesticated, and as human food in times of scarcity. A desirable animal feed in Europe and the British Isles for the flavour imparted to meats, the irregularity of mast imparted even greater value to meat produced from mast-fed animals (Davidson 1999).

The irregularity of the food resource provided by masting schedules has parallel effects in more natural systems, with ramifications for interannual variation in population growth and individual survival at multiple levels of the community foodweb (Jones *et al.* 1998, Vander Wall 2001). Such effects have generated attention on the part of animal ecologists needing to know what keeps animals alive, and from plant ecologists wanting to understand what value this distinctive behaviour might have for the plant (Janzen 1971, Kelly 1994).

One of the earliest explanations for the intermittent character of masting was very straightforward: fruit is produced under favourable weather conditions and not produced when conditions are unfavourable; synchrony of co-occurring individuals and species followed from experience of the same general weather conditions (Büsgen and Münch 1929). Unfortunately, such weather tracking is not sufficient to produce observed patterns of masting for the majority of species (Koenig and Knops 2000). Furthermore, allocation to reproduction is much more complex than the simple use of resources surplus to need as produced by a year of good weather tracking (e.g. Morris 1951, Norton and Kelly 1988, Fenner 1991, Kelly and Sork 2002).

How, then, is masting an example of internally induced oscillations? When an individual tree depletes its resources severely in an episode of fruiting, it will go through a period of recovery of one to several years before it is again able to produce a large fruit/seed crop (Sork *et al.* 1993). Allowed to go on at this point, individual variation would lead one tree to recover sooner than another, and abetted by year-to-year differences in general climate, the number of trees flowering in any one year eventually would become temporally unpredictable (Satake and Iwasa 2002, Iwasa and Satake 2004). The final two chapters of this volume therefore deal with the factors that favour the evolution of temporal coupling of reproduction in forest trees.

The temporal dynamics of masting are not directly related to species coexistence, but are an indirect link between the resource fluctuations that mast fruiting presents and the coexistence of consumer species that depend on masting. The tools to put these two areas together are applied and the two contributions on the subject here offer examples of how those joint processes may regulate an ecologically important phenomenon.

REFERENCES

Adler, P. B., HilleRisLambers, J., Kyriakidis, P. C., Guan, Q. and Levine, J. M. (2006). Climate variability has a stabilizing effect on the coexistence of prairie grasses. *Proceedings of the National Academy of Sciences, USA* **103**, 12793–12798.

Armstrong, R. A. and McGehee, R. (1976a). Coexistence of species competing for shared resources. *Theoretical Population Biology* **9**, 317–328.

Armstrong, R. A. and McGehee, R. (1976b). Coexistence of two competitors on one resource. *Journal of Theoretical Biology* **56**, 499–502.

Armstrong, R. A. and McGehee, R. (1980). Competitive exclusion. *American Naturalist* **115**, 151–170.

Benincá, E., Huisman, J., Heerkloss, R. *et al.* (2008). Chaos in a long-term experiment with a plankton community. *Nature* **451**, 822–825.

Büsgen, M. and Münch, E. (1929). *The Structure and Life of Forest Trees*. London: Chapman and Hall.

Cáceres, C. E. (1997). Temporal variation, dormancy, and coexistence: a field test of the storage effect. *Proceedings of the National Academy of Sciences, USA* **94**, 9171–9175.

Chesson, P. and Kuang, J. J. (2008). The interaction between predation and competition. *Nature* **456**, 235–238.

Chesson, P. L. and Warner, R. R. (1981). Environmental variability promotes coexistence in lottery competitive systems. *American Naturalist* **117**, 923–943.

Davidson, A. (1999). *The Oxford Companion to Food*. Oxford: Oxford University Press.

Fenner, M. (1991). Irregular seed crops in forest trees. *Quarterly Journal of Forestry* **85**, 166–172.

Huisman, J. and Weissing, F. J. (1999). Biodiversity of plankton by species oscillations and chaos. *Nature* **402**, 407–410.

Huisman, J., Pham Thi, N. N., Karl, D. M. and Sommeijer, B. (2006). Reduced mixing generates oscillations and chaos in the oceanic deep chlorophyll maximum. *Nature* **439**, 322–325.

Hutchinson, G. E. (1961). The paradox of the plankton. *American Naturalist* **95**, 137–145.

Iwasa, Y. and Satake, A. (2004). Mechanisms inducing spatially extended synchrony in mast seeding: the role of pollen coupling and environmental fluctuation. *Ecological Research* **19**, 13020.

Janzen, D. H. (1971). Seed predation by animals. *Annual Review of Ecology and Systematics* **2**, 465–470.

Jones, C. G., Ostfeld, R. S., Richard, M. P., Schauber, E. M. and Wolff, J. O. (1998). Chain reactions linking acorns to gypsy moth outbreaks and Lyme disease risk. *Science* **279**, 1023–1026.

Kelly, C. K. and Bowler, M. G. (2002). Coexistence and relative abundance in forest tree species. *Nature* **417**, 437–440.

Kelly, C. K. and Bowler, M. G. (2009). Temporal niche dynamics, relative abundance and phylogenetic signal of coexisting species. *Theoretical Ecology* **2**(3), 161–169.

Kelly, D. (1994). The evolutionary ecology of mast seeding. *Trends in Ecology and Evolution* **9**, 465–470.

Kelly, D. and Sork, V. L. (2002). Mast seeding in perennial plants: why, how, where? *Annual Review of Ecology and Systematics* **33**, 427–447.

Koch, A. L. (1974a). Coexistence resulting from an alternation of density dependent and density independent growth. *Journal of Theoretical Biology* **44**, 373–386.

Koch, A. L. (1974b). Competitive coexistence of two predators utilizing the same prey under constant environmental conditions. *Journal of Theoretical Biology* **44**, 387–395.

Koenig, W. D. and Knops, J. M. H. (2000). Patterns of annual seed production by northern hemisphere trees: a global perspective. *American Naturalist* **155**, 59–69.

McGehee, R. and Armstrong, R. A. (1977). Some mathematical problems concerning the ecological principle of competitive exclusion. *Journal of Differential Equations* **23**, 30–52.

Morris, R. F. (1951). The effects of flowering on the foliage production and growth of balsam fir. *Forestry Chronicle* **27**, 40–57.

Norton, D. A. and Kelly, D. (1988). Mast seeding over 33 years by *Dacrydium cupressinum* Lamb. (rimu)(Podocarpaceae) in New Zealand: the importance of economies of scale. *Functional Ecology* **2**, 399–408.

Sale, P. F. (1977). Maintenance of high diversity in coral reef fish communities. *American Naturalist* **111**, 337–359.

Sale, P. F. (1978). Coexistence of coral reef fishes: a lottery for living space. *Environmental Biology of Fishes* **3**, 85–102.

Satake, A. and Iwasa, Y. (2002). The synchronized and intermittent reproduction of forest trees is mediated by the Moran effect, only in association with pollen coupling. *Journal of Ecology* **90**, 830–838.

Snyder, R. E. (2007). Spatiotemporal population distributions and their implications for species coexistence in a variable environment. *Theoretical Population Biology* **72**, 7–20.

Sork, V. L., Bramble, J. and Sexton, O. (1993). Ecology of mast fruiting in three species of North American deciduous oaks. *Ecology* **83**, 528–541.

Vander Wall, S. B. (2001). The evolutionary ecology of nut dispersal. *Botanical Review* **67**, 74–117.

Warner, R. R. and Chesson, P. L. (1985). Coexistence mediated by recruitment fluctuations: a field guide to the storage effect. *American Naturalist* **125**, 769–787.

Part I Observing temporal processes in nature

The chapters in the first section of this book are aimed at documenting the action of temporal dynamics in natural systems. We include here an assortment of life-history types as well as both experimental and observational evidence of temporal processes.

The desert annual community of the American southwest was and is a continuing resource for investigating temporal dynamics, and two of our chapters make use of it. One advantage to this community is the short generation times of the component species – up to a year aboveground, and a few years to (at most) a few decades in the seed bank. Peter Chesson is best known for his theoretical work on the ecology of diversity, in particular the mechanism known as the storage effect, dating from Chesson and Warner (1981; citation in most chapters). Here, he and his colleagues use desert annuals and clonal perennials to apply two different approaches to testing the storage hypothesis; each is drawn largely from field observations but also includes experimental elements. Larry Venable and Sarah Kimball provide an exhaustive yet concise summary of the studies carried out by the Venable group, also in the desert annual system, over a period of some 28 years. These studies are not limited to demography but include a strong ecophysiological component, which, among other things, has allowed them to extract the effects of climate change over this period.

The chapter by Peter Adler and that by Norma Fowler and Craig Pease also exploit long-term studies. The first reviews work extracting information from two historical data sets, long-term censuses from the last century of grassland areas in Kansas (prairie) and Idaho (steppe-sagebrush). Analyses of changing patterns of abundance in several common grasses in each result in the conclusion that the impact of temporal dynamics differs between sites. On the other hand, Fowler and Pease extract data from Fowler's 16-year study of herbaceous perennials in a Texas shortgrass community to find a strong potential for temporal dynamics promoting coexistence among these eight species.

The remaining three chapters of this section all address the action of temporal dynamics using congeneric species pairs. Colleen Kelly and collaborators review earlier work with temporal niche dynamics in trees of the hyperdiverse tropical dry forest of México, and present evidence, also from the Mexican dry forest, on the role temporal dynamics may play in speciation and in landscape-level patterns of species coexistence. Mick Hanley and Rebecca Sykes present a strong case for herbivory as a possible provider of the fluctuating environmental variability behind temporal dynamics, combining a review of the literature with experimental work on the tradeoff between growth rate and palatability (i.e. between sensitive and resistant species) in congeneric pairs of herbaceous temperate zone plant species. The section is brought to a close by Susanne Schwinning and colleagues, with an investigation into the potential significance of temporal niche processes in current ecosystem function. They find that, although temporal niches are still poorly understood by ecologists in general, too often overlooked in physiological ecology and often confounded with the climate responses and climate brackets of plant functional types, a consideration of the separate roles of seedling and adult traits in recruitment will not only inform diversity–productivity relationships, but may strike at the core of understanding climate change response from local to global scales.

Samples, organisms and techniques vary greatly among these researchers and projects. Some have gone looking for evidence of temporal dynamics; for others, the insights gained may be better attributed to serendipity. In summary, we restrict any directions for how to obtain a better understanding of temporal niche dynamics to the eternally useful ecological directive to keep your minds and eyes open.

PETER CHESSON, NANCY J. HUNTLY, STEPHEN H. ROXBURGH, MARISSA PANTASTICO-CALDAS AND JOSÉ M. FACELLI

2

The storage effect: definition and tests in two plant communities

2.1 INTRODUCTION

Nature is pervaded by variation: the physical environment is ever changing in time and in space, populations fluctuate, and no two organisms are the same. To explore natural environments is to be confronted by variation, and the science of ecology is challenged by the persistent question: is this variation more than variation itself? Environmental variation can cause population fluctuations (Ripa *et al.* 1998), but can it do more than this? Does it affect how organisms interact with one another? Does it shape populations and communities? How and in what ways? Biologists firmly accept that variation shapes the organisms. Heritable variation is the engine of evolution, which is fuelled by environmental change. In life-history theory, it is widely accepted that organisms show adaptations to variation in the physical environment, exemplified by evolutionary theories of iteroparity and seed dormancy (Cohen 1966, Bulmer 1985, Ellner 1985a, Real and Ellner 1992). Fundamentally, these adaptations allow species to take advantage of favourable environmental conditions without being too vulnerable to unfavourable environmental conditions.

Community ecologists have had a variety of attitudes to variation, especially variation in the physical environment (Chesson and Case 1986). Successional change after disturbance had a prominent role in the early development of plant and ecosystem ecology (Clements 1916) and now has an important role in diversity maintenance theory relying on competition–colonisation tradeoffs (Hastings 1980). Spatial variation is often assumed to provide for, and should therefore promote, species

Temporal Dynamics and Ecological Process, ed. C. K. Kelly, M. G. Bowler, G. A. Fox. Published by Cambridge University Press.

diversity (Pacala and Tilman 1994, Amarasekare and Nisbet 2001, Snyder and Chesson 2004). Although it is often assumed that regular temporal variation, such as seasonal and diurnal variation, provides for temporal niches (Armstrong and McGehee 1976, Levins 1979, Brown 1989a, b, Chesson *et al.* 2001), there is also much unpredictable temporal variation, such as deviations of weather and climate from seasonal averages (Davis 1986) and disturbances such as fire (Connell 1978, Bond and Keeley 2005). Should we think of this unpredictable temporal variation as disruptive to ecological processes (May 1974)? Do organisms fail to adapt to unpredictable temporal variation? Are they merely jerked around by it? Life-history theory suggests otherwise (Bulmer 1985, Real and Ellner 1992), yet conclusions are often drawn from models that reflect no such adaptations, for example Lotka–Volterra models with unpredictable environmental variation added arbitrarily (Turelli 1981, Kilpatrick and Ives 2003). Moreover, conclusions are often coloured by the false expectation that unpredictable environmental variation merely disrupts outcomes dictated independently by interactions between organisms. For example, one view of disturbance is that it prevents equilibrium from occurring (Rohde 2005). This outcome is seen as promoting species diversity if few species could coexist at equilibrium (Hutchinson 1961). On the other hand, disturbance disrupting a high diversity equilibrium is seen as lowering diversity (May 1974). It is important to realise that views like these treat the organisms and their interactions as shaped separately from the temporal variation they experience. Why should we regard organisms, their interactions and the variable milieu that they inhabit as other than intrinsically linked?

When adaptations to environmental variation are included in models of species interactions, the conclusions change dramatically (Chesson and Huntly 1988). A modern view of environmental variation treats it as part of an ecosystem, not as something imposed upon it. There is every reason to believe that ecological communities are shaped by environmental variation, be it spatial or temporal, predictable or unpredictable. The effects of environmental variation should be seen in the adaptations of individual species and their relationships with each other. An early articulation of this perspective is the regeneration niche concept of Grubb (1977), who suggested extending the idea of a species niche to include regeneration processes, which are often affected substantially by temporal environmental fluctuations. The question that we address here is how adaptations to a temporally varying physical environment affect the structure of an ecological community. Although our focus will be on temporal variation, there is a parallel theory for spatial variation (Chesson 2000, Snyder and Chesson 2004). We are specifically concerned with coexistence of species in the same guild. By the term 'guild', we mean a subset of a community consisting of species with similar ecology. How such species coexist with one another is a central question in community ecology.

Striking temporal fluctuation is a common feature of ecological processes. There has been a tendency to think of some environments as harsh and fluctuating (e.g. deserts), and others as benign and stable (e.g. tropical regions). However, major variation in biological processes occurs even in seemingly benign systems. For example, tropical forests show enormous variation in seedling recruitment (Wright *et al.* 2005), and coral reefs likewise show high recruitment variation (Soong *et al.* 2003). Although the environment may seem benign to humans, the organisms in that environment can be sensitive, at least at some stages of the life cycle, to environmental variation of the magnitude that does occur. As we shall see, it is variation at critical life stages that leads to effects of variation on coexistence of species in the same guild.

2.2 MODELLING TEMPORAL ENVIRONMENTAL VARIATION

Early approaches to understanding the role of environmental variation in community dynamics treated variation in the physical environment as generic fluctuations that could be added either to the per capita growth rate or to the carrying capacity, often in Lotka–Volterra models or their discrete-time counterparts (Turelli 1980, 1981). The underlying models often had stable equilibrium points, and these ways of adding environmental variation do little more than push populations away from equilibrium. It is not surprising that the initial conclusions of such models were that environmental variation is unfavourable to species persistence and unfavourable to species coexistence (May 1974). Moreover, more careful analyses revealed much overzealousness in the conclusion that such environmental fluctuations have appreciable effects on coexistence (Turelli 1980, 1981).

The key problem with simply adding fluctuations to population growth rates or to the carrying capacity is that it fails to consider how organisms may be adapted to varying environmental conditions. Just adding environmental fluctuations to the per capita growth rate fails to consider which aspects of the biology of organisms are sensitive to environmental fluctuations, and which are not. Adding fluctuations to the carrying capacity is really no better, and it can be argued that it is in fact worse because the carrying capacity is an outcome of dynamics, not a basic biological trait. Thus, the implications of varying the carrying capacity alone are unclear. Most important, quite different results are obtained from incorporating environmental variation in ways that do respect the biological processes at work (Chesson and Huntly 1988). We are going to consider one such approach here.

The approach we consider, although making specific postulates about the biological effects of environmental variation, nevertheless does not need to be highly specific because in fact biology of a generic nature distinguishes the important cases (Chesson 1994). This approach postulates that the per capita population growth rates, $r_j(t)$, of the various species in a guild can be represented as functions of direct responses, $E_j(t)$, of species to the physical environment, and to competition, $C_j(t)$. Thus,

$$r_j(t) = g_j\left(E_j(t), C_j(t)\right), \tag{2.1}$$

where g_j is an arbitrary function of two variables (Chesson 1994), and the responses $E_j(t)$ and $C_j(t)$ represent distinct biological process, but often will not be statistically independent of one another. Under this quite general and nonspecific formulation, two general mechanisms of species coexistence emerge. The first of these is the storage effect, which might be thought of as the mechanism behind coexistence by temporal niche differences. It involves an interaction between the environmental response and the response to competition (Chesson 1994). The second is termed relative nonlinearity (Chesson 1994). It involves fluctuations in competition and emerges from species differences in how sensitivity to competition changes with the magnitude of competition. Although other sorts of mechanisms might arise from temporal fluctuations, analysis of the model (2.1) under reasonable assumptions all but rules them out (Chesson 1994). Both the storage effect and relative nonlinearity have the potential for positive or negative effects on species coexistence, i.e. rather than promote diversity, under certain assumptions, either of these mechanisms might promote competitive exclusion. An interesting finding, however, is that under the assumptions that seem most biologically realistic, these mechanisms have positive effects and so promote species coexistence. Of these two mechanisms, the storage effect is by far the stronger mechanism (Chesson 1994), and so the focus here is on the storage effect.

The storage effect

The storage effect is the formalisation of temporal niche differences (Armstrong and McGehee 1976, Levins 1979, Brown 1989b, Chesson *et al.* 2001). In simplest terms, species vary in resource consumption over time, and different species have different patterns of variation. Different patterns of variation do not in themselves lead to stable coexistence. Critical to coexistence is the way these patterns of variation are related to the mediators of species interactions. However, we begin our discussion focusing on these patterns of variation. We then go on to consider additional requirements necessary for such patterns of variation to promote coexistence.

We consider in detail examples of two guilds of plant species and provide data suggesting that the storage effect is important for each. The first guild is desert annual plants, which have temporally fluctuating germination patterns and persist over time as dormant seeds. Different species of desert annual plants living in the same environment are known to germinate at different times (Juhren *et al.* 1956, Adondakis and Venable 2004, Facelli *et al.* 2005). The seeds of desert annual species are sensitive in different ways to the weather, most notably temperature at the time of rainfall (Figure 2.1), though it is also clear that weather conditions experienced by a seed prior to any given rainfall event are important too. The simplest way of thinking about these patterns is that in different years sufficient rain for germination occurs at different times during the fall and early winter germination seasons. If the rains come late, then those species preferring cooler conditions predominate in that year's crop, while if the rains are early, species preferring warmer conditions predominate. However, a complex of weather factors contribute to which species will predominate in any year (Baskin and Baskin 1998, Facelli *et al.* 2005). The outcome is that the abundances of different species in a guild of annual plants at a site vary greatly from year to year, with substantial variation also in their relative abundances (Figure 2.2).

Perennial organisms also have figured prominently in development of ideas on the storage effect. Indeed, the idea was originally formulated for coral reef fishes, recognising that recruitment (establishment of young fish on a reef) varies dramatically over time, with temporal patterns differing between species (Chesson and Warner 1981, Chesson 1997). Plants provide many examples of guilds of perennials, with notable applications of the storage effect to forest trees, especially in the tropics (Comins and Noble 1985, Warner and Chesson 1985, Kelly and Bowler 2002, 2005). Despite the fact that tropical environments are frequently considered to be environmentally stable, tropical trees show very large

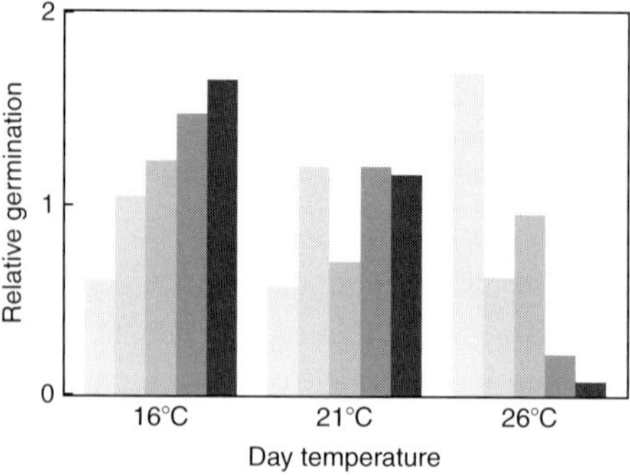

Figure 2.1 Germination of winter annual plants from the soil seedbank at the field site near Portal, Arizona, at different temperatures in growth chambers. The bar heights are emergence of a species at the given temperature relative to its mean emergence over all three temperatures. Species are arranged in order of relative germination at 16°C: from left to right, *Baileya multiradiata*, *Descurainea pinnata*, *Spermolepis echinata*, *Eriastrum diffusum* and *Eriogonum abertianum*. Note that the pattern at 26°C is essentially the reverse of that at 16°C. These changes in germination pattern with temperature are statistically significant ($p < 10^{-6}$). Growth chamber conditions were 13L:11D, with day/night temperatures 16°C/6°C, 21°C/11°C and 26°C/16°C. Twenty 10-cm diameter by 2-cm deep undisturbed samples of the surface soil from the field site were used per chamber. Samples were collected November 1986, stored dry at 5°C until used in the experiment in April 1987, where they were kept moist over wet sand. The data show total emergence at 35 days.

recruitment variation (Wright *et al.* 2005). Many herbaceous perennials also have strong recruitment variation (Chesson *et al.* 2001, Tilman 2004). We have studied herbaceous perennial plants in the understorey of a wet sclerophyll forest in eastern Australia. There one finds notable growth differences between species. We studied two species that tend to co-occur on the scale of a decimetre, the grass *Poa sebieriana* and the small aster *Lagenifera stipitata* (= *Lagenophora stipitata*) (Roxburgh and Chesson 1998); indeed, they are often intermingled. Although such tight co-occurrence might be taken to indicate facilitation of one by the other, and indeed to indicate that the species with somewhat larger local biomass, *Poa sebieriana*, might facilitate *Lagenifera* in some way, we found that removal of *Poa* led to expansion of *Lagenifera* over time. Regardless of the factors that cause them to co-occur, the net effect of *Poa* on *Lagenifera* is negative.

The two species, *Poa* and *Lagenifera*, have very different temporal patterns of growth over time. *Poa* tends to be green and growing in the cooler months of the year, while *Lagenifera* puts on new shoots and

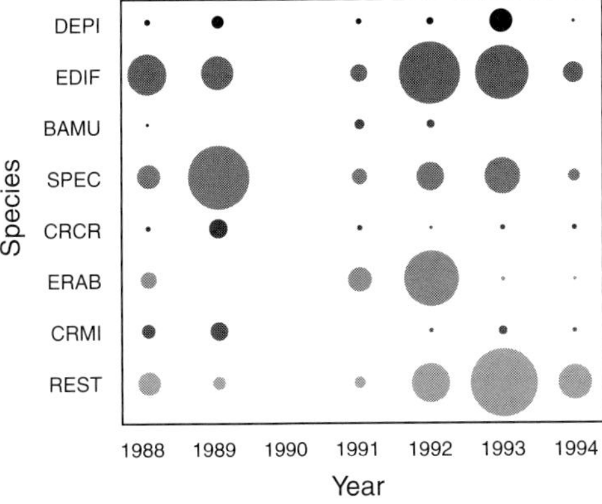

Figure 2.2 Total abundances of winter annual plant species flowering at the field site near Portal, Arizona, counted annually on 12 50cm × 80 cm permanent plots. Each row corresponds to a species, and each column corresponds to the year indicated. The areas of the circles are proportional to abundances, with the largest circle (1993, REST) exactly equal to a density of 700 plants per m^2. The species are *Descurainea pinnata* (DEPI), *Eriastrum diffusum* (EDIF), *Baileya multiradiata* (BAMU), *Spermolepis echinata* (SPEC), *Cryptantha crassisepala* (CRCR) and *Eriogonum abertianum* (ERAB), *C. micrantha* (CRMI), and the total of all other annual plant species (REST). The absence of circles for 1990 indicates that no plants survived to flowering, due to severe drought. The census continued after 1994, but from 1995 the invasive species *Erodium cicutarium* began establishing and now forms most of the biomass of the winter annual flora.

expands its leaves in the spring and early summer, dying back in the autumn. Variation in growth of *Lagenifera* is shown in its seasonally fluctuating incidence over time (Figure 2.3a), whereas *Poa* maintains steady aboveground biomass in all seasons, and indeed the abundance of *Poa* over time is aseasonal (Figure 2.3b).

Environmental responses

Both guilds illustrated here have characteristics that would seem to indicate temporal resource partitioning, meaning different patterns of resource use over time. In the case of the desert annuals, germination is the precursor to resource use, and higher germination leads to higher demand for resources. In the case of herbaceous understorey plants, the expansion of aboveground biomass is the activity most closely associated with resource demand. The environmental response, $E_j(t)$, required by the theory is thus the germination fraction in the case of the desert annual

(a)

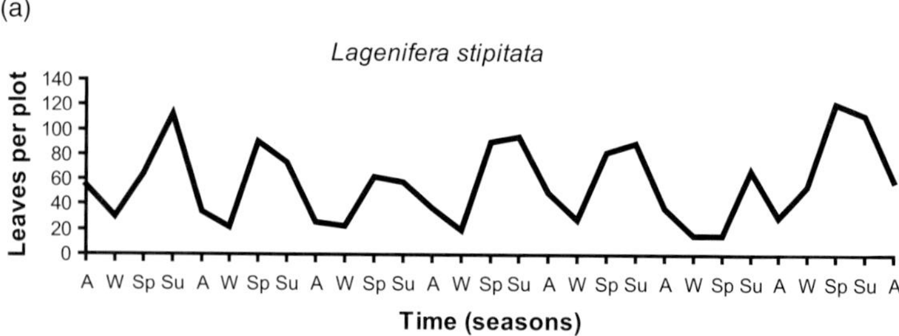

(b)

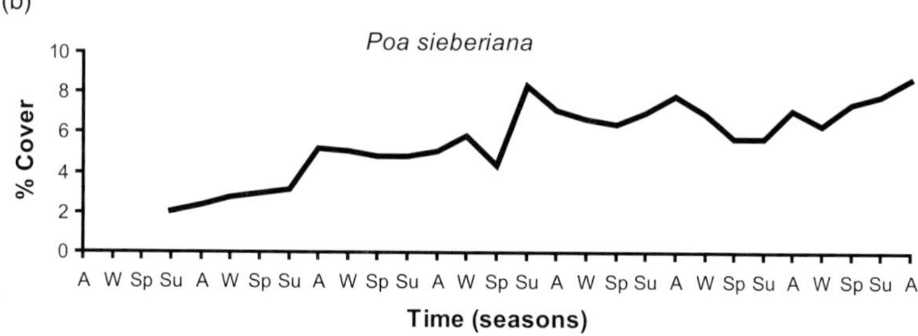

Figure 2.3 Changes in abundance of *Lagenifera stipitata* and *Poa sieberiana* measured every 3 months (A, autumn; W, winter; Sp, spring; Su, summer) on 30 1-m^2 permanent plots at the field site in the Namadgi National Park, Australian Capital Territory. **(a)** Mean number of *L. stipitata* leaves per plot. **(b)** Percent cover of *P. sieberiana*. Different measurements were used for the abundance of the two species because of their large difference in morphology.

plants (Chesson and Huntly 1989), and biomass growth for understorey perennials (Chesson *et al.* 2001). Extensions of the theory allow multiple environmental responses; for example, in annual plants germination and subsequent growth have been studied as two distinct environmental responses (Angert *et al.* 2009), but the overall message changes very little.

Responses to competition

For coexistence by the storage effect, it is important that responses to the physical environment can be directly linked to the way species interact with one another. The notion of temporal resource partitioning suggests this at the outset. It is not enough for species to have different patterns of fluctuation over time. If these patterns do not have any consequences for the way species interact with one another, then they will not affect species coexistence (Chesson 2008). In storage effect theory, we look to the quantity called the response to competition

($C_j(t)$ of Equation (2.1)) to summarise species interactions. Competition is detected through the effects of density on average performance of individuals in a population. Ultimately this is performance in per capita population growth, but it is much more informative to look closely at how competition affects the species biologically. In desert annuals, for instance, population growth is made of up to two components: persistence of dormant seed in the seedbank and production of new seed. The process of production of new seed is where competition acts, not persistence of dormant seed in the seedbank. Production of new seed involves germination, growth of the plant, and fruiting, with competition presumably mostly affecting growth. Seed production directly reflects that growth, and competition is measured as the reduction in individual seed production due to the presence of neighbouring plants in the same guild. This measurement thus includes both intraspecific and interspecific competition. In general competition is quantified by log seed production without neighbours minus log seed production with neighbours. This quantity is $C_j(t)$ in Equation (2.1), but has independently emerged as a measure of the intensity of competition in the field of study, focusing on variation in the intensity of competition along environmental gradients (Choler *et al.* 2001).

For herbaceous understorey perennial plants, individual plants are clonal, and much of the growth of a population reflects clonal growth of individual genets (Harper 1985). Population density is appropriately summarised as biomass per unit area. The main productive process is accumulation of photosynthate. Competition reduces this production. Precisely measuring competition with clonal plants presents challenges in the field, although it is certainly amenable to definition theoretically, which is a prerequisite to devising approaches empirically. In the work reported here, we bypass the problem of a complete measurement of $C_j(t)$, studying directly only that component due to interspecific competition from *P. sebieriana*, which is revealed by complete removal of *Poa*.

How is competition linked to environmental responses?

For annual plants, higher germination of a species leads to more individuals of that species placing demands on resources, and so we expect the competitive response of any species to be directly affected by the environmental responses of all species, assuming that they compete for common resources whenever they are growing. More generally, we can think of a more favourable physical environment for a species as directly increasing its environmental response, with a higher environmental response leading to higher demand for resources. For instance, in clonal perennials we think of favourable conditions as directly promoting growth, as warm spring conditions do for *L. stipitata*, leading to demand for resources, such as water, nutrients and light, that allow growth. In previous work on recruitment variation in perennial organisms, a favourable environment leading to higher numbers of juvenile organisms (plants or animals) was postulated to lead to higher demand for resources for those juveniles to establish or recruit to the adult stage (Chesson 2003). When higher demand for resources leads to reduction in resource availability, it leads to increases in competition. Thus, the overall favourability of the environment may be linked to competition. This linkage is measured in the theory by a quantity called covariance between environment and competition. It is the statistical covariance over time between the environmental response and the competitive response of a species.

Covariance between environment and competition is measured in general in two different states (Chesson 2008). In the first state, a species is depressed to low density experimentally. This is called the *invader state*. In this state, its competitive response reflects interspecific competition. The second state, the *resident state*, is where the species is not perturbed to low density, but instead fluctuates unconstrained over time. In a two-species system, the competitive response of a resident is measured with the other species in the invader state. Thus, a resident is effectively studied in monoculture, and its competitive response reflects intraspecific competition. With more than two species, the competitive response of a resident is studied with exactly one other species in the invader state at any given time. Thus, with more than two species, the competitive response of a resident reflects both intraspecific and interspecific competition, but the competitive response of an invader continues to reflect interspecific competition alone. Regardless of the number of species, a comparison of the competitive responses in these two different states compares intraspecific competition with interspecific competition.

The chief interest in comparing performance between invader and resident states is to see if species coexistence is stabilised, i.e. whether a species has an overall tendency to resist extinction (Chesson and Huntly 1989). An advantage in per capita growth in the invader state means it does resist extinction. If we were examining standard equilibrium hypotheses on coexistence, our concern would be with the difference between mean competitive responses in invader and resident states

(Chesson 1994). To understand the effects of temporal resource partitioning on coexistence, however, we do not compare the competitive responses between invader and resident states; rather, we compare the environment–competition covariances between resident and invader states (Chesson 1994, 2008).

The linkage between the response to the environment and the response to competition generates a fundamental expectation. As the response to the environment, i.e. the direct environment-dependent component of growth, fluctuates over time, so does intraspecific competition for a species in its resident state. The positive relationship between a species' environmental response and intraspecific competition makes a positive contribution to covariance between environment and competition for a resident. Indeed, in a two-species system, resident covariance between environment and competition must be positive. With a multispecies system, resident covariance will depend on the environmental responses of other resident species. If the environmental responses of different residents are independent, then the covariance for any given resident remains positive. Positively correlated environmental responses boost resident environment–competition covariances, while negatively correlated ones diminish them.

For the invader, covariance between environment and competition is simpler (Chesson 2008), as summarised in Table 2.1. The invader's environmental response has no direct effect on the competition that it experiences. Thus, if the environmental responses of the invader and residents are independent, invader covariance between environment and competition is zero. Should the invader environmental response be negatively correlated with resident environmental responses, then the invader covariance will be negative. In the case where invader environmental response is positively correlated with resident environmental responses, invader covariance will be positive (Table 2.1 columns 1 and 2). Nevertheless, the general expectation is that invader covariance will be less than resident covariance (Table 2.1, column 3), unless all species have identical patterns of fluctuation in their environmental responses. This difference in covariance of environmental and competitive responses for species in resident versus invader states measures a critical aspect of temporal partitioning. It measures how much a shift between intraspecific and interspecific competition changes the relationship between environmental favourability and total competition experienced by a species. The concern is with *total* competition, as the sum of the effects of all individual organisms from all species in the guild, rather than

Table 2.1 *Simplified scheme showing what environmental response correlations mean for variance in the invader's growth rate and the magnitude of the storage effect. Covariance between environment and competition for the resident is positive and unaffected by the correlation between invader and resident environmental responses. Note that the terms are qualitative and relative. Equation (2.4) shows that the quantitative relationship between the first and last columns of this table is approximately linear.*

Correlation between invader and resident environmental responses	Invader covariance between environment and competition	Resident–invader covariance difference	Variance of the invader growth rate	Magnitude of the storage effect
Positive	Positive	Small positive	Low	Weak
Zero	Zero	Intermediate positive	Intermediate	Moderate
Negative	Negative	Large positive	High	Strong

with a competition coefficient, which measures only the effect of one individual organism (or one unit of the organism) on another.

Interactions between environment and competition

The next task is to ask what the covariances between environmental and competitive responses mean. Positive covariance per se means that favourable environmental conditions are accompanied by higher competition, which reduces the ability of a species to take advantage of those favourable conditions. Conversely, unfavourable environmental conditions are accompanied by a reduction in competition. Thus, positive covariance tends to reduce fluctuations in population growth rates (Chesson and Huntly 1989, Chesson 2003). Negative covariance, by contrast, means that favourable environmental conditions tend to coincide with low competition. Such conditions potentially lead to strong population growth (Chesson and Huntly 1989). However, negative covariance also means that unfavourable conditions tend to occur simultaneously with high competition, which is a double dose of negative conditions and hence unfavourable to population growth. Negative covariance, therefore, tends to maximise variance in population growth rates (Table 2.1, column 4). As covariances for invaders tend to be more negative than covariances for residents, it is apparent that environmental conditions favourable to positive growth are thwarted by competition less often and less severely for invaders than for residents. However, unfavourable environmental conditions should be made worse by higher competition more often and more severely for invaders than for residents. Thus, temporal variance in growth rates should be higher for invaders than for residents (Chesson and Huntly 1989).

In ecology, higher variance in population growth has traditionally been seen as a disadvantage (May 1974, Lande 1993). Here, we see that, in fact, it can reflect a situation in which invaders have advantages over residents, allowing their recovery from low density. Long-term trends in population growth depend on the average over time of the population growth rate. So how can the variance be involved? This is where buffered population growth comes in. Buffered population growth means that when unfavourable conditions occur, the magnitude of the negative growth during such times is limited, but when favourable conditions occur, positive growth is unrestricted. Under these circumstances, a higher variance in growth leads to an increase in mean population growth (Table 2.1, Chesson and Huntly 1989, Chesson 2003).

Buffered population growth has been assumed to arise in perennial organisms, such as woody plants and fishes, because adult survival is relatively insensitive to varying environmental factors and competition, with most sensitivity to these factors occurring in reproduction and juvenile survival (Chesson 2003). Persistence of adults limits the damage from unfavourable conditions, but does not prevent strong growth at other times. Thus, adult survival buffers population growth, converting higher growth variance into higher mean growth. Similarly, the dormant seeds of annual plants are relatively insensitive to environmental factors and competition in comparison with the actively growing plants. For annual plants with germination as the environmental response, poor environmental conditions mean that competition affects only a small fraction of the population, because only a small fraction of the population germinates and is exposed to competition when environmental conditions are poor. The effects of competition on population growth are then quite limited. In particular, the double dose of negative effects that accompany poor environmental conditions and high competition leads to only modestly negative population growth unless survival of dormant seeds is low (Chesson and Huntly 1989). On the other hand, high germination with low competition leads to strong positive population growth. These observations imply that negative covariance between environment and competition increases not just the variance in growth, but also the average population growth rate (Chesson and Huntly 1989).

The issues are similar for clonal herbaceous plants. In the simplest terms, persistence of biomass under poor conditions for photosynthesis buffers these unfavourable conditions. Moreover, clonal perennial plants often have adaptations that protect accumulated photosynthate during unfavourable conditions. These include storage of photosynthate in storage structures such as rhizomes, leaf bases, bulbs, corms and tubers (Chapin *et al.* 1990). The net effect once again is that negative conditions are buffered, while positive growth is not restricted, converting more variable conditions into higher mean growth. Here, 'more variable conditions' means lower covariance between environment and competition, taking account of sign. Thus, a strong negative covariance, zero covariance, weak positive covariance and moderate positive covariance are all more variable conditions than a strong positive covariance. To see if the storage effect promotes coexistence, when we know that population growth is buffered, we look to see if covariance between environment and competition is smaller for the invader than the resident state (Table 2.1, column 3). It then promotes recovery of a species from low density, preventing competitive exclusion. The ability of all species to recover from low density means that they coexist stably.

These conditions for a positive storage effect can be equated with the conditions necessary for temporal resource partitioning to promote coexistence or, more generally, for temporally fluctuating environmental responses to promote coexistence. We shall see this more precisely when we consider the formula defining the magnitude of the storage effect in the next section. Fundamentally, without the linkage between competition and the environment discussed here, environmental fluctuations have no effect on the way species interact with one another. Second, without buffered population growth, the patterns of fluctuating interactions that arise when there is covariance between the environmental and competitive responses of a species are just fluctuations and have no significance for long-term outcomes.

The storage effect has often been described as having three ingredients: species-specific responses to the environment, covariance between environment and competition and buffered population growth. We now think it is preferable to state that what is needed is invader–resident differences between environment and competition and buffered population growth. In the list of three ingredients, species-specific responses to the environment, which can be measured by the correlation between the environmental responses of the species, create invader–resident covariance differences (Table 2.1).

Quantifying the storage effect

The above qualitative description of the requirements for the storage effect can be quantified as a measure of its overall magnitude. We give this magnitude here in its simplest and most useful form, signified by the notation $\overline{\Delta I}$, as an average for all species in a guild:

$$\overline{\Delta I} = \overline{s(\chi_{res} - \chi_{inv})}^{spp}. \tag{2.2}$$

Here $\chi_{res} - \chi_{inv}$ measures the difference between resident and invader covariance between environment and competition for a species, and the constant s measures how much population growth is buffered. This buffering constant s is introduced here to simplify the notation, and is equal to the quantity $-\gamma/d$ of Chesson (2003) and $-\gamma/b$ of Chesson (2008). In contrast to γ, where a negative value indicates buffered population growth, a positive value of s indicates that population growth is buffered. The bar on the right-hand side of Equation (2.2), with attached species abbreviation, 'spp', indicates that this quantity is averaged over species. Here, the average is over all pairs of species in their invader and resident states (Chesson 2008). The form of Equation (2.2) involves the assumption that the species compete for just one resource, which might be a composite resource on which all depend. An important resource for herbaceous plants is soil water, but space to establish, which allows an individual to monopolise resources on that space, is in effect a composite resource assumed in discussions of coexistence of organisms like coral reef fishes or forest trees. When this assumption of a single resource or composite resource is not valid, analogous but more complex formulae apply (Chesson 2003, 2008).

Under the assumption that the storage effect is the only mechanism present, the population growth rate, $\bar{r}_i$, for a given species i in the invader state, takes the form,

$$\bar{r}_i = \mu_i - \overline{\mu}^{spp} + \overline{\Delta I} \tag{2.3}$$

(Chesson 2003, Chesson 2008). Here, μ_i is a measure of the performance, or fitness, of species i as a whole in its environment, and $\overline{\mu}^{spp}$ is the average of this quantity for the competitors of species i. Thus, $\bar{r}_i$ consists of a fitness comparison, plus the storage effect. It is the long-term rate at which an invader grows at low density, and thus the bar refers to a time average of the per capita population growth rate, $r_i(t)$, at low density. The unit of time used here is the so-called natural time unit (Chesson 2008), which is often the generation time, as it is in the model for the annual plant guild discussed below. Then the growth rate, Equation (2.3), may be converted to growth per year by dividing by the generation time in years.

More generally, a term due to relative nonlinearity is added to Equation (2.3), but, as this is often of much smaller magnitude than the storage effect (Yuan and Chesson, unpublished manuscript), we do not consider it. Without the storage effect term in Equation (2.3), a species could only recover from low density if it had a value of μ_i greater than the average of its competitors. As a consequence of this form, in fact only one species would persist in the system: that species with the maximum μ would displace all others. With a positive storage effect term present, i.e. a positive value of $\overline{\Delta I}$, it is also possible for species at a disadvantage in average fitness to increase from low density. Thus, we can think of $\overline{\Delta I}$ in two different ways. It can be thought of as the average boost to the growth from low density provided by the storage effect, and it can be thought of as defining the degree of fitness inequality that is compatible with coexistence under this mechanism.

In simple circumstances where the different species have the same variance in fluctuations over time and their similar life histories dictate the same buffering constant, then Equation (2.2) takes the very simple form,

$$\overline{\Delta I} = \frac{s\sigma^2(1-\rho)B}{n-1}, \tag{2.4}$$

where σ^2 specifies the variance in the environmental response, ρ specifies the average correlation between species in their environmental responses, n is the number of species in the guild under consideration, and B is a positive number specifying the relationship between the average intensity of competition and the environmental response (Chesson 2003). We can specify the constant B in a number of important situations. For lottery competition where it is assumed that the guild in question monopolises the resources present in an area, and always manages to consume them all, B is simply the number one (Chesson 1994, 2003). If we assume a slightly more sophisticated model of plant competition, then B is a number between zero and one, with closeness to one indicating the degree to which the guild uses those resources.

For the desert annual guild described here, it appears that increasing the germination fraction does not increase total plant biomass, which means that a model close to the lottery case seems reasonable (Inouye 1980), and so we base our analysis on the annual plant model with lottery competition, studied in Chesson (1994). For the desert annual guild, we estimated σ^2 $(1{-}\rho)$ from the raw data behind Figure 2.2. The method is explained in Box 2.1. Along with estimates of the survival rate of dormant seeds derived from buried seed bags, this gives an estimate of the storage effect, $\overline{\Delta I} = 0.06$. This result means that the recovery rate from low density is boosted by 0.06, i.e. a species at average fitness would increase from low density at a rate of 6% per seed generation, or 3–5% per year (Box 2.1). Translated into yield differences compatible with coexistence, this means that the least fit species cannot deviate more than 6% below the average yield of its competitors if it is to persist in the group of interacting species.

A 6% yield difference does not seem large, but note that the denominator $(n - 1)$ in the formula for the storage effect implies that the storage effect is stronger with fewer species. Thus, in a four-species system, a species with yield 13% less than the average of other species would be tolerated, and in a two-species system a yield disadvantage of 40% is compatible with coexistence. Thus, the storage effect can quite powerfully promote coexistence of a few species, but, with as many as eight species, the guild is only weakly stabilised by the storage effect. It thus seems likely that additional mechanisms play a role in maintaining diversity of the annual plant guild at our study site. For example, there is evidence that spatial variation (and the storage effect as applied to spatial variation) contributes to diversity maintenance in this guild (Sears and Chesson 2007). Predation may also have a role, as much seed is taken by seed predators. When predation suitably differentiates species, it can help maintain diversity (Kuang and Chesson 2010). While seed predation may also undermine the storage effect (Kuang and Chesson 2009), this tendency is limited when competition is of the lottery form, as suggested here (Kuang and Chesson, unpublished).

Testing the storage effect hypothesis

Above we were able to fit a plausible model and examine its power in maintaining diversity. In doing so, the mechanism passes a plausibility test, but is not shown definitively to be at work maintaining diversity (Chesson 2008). Ideally, we would like to show that the environment–

Box 2.1 Estimating the storage effect for desert annual plants.

To estimate the storage effect for desert annual plants, we used the following model of the dynamics of the guild:

$$N_j(t+1) = s_j(1 - G_j(t))N_j(t) + \frac{Y_j V_j G_j(t) N_j(t)}{C'(t)}. \quad (2.5)$$

Here $N_j(t)$ is the density of seeds of species j in the seedbank before germination in year t, $G_j(t)$ is the fraction of the seedbank of species j that germinates in year t, s_j is the survival rate of ungerminated seeds, V_j is the seedling survival rate, Y_j is the average yield of a surviving seedling in the absence of competition, and $C'(t)$ is the magnitude of competition. Competition is assumed to be of the lottery form, which means that it is a linear function of the seedling densities,

$$C'(t) = \sum_{k=1}^{n} a_k V_k G_k(t) N_k(t), \quad (2.6)$$

where the a_k are competition coefficients giving the competitive effect of species k. We used the germination fraction $G_j(t)$ as the environmental response. More generally, one might consider multiple environmental responses including seedling mortality and seedling growth (Chesson *et al.* 2005), each of which has the potential to add to the storage effect (Angert *et al.* 2009). Here we consider a simpler and likely conservative approach focusing on germination fraction alone.

To estimate the storage effect, we use the log scale for both the germination fraction and competition because the log scale has proved to be a suitable scale for model analysis and also statistical analysis of field data (Chesson 2008). Thus, the environmental response is formally defined as $E_j(t) = \ln G_j(t)$, and the competitive response is defined as $C(t) = \ln C'(t)$. With these definitions, the growth rate function, $g_j(E_j(t), C_j(t))$, takes the form

$$r_j(t) = g_j\left(E_j(t), C_j(t)\right) = \ln\left\{s_j\left(1 - e^{E_j(t)}\right) + Y_j V_j e^{E_j(t) - C(t)}\right\}. \quad (2.7)$$

Note that here the competitive response is the same for all species, which means that stable coexistence is impossible if the environment does not vary over time.

With the data from Figure 2.2, germination fraction is not directly observable, but what we need is not germination fraction,

Box 2.1 (cont.)

but relative variation in ln germination fraction between species. In Equation (2.4), this is given as $\sigma^2(1-\rho)$, but if variance in ln germination fraction, σ^2, differs between species, $\sigma^2(1-\rho)$ is replaced by the more general quantity, the time by species interaction variance component of ln germination fraction (Angert *et al.* 2009, supplementary appendix). In any case, $\sigma^2(1-\rho)$ would be estimated by this variance component according to standard techniques (Searle *et al.* 1992). This variance component is best understood as that part of the variance over time that is unique to a species. Thus, it comes from the deviation of an individual species' temporal germination pattern from the average temporal pattern for all species.

We estimated this time by species variance in ln germination fraction from the time by species variance component of ln abundance of plants at flowering. However, as ln abundance at flowering equals $\ln N_j(t) + \ln V_j + \ln G_j(t)$, this variance component contains two sources of contamination, viz. variation in survival of seedlings, V_j, and variation in abundance, $N_j(t)$, in the seedbank. However, we found from simulation of the model (2.5) that the time by species ln abundance variance for successive pairs of years is minimally affected by seedbank variation. Indeed, contrary to intuition, seedbank variation causes a slight underestimate of the time by species variance in germination fraction alone. Hence, we used the time by species variance component from successive pairs of years, averaged over successive pairs of years, to estimate the needed time by species variance in ln germination fraction. Contamination due to survival variance is not removed in this way, but field data indicated that species-specific variation in germination fraction, as depicted in Figure 2.1, is much more important than species-specific variation in survival. If this observation were not valid, no large error would be introduced because survival variation leads to an additional and similar storage effect contribution, as does germination fraction (Angert *et al.* 2009, supplementary appendix).

For our data, this method yields a value of time by species variance of 0.843 (95% confidence interval 0.597–0.946, hybrid bootstrap (Shao and Tu 1995) resampling plots). The final quantity needed to estimate the storage effect is the buffering constant s,

Box 2.1 (cont.)

which is equal to the survival rate of dormant seeds in the seedbank. To estimate this value, we followed survival of seed over time in mesh bags buried at a depth of 1–2 cm in soil at our field site. With an estimated value of 0.67, and eight species, this gives an estimated value from Equation (2.4) of 0.08. However, Equation (2.4) is based on an approximation (Chesson 1994, Angert *et al.* 2009) that can be inaccurate for environmental variation as large as observed here. We corrected this estimate by simulation of model (2.5) using the time by species variance as a model parameter. The technique is to adjust the mean difference, $\mu_i - \bar{\mu}^{spp}$, by adjusting the ratio of resident and invader seed yield parameters, until the simulated $\bar{r}_i$ value is zero. At this point, $\bar{\mu}^{spp} - \mu_i = \overline{\Delta I}$, allowing $\overline{\Delta I}$ to be estimated as $\bar{\mu}^{spp} - \mu_i$. This technique gives the value of $\overline{\Delta I} = 0.06$ reported in the text, which means that the recovery rate from low density is boosted by 0.06, i.e. a species at average fitness would increase from low density at a rate of 6% per seed generation. A seed generation is the reciprocal of the mean rate of seed loss from the seedbank, viz. $1/[1 - s(1 - \bar{G})]$. If the mean germination fraction over time and species is in the reasonable range 0.3–0.7, the seed generation time is 1.25 to 2 years, giving recovery rates per year in the range 3–5%.

competition covariance changes between invader and resident states, with residents having higher covariance than invaders. A natural experiment allowed the presence of the spatial storage effect to be tested by this method at our study site near Portal, Arizona (Sears and Chesson 2007). In general, experimentally preparing resident and invader states in the field would allow the covariances associated with temporal variation to be detected, potentially yielding very strong tests of the storage effect. Here we wish to illustrate a variation on this idea that is simpler to carry out, using data from the herbaceous understorey guild. A similar test might be conducted with desert annual plants, but the seasonal variation present in the understorey system means that the full range of relevant temporal variation can be observed over the space of a single year. That makes such tests simpler to perform, although clonality of the understorey species presents other challenges not found with annual plants.

The idea behind the test is as follows. Rather than directly measure Equation (2.2) for the storage effect, which is an invader–resident comparison, we might make a comparison between the state where all species are present together and the state where one species is present in monoculture. The idea is to check for changes in covariance between environment and competition, comparing monoculture and polyculture states. As explained above, a resident in monoculture shows maximal covariance between environment and competition. The unmanipulated situation with two species present should have a lower covariance if the storage effect is functioning, because it compares a covariance based on intraspecific and interspecific competition with one based on intraspecific competition alone. Moreover, we should be able to detect these covariance changes indirectly through changes in variance of growth rather than by measuring changes in the covariance directly. Thus, if the covariance behaves the way it needs to behave to generate the storage effect, we should see higher variation in polyculture than in monoculture. This approach deals with the serious difficulties of preparing invader and resident states with clonal plant species. Thus, we followed this approach in testing whether the understorey herb, *L. stipitata*, shows the changes in variance expected if the storage effect allows it to coexist with *P. sebieriana*.

We performed the same monoculture–polyculture comparison on two separate occasions and got the same results each time. The method was similar on each occasion. We located our study plots in areas of high *Poa* abundance to identify the appropriate habitat for the manipulations. In these areas, we set up paired plots, where all *Poa* was removed from one member of the pair and the other member of the pair was left unmanipulated. In the first experiment (labelled (a) in the figures), all other herbaceous plants except *Lagenifera* were removed along with *Poa*. In the second experiment, (b), only *Poa* was removed. However, other species were minor in comparison with the two focal species in these areas of high *Poa* abundance. In both cases, the effects were marked, with removals showing lower variation in aboveground biomass than controls (Figure 2.4). We tested for this reduced variation in recruitment of leaves by converting the data into change in log-transformed leaf abundance from one census time to the next, and separating these into positive and negative changes on each plot (Figure 2.5). In both experiments, an initial period of data was discarded to allow *Lagenifera* in removal plots to adjust to the absence of *Poa*.

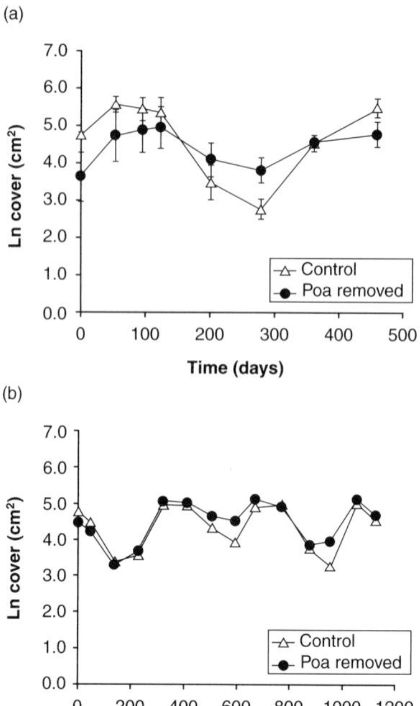

Figure 2.4 Changes in abundance of *Lagenifera stipitata* measured on experimental plots at the field site in the Namadgi National Park, Australian Capital Territory. The controls are unmanipulated, but the aboveground biomass of *Poa sieberiana* was completely removed from all removal plots by clipping *Poa* tussocks at the soil surface. Regular visits confirmed the absence of regrowth. **(a)** and **(b)** show the outcome of this experiment in two different time periods (1992–93, and 1995–97), and two different methods: (**a**) direct estimates of leaf area, removal of all species except *Lagenifera* along with *Poa* in removal plots; (**b**) incidence of leaves in 1.5-cm square areas). The symbols in (**a**) at approx. 100 days are not measurements but interpolated values exactly 1 year from the end of the experiment.

Our hypothesis of greater variation in polyculture versus monoculture was a test of whether both positive growth and negative growth were greater in polyculture than monoculture. This prediction was borne out in both cases (Figure 2.5), even though slightly different methods were used, suggesting indeed that covariance between environment and competition is changing in the directions predicted by the storage effect.

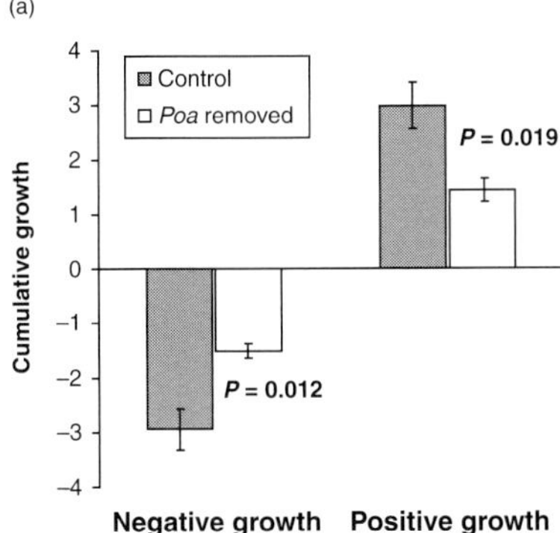

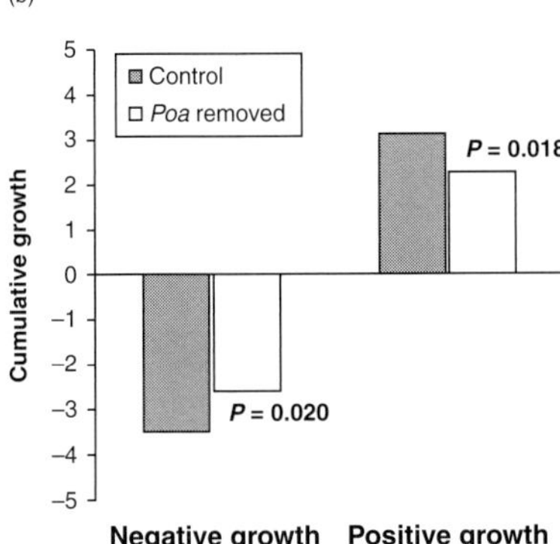

Figure 2.5 Variation in *Lagenifera stipitata* growth in monoculture versus polyculture assessed by cumulative positive and negative growth. **(a)** and **(b)** are derived from the same data as **(a)** and **(b)** of Figure 2.4, but with initial data excluded from the calculation of growth. The excluded data are from the initial 3 months in **(a)** and the initial 13 months in **(b)**. These time periods were treated as times when *Lagenifera* was adjusting to the absence of *Poa*. The remaining data represent whole years. Interpolation, as illustrated in Figure 2.4a, was used to obtain initial values representing exact multiples of a year from termination of the experiment.

2.3 DISCUSSION

Variation in the physical environment is a pervasive feature of nature. We now have the theory providing hypotheses about how it should affect community-level outcomes. The ideas presented here depend greatly on life-history traits that buffer population growth. Such traits are found in one form or another in most organisms, but are commonly ignored in most other approaches to studying the effects of environmental variation on community processes. Moreover, mechanisms of coexistence do not result from more common approaches that use Lotka–Volterra competition models with environmental fluctuations in the carrying capacity alone, or directly in the per capita growth rates (Turelli 1981, Chesson 1994, Chesson and Huntly 1997). One may well ask why coexistence mechanisms do not emerge in these circumstances. The technical answer is that these approaches incorporate no buffering of population growth. Hence the storage effect is precluded. Moreover, competition is linearly related between species, precluding the coexistence mechanism termed relative nonlinearity (Chesson 1994). However, the biological answer is that the Lotka–Volterra model with fluctuations added in these arbitrary ways fails to reflect the adaptations that most organisms have evolved to turn varying conditions to their advantage. The presence of a seedbank in our annual plant guild is predicted by life-history theory to arise by natural selection in a variable environment (Ellner 1985a, b, Real and Ellner 1992). In the context considered here, it provides buffered population growth, which has the effect of protecting individuals from serious losses, while allowing strong performance when conditions are favourable. Similarly, the mechanisms that clonal herbaceous plants have to store photosynthate and protect their bodies from herbivory (Chapin *et al.* 1990) also buffer population growth in the sense studied here. These mechanisms greatly improve the functioning of organisms in the variable world that they inhabit, and, as illustrated by analysis of the models presented here, they have profound consequences for species interactions.

We have discussed here the storage effect as it applies to competitive interactions with temporal variation. But there are spatial versions of the storage effect (Chesson 2000), and also versions involving both temporal (Chesson and Kuang 2010, Kuang and Chesson 2010) and spatial variation (Chesson and Stump, unpublished manuscript) with apparent competition. The necessary description of the storage effect in terms of statistical concepts has perhaps limited general understanding of these ideas. However, we are now in a strong position to move beyond past simplistic notions of the roles that variation in nature has in natural systems. A number of different research groups have set about testing the storage effect using various approaches (Kelly and Bowler 2002, Adler *et al.* 2006, Sears and Chesson 2007, Angert *et al.* 2009) and indeed this volume provides several examples. A common perception is that ideas involving temporal variation, especially of the stochastic sort, are difficult to test. However, the methodology that has been developed and implemented to test these ideas shows instead that they are not appreciably more difficult to test than other diversity maintenance hypotheses. Hypotheses for diversity maintenance are intrinsically difficult to test in general, and strong tests of these hypotheses are uncommon (Chesson 2008). Most analyses do not extend beyond study of patterns that are consistent with a hypothesis.

We have illustrated here two ways in which we think the study of diversity maintenance mechanisms can be improved. In the first instance, we estimated the overall magnitude of a mechanism from data from our study system. This magnitude could be interpreted in two different ways. First, it gives the overall contribution of the mechanism to the tendency to recover from low population density. Fundamentally, species coexist stably with their guild members if when perturbed to low density, for example by a run of years that are unfavourable to them, they are able to recover, in other words, to have a positive long-term low-density growth rate, $\bar{r}_i$. The greater the magnitude of this long-term low-density growth rate, the greater the stability of coexistence. We were able to estimate directly the average contribution of the storage effect to recovery in the annual plant guild. This number is an average over the group of species considered, and it is equal to the actual average recovery rate if no other mechanism is present. Second, this measure of the magnitude of the mechanism can be interpreted in terms of fitness inequalities compatible with coexistence, and so we were able to convert the predicted recovery rate into a prediction about seed yield differences that would be compatible with coexistence. By understanding how the magnitude of the mechanism changes with the number of species, we were able to predict that this mechanism alone is able strongly to stabilise coexistence of several species, but is unlikely to be the sole explanation of coexistence for the entire group of species that were studied. Looking directly at the magnitude of a mechanism, as we have done, allows informative conclusions of this sort.

The second illustration of how we think the study of diversity maintenance can be improved involves looking for signatures of the functioning of the mechanism. For the storage effect, the change between the invader and resident states of the environment–competition covariance is critical to the functioning of the mechanism. Buffered population growth is essential for this change to be at all meaningful, but buffering of population growth is relatively easy to check as it is basic to the life history of the organisms. Being satisfied that buffered population growth is present, a strong test of species coexistence resides in the behaviour of covariance between environment and competition. Experiments directly targeting covariance between environment and competition are discussed elsewhere (Sears and Chesson 2007, Chesson 2008) and must be regarded as the Holy Grail for definitive tests of the storage effect. However, covariance between environment and competition is reflected by variance of population growth and by variance in life-history parameters with a dominant role in population growth (Chesson and Huntly 1989, Chesson 2003). This means that strong tests of the storage effect might be found by studying such variances.

In many cases, a life-history parameter with a dominant role in population growth might be described as 'recruitment'. Indeed, as explained elsewhere (Chesson 2003, Chesson 2008), it is variance in recruitment that is most sensitive to the changes in covariance between environment and competition that drive the storage effect. This variance can and has been looked at in several ways. These include between-species comparisons, for species naturally at high and low density (Kelly and Bowler 2002), but within-species comparison is more reliable if appropriate within-species resident and invader states can be prepared (Chesson 2008). Here, we illustrate an approach to within-species comparison that involves not direct invader and resident state detection, but a monoculture–polyculture comparison. This approach is incapable of giving the direct measure of the strength of the storage effect that a resident–invader comparison reveals, but it is capable of detecting the underlying phenomenon of change in covariance as population densities are manipulated. As the procedure involves simple removals, it is likely to be much more easily implemented than invader–resident comparisons.

Understanding the mechanisms that underlie diversity maintenance is a central goal of community ecology. It has been hampered on many fronts, both on theoretical and empirical levels. An enormous problem in the theory has been adequately representing the dynamics of communities in mathematical form and then drawing conclusions from them. Although there is much more to do, we now have the basic concepts with which to understand the very serious problem of the role of environmental variation in species coexistence. Understanding is now moving beyond a focus on competition to a fuller understanding of the role of natural enemies in the maintenance of diversity (Holt 1984, Chesson and Kuang 2008). Indeed, an emerging theory is extending ideas on competitive interactions to include Holt's concept of apparent competition (Holt 1977), which holds that joint limitation of two or more species by a predator is analogous to joint limitation by a resource. Importantly, this emerging theory fully extends the ideas on environmental variation that have been central to the discussion here (Chesson and Kuang 2010, Kuang and Chesson 2010).

On the empirical front, there is the huge challenge of conducting rigorous tests of coexistence mechanisms. Much community ecology generates information relevant to understanding species coexistence mechanisms, but stops short of actually testing them. The ability to quantify mechanisms in functional form, as discussed here, holds much promise for rigorous testing in nature. We have presented the beginnings of such tests by estimating the magnitude of a mechanism, which gives a good indication of its sufficiency, and by seeking the signature of a mechanism through patterns found experimentally. However, there is a need to go much further. We have shown how a mechanism can be represented quantitatively in functional form. This provides the means to determine convincingly if a mechanism is in fact functioning in a given setting (Chesson 2008). The experiments needed to do this may be difficult, but they are well within the realms of possibility in community ecology. Doing these experiments should teach us much.

ACKNOWLEDGEMENTS

We are grateful for help with the fieldwork described here from Stefan Sommer, Sandra Berry, Natalie Hocking and Suzie Richmond. This work was supported by National Science Foundation grants BSR-8615028, DEB-9307568, DEB-0816231 and DEB-0542991 and while NJH was working at the Foundation. Any opinion, finding and conclusions or recommendations expressed in this material are those of the authors and do not necessarily reflect the views of the National Science Foundation.

REFERENCES

Adler, P. B., HilleRisLambers, J., Kyriakidis, P. C., Guan, Q. F. and Levine, J. M. (2006). Climate variability has a stabilizing effect on the coexistence of prairie grasses. *Proceedings of the National Academy of Sciences, USA* **103**, 12793–12798.

Adondakis, S. and Venable, D. L. (2004). Dormancy and germination in a guild of Sonoran Desert annuals. *Ecology* **85**, 2582–2590.

Amarasekare, P. and Nisbet, R. M. (2001). Spatial heterogeneity, source-sink dynamics, and the local coexistence of competing species. *American Naturalist* **158**, 572–584.

Angert, A. L., Huxman, T. E., Chesson, P. and Venable, D. L. (2009). Functional tradeoffs determine species coexistence via the storage effect. *Proceedings of the National Academy of Sciences, USA* **106**, 11641–11645.

Armstrong, R. A. and McGehee, R. (1976). Coexistence of species competing for shared resources. *Theoretical Population Biology* **9**, 317–328.

Baskin, C. C. and Baskin, J. M. (1998). *Seeds: Ecology, Biogeography, and Evolution of Dormancy and Germination*. San Diego, CA: Academic Press.

Bond, W. J. and Keeley, J. E. (2005). Fire as a global 'herbivore': the ecology and evolution of flammable ecosystems. *Trends in Ecology and Evolution* **20**, 387–394.

Brown, J. S. (1989a). Coexistence on a seasonal resource. *American Naturalist* **133**, 168–182.

Brown, J. S. (1989b). Desert rodent community structure: a test of four mechanisms of coexistence. *Ecological Monographs* **59**, 1–20.

Bulmer, M. G. (1985). Selection for iteroparity in a variable environment. *American Naturalist* **126**, 63–71.

Chapin, F. S., Schulze, E. and Mooney, H. A. (1990). The ecology and economics of storage in plants. *Annual Review of Ecology and Systematics* **21**, 423–447.

Chesson, P. (1994). Multispecies competition in variable environments. *Theoretical Population Biology* **45**, 227–276.

Chesson, P. (1997). Diversity maintenance by integration of mechanisms over various scales. In *Proceedings of the Eighth International Coral Reef Symposium*. Balboa, Panama City, Panama: Smithsonian Tropical Research Institute, pp. 405–410.

Chesson, P. (2000). General theory of competitive coexistence in spatially-varying environments. *Theoretical Population Biology* **58**, 211–237.

Chesson, P. (2003). Quantifying and testing coexistence mechanisms arising from recruitment fluctuations. *Theoretical Population Biology* **64**, 345–357.

Chesson, P. (2008). Quantifying and testing species coexistence mechanisms. In F. Valladares, A. Camacho, A. Elosegui *et al.* (eds), *Unity in Diversity: Reflections on Ecology after the Legacy of Ramon Margalef*. Bilbao, Spain: Fundación BBVA, pp. 119–164.

Chesson, P. and Huntly, N. (1989). Short-term instabilities and long-term community dynamics. *Trends in Ecology and Evolution* **4**, 293–298.

Chesson, P. and Huntly, N. (1997). The roles of harsh and fluctuating conditions in the dynamics of ecological communities. *American Naturalist* **150**, 519–553.

Chesson, P. and Kuang, J. J. (2008). The interaction between predation and competition. *Nature* **456**, 235–238.

Chesson, P. and Kuang, J. J. (2010). The storage effect due to frequency-dependent predation in multispecies plant communities. *Theoretical Population Biology* **78**, 148–164.

Chesson, P., Donahue, M. J., Melbourne, B. A., and Sears, A. L. W. (2005). Scale transition theory for understanding mechanisms in metacommunities. In M. Holyoak, A. M. Leibhold and R. D. Holt (eds), *Metacommunities: Spatial Dynamics and Ecological Communities*. Chicago, IL: University of Chicago Press, pp. 279–306.

Chesson, P., Pacala, S. and Neuhauser, C. (2001). Environmental niches and ecosystem functioning. In A. P. Kinzig, S. W. Pacala and D. Tilman (eds), *The Functional Consequences of Biodiversity*. Princeton, NJ: Princeton University Press, pp. 213–245.

Chesson, P. L. and Case, T. J. (1986). Overview: nonequilibrium community theories: chance, variability, history, and coexistence. In J. Diamond and T. J. Case (eds), *Community Ecology*. New York: Harper and Row, pp. 229–239.

Chesson, P. L. and Huntly, N. (1988). Community consequences of life-history traits in a variable environment. *Annales Zoologici Fennici* **25**, 5–16.

Chesson, P. L. and Warner, R. R. (1981). Environmental variability promotes coexistence in lottery competitive systems. *American Naturalist* **117**, 923–943.

Choler, P., Michalet, R. and Callaway, R. M. (2001). Facilitation and competition on gradients in alpine plant communities. *Ecology* **82**, 3295–3308.

Clements, F. E. (1916). *Plant Succession: An Analysis of the Development of Vegetation*. Washington DC: Carnegie Institution of Washington.

Cohen, D. (1966). Optimizing reproduction in a randomly varying environment. *Journal of Theoretical Biology* **12**, 119–129.

Comins, H. N. and Noble, I. R. (1985). Dispersal, variability, and transient niches: species coexistence in a uniformly variable environment. *American Naturalist* **126**, 706–723.

Connell, J. H. (1978). Diversity in tropical rain forests and coral reefs. *Science* **199**, 1302–1310.

Davis, M. B. (1986). Climatic instability, time-lags and community disequilibrium. J. Diamond and T. J. Case (eds), *Community Ecology*. New York: Harper and Row, pp. 269–284.

Ellner, S. (1985a). ESS germination strategies in randomly varying environments. I. Logistic-type models. *Theoretical Population Biology* **28**, 50–79.

Ellner, S. (1985b). ESS germination strategies in randomly varying environments. II Reciprocal yield-law models. *Theoretical Population Biology* **28**, 80–116.

Facelli, J. M., Chesson, P. and Barnes, N. (2005). Differences in seed biology of annual plants in arid lands: a key ingredient of the storage effect. *Ecology* **86**, 2998–3006.

Grubb, P. J. (1977). The maintenance of species-richness in plant communities: the importance of the regeneration niche. *Biological Review* **52**, 107–145.

Harper, J. L. (1985). Modules, branches, and the capture of resources. In J. B. C. Jackson, L. W. Buss and R. E. Cook (eds), *Population Biology and Evolution of Clonal Organisms*. New Haven, CT: Yale University Press, pp. 1–33.

Hastings, A. (1980). Disturbance, coexistence, history, and competition for space. *Theoretical Population Biology* **18**, 363–373.

Holt, R. D. (1977). Predation, apparent competition, and the structure of prey communities. *Theoretical Population Biology* **12**, 197–229.

Holt, R. D. (1984). Spatial heterogeneity, indirect interactions, and the coexistence of prey species. *American Naturalist* **124**, 377–406.

Hutchinson, G. E. (1961). The paradox of the plankton. *American Naturalist* **95**, 137–145.

Inouye, R. S. (1980). Density dependent germination response by seeds of desert annuals. *Oecologia* **46**, 235–238.

Juhren, M., Went, F. W. and Phillips, E. (1956). Ecology of desert plants. IV. Combined field and laboratory work on germination in the Joshua Tree National Monument, California. *Ecology* **37**, 318–330.

Kelly, C. K. and Bowler, M. G. (2002). Coexistence and relative abundance in forest trees. *Nature* **417**, 437–440.

Kelly, C. K. and Bowler, M. G. (2005). A new application of storage dynamics: differential sensitivity, diffuse competition, and temporal niches. *Ecology* **86**, 1012–1022.

Kilpatrick, A. M. and Ives, A. R. (2003). Species interactions can explain Taylor's power law for ecological time series. *Nature* **422**, 65–68.

Kuang, J. J. and Chesson, P. (2009). Coexistence of annual plants: generalist seed predation weakens the storage effect. *Ecology* **90**, 170–182.

Kuang, J. J. and Chesson, P. (2010). Interacting coexistence mechanisms in annual plant communities: frequency-dependent predation and the storage effect. *Theoretical Population Biology* **77**, 56–70.

Lande, R. (1993). Risks of population extinction from demographic and environmental stochasticity and random catastrophes. *American Naturalist* **142**, 911–927.

Levins, R. (1979). Coexistence in a variable environment. *American Naturalist* **114**, 765–783.

May, R. M. (1974). *Stability and Complexity in Model Ecosystems*. 2nd edn. Princeton, NJ: Princeton University Press.

Pacala, S. W. and Tilman, D. (1994). Limiting similarity in mechanistic and spatial models of plant competition in heterogeneous environments. *American Naturalist* **143**, 222–257.

Real, L. A. and Ellner, S. (1992). Life-history evolution in stochastic environments: a graphical mean variance approach. *Ecology* **73**, 1227–1236.

Ripa, J., Lundberg, P. and Kaitala, V. (1998). A general theory of environmental noise in ecological food webs. *American Naturalist* **151**, 256–263.

Rohde, K. (2005). *Nonequilibrium Ecology*. New York: Cambridge University Press.

Roxburgh, S. and Chesson, P. (1998). A new method for detecting species associations with spatially autocorrelated data. *Ecology* **79**, 2180–2192.

Searle, S. R., Casella, G. and McCulloch, C. E. (1992). *Variance Components*. New York: John Wiley and Sons.

Sears, A. L. W. and Chesson, P. (2007). New methods for quantifying the spatial storage effect: an illustration with desert annuals. *Ecology* **88**, 2240–2247.

Shao, J. and Tu, D. (1995). *The Jackknife and Bootstrap*. New York: Springer-Verlag.

Snyder, R. E. and Chesson, P. (2004). How the spatial scales of dispersal, competition, and environmental heterogeneity interact to affect coexistence. *American Naturalist* **164**, 633–650.

Soong, K., Chen, M.-h., Chen, C.-l. *et al.* (2003). Spatial and temporal variation of coral recruitment in Taiwan. *Coral Reefs* **22**, 224–228.

Tilman, D. (2004). Niche tradeoffs, neutrality, and community structure: a stochastic theory of resource competition, invasion, and community assembly. *Proceedings of the National Academy of Sciences, USA* **101**, 10854–10861.

Turelli, M. (1980). Niche overlap and invasion of competitors in random environments II. The effects of demographic stochasticity. In W. Jager, H. Rost and P. Tautu (eds), *Biological Growth and Spread: Mathematical Theories and Applications*. Berlin: Springer-Verlag, pp. 119–129.

Turelli, M. (1981). Niche overlap and invasion of competitors in random environments I: models without demographic stochasticity. *Theoretical Population Biology* **20**, 1–56.

Warner, R. R. and Chesson, P. L. (1985). Coexistence mediated by recruitment fluctuations: a field guide to the storage effect. *American Naturalist* **125**, 769–787.

Wright, S. J., Muller-Landau, H. C., Calderon, O. and Hernandez, A. (2005). Annual and spatial variation in seedfall and seedling recruitment in a neotropical forest. *Ecology* **86**, 848–860.

COLLEEN K. KELLY, MICHAEL G. BOWLER, GORDON A. FOX,
J. ARTURO SOLÍS-MAGALLANES, J. MARCELA RAMOS-TAPIA,
PILAR LOPERA BLAIR, SUSANNE SCHWINNING,
JOHN N. WILLIAMS AND JEFFREY B. JOY

3

What temporal processes in trees tell us about competition, community structure and speciation

3.1 INTRODUCTION

This chapter reviews evidence concerning the vital role that temporal dynamics can have in the ecology of trees and other long-lived species in the assembly and maintenance of natural communities. The research synthesised here was stimulated by a desire to determine the action of temporal dynamics in nature, and its implications for the nature of competition, community structure and assembly on multiple scales and across a range of climatic conditions. For the most part, the results discussed concern tropical forests, but we think they provide strong support for a more general view that can be applied across biomes. Finally, we ask if there may be a potential role for temporal dynamics in speciation, in light of what we have learned from the tropical trees.

A field programme begun in the late '90s in the tropical dry forest of México was consciously designed to study the coexistence of closely related species in a very speciose community, but the role of temporal dynamics had not been suspected and its finding was serendipitous. With centuries-long lifespans, decades-long juvenile stages and low population turnover rates, trees are problematic candidates for demographic analyses, either observational or experimental. Unless instant death is involved, the particular hurdle with trees, as with any long-lived organism, is directly connecting any specific response in the early

Temporal Dynamics and Ecological Process, ed. C. K. Kelly, M. G. Bowler, G. A. Fox.
Published by Cambridge University Press.

life of the individual with the long-term individual persistence or character of the standing population. However, trees differ from many long-lived organisms in carrying their history in their structure at both the individual and population levels. Thus, a tree population itself documents individual success over the history of the population (Parker *et al.* 1997, Cole *et al.* 2011). The distribution of a population with regard to physical conditions, size and age structure and relative to other woody species all contain information on the ecology and interactions of species (e.g. Veblen 1989, 1992, Villalba and Veblen 1998, Kelly *et al.* 2001) and it was the age structure of populations that revealed the action of temporal dynamics at Chamela Biological Station.

Although not exclusively so, the majority of studies that we discuss in this chapter have focused on forest trees in the tropics, in the interest of acquiring sample sizes sufficient to generalise our inferences. A substantial number of these studies examine the relationships between congeneric species forming terminal dichotomies in the phylogenetic tree of the examined communities. This design utilises the a priori similarity of close relatedness unbiased by pre-existing ecological expectations. In this way, we have been able to identify a general role of similarity in determining species interaction and community assembly. Congeneric comparison is also an analysis that has been well explored as a natural 'all else being equal' comparison of independent contrasts, strengthening the inference that an observed pattern of difference between taxa is attributable to the target factor (Felsenstein 1985, Harvey and Pagel 1991, Harvey 1996).

Congener comparisons also suggest evolutionary processes, and we present results and theory on temporal niche dynamics as a route for speciation, and a potential source of plant 'supergenera'. At the landscape level, supergenera comprise a small proportion of genera present, but a disproportionate amount of standing mature individuals (Williams and Kelly 2013). From the patterns of change seen across a gradient of temperature and rainfall, we can better understand directional evolutionary change resulting from ecological pressures, and the response of the community to environmental variation.

Finally, we discuss the ramifications of these results. In this, we touch upon the differences and similarities between temporal processes for congeneric and non-congeneric temporal niche partners. We also consider the problem of assessing competitive relationships in light of the effects and increasingly apparent ubiquity of both temporal and spatial storage dynamics in which competitive interaction is condition dependent rather than definitive. Intermingled congeneric species form a significant proportion of woody communities worldwide (Gentry 1991, Kelly and Bowler 2005); in itself, this number represents a significant proportion of species potentially able to support temporal dynamics, but temporal dynamics may also occur between unrelated species, making its potential impact on ecological process of even greater significance (Mason 2005, Ellis 2007, Chesson *et al.* this volume, Adler this volume, Fowler this volume, Ellis *et al.* unpublished manuscript.)

3.2 POPULATIONS AND COMMUNITIES

Population structure

For mature trees, growth and mortality at any one point in time will vary only gradually in a regular manner across size classes (Johnson and Frier 1989, Hara *et al.* 1991). Thus, if adult mortality rates were the only determinant of age structure in mature trees then the age distribution of adult trees of each species should be exponential and related measures should show smooth curves, with regular transitions from one age class to the next when a population is represented as sequential age classes. Substantial deviations from this simple expectation are observed among the species resident at Chamela (Figure 3.1). The most likely causes for deviations are effects taking place prior to an individual reaching the canopy, in particular as variation either in seedling/sapling mortality or in availability of propagules. If recruitment to the adult stage were cut off for a number of years, the result would be an exponential punctuated with gaps. To infer other than recruitment depression would require assuming that at any one time, size-dependent growth or mortality of mature trees is low in one size class, high in the next largest size class, low again in the following size class, etc. Such a pattern of 'pick-and-choose' growth and death rates directly contradicts what is well known and documented about size-dependent patterns of vital rates of mature trees. Consistent with this explanation, a summary of a 32-year-long study of a single species (Connell and Green 2000) showed that seedling cohort size, i.e. recruitment, could predict the relative frequency of the corresponding adult size class.

The premise that such variability in recruitment over time may allow species to coexist produces an expectation of a specific relationship between age classes of those coexisting species: a short-term loss in one species offsets gain in the other(s) in the dynamic, and a long-term stability of numbers at any one site. From this, the rarer of the

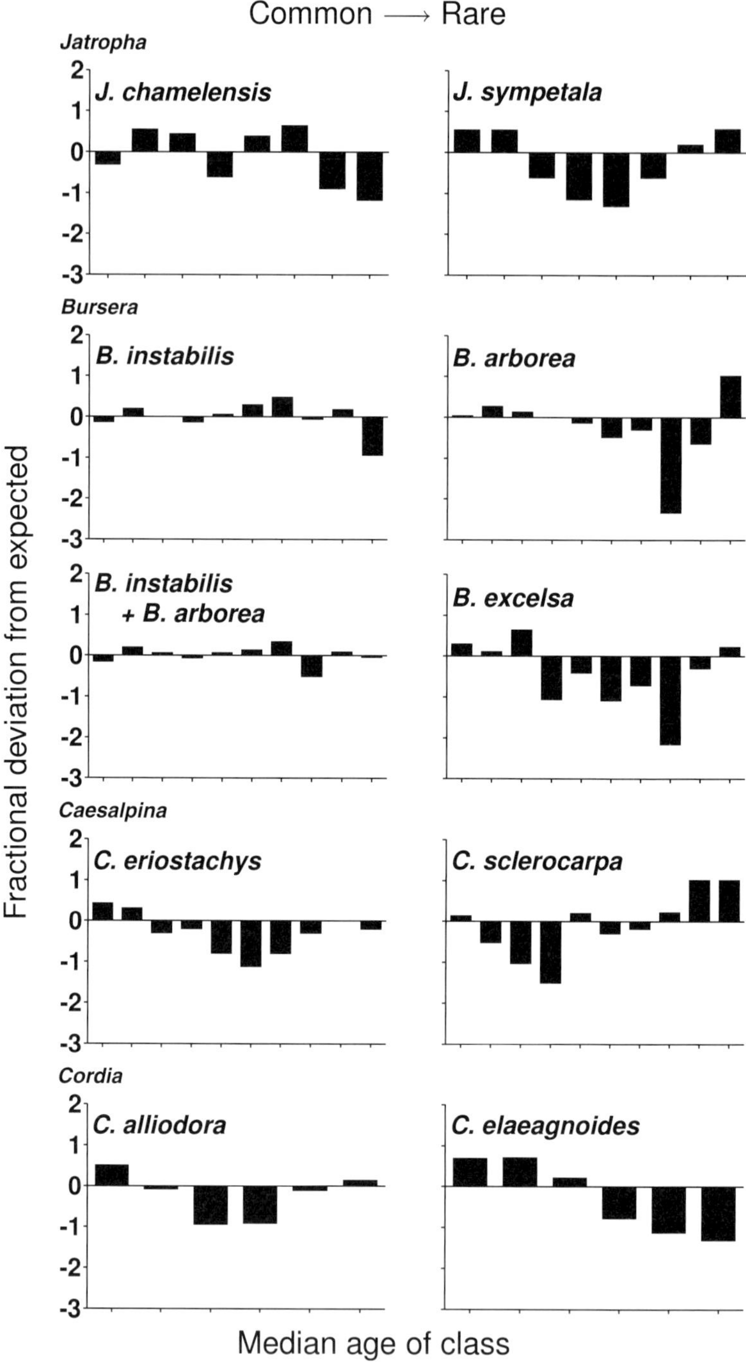

Figure 3.1 Fractional deviation of age classes from an expected exponential distribution. The deviations were calculated as (Expected – Observed)/ Expected. Comparisons (contrasts) were made within, not between rows of

interacting taxa should show greater deviation than the commoner taxon does in the transition from one age class to the next: trading off approximately the same number of individuals would produce a greater fractional change in the less abundant species. Applying a quantitative model of a temporal dynamic to this interaction also generates a testable prediction at the population level that the rarer species will be the more competitive/more responsive to environmental variation, and the commoner species will be the more tolerant/less responsive to environmental variation (Kelly and Bowler 2002).

Support for these expectations was found in the highly diverse ecosystem of the tropical dry forest of México. At Chamela Biological Station in the western state of Jalisco, the high percentage of congeneric tree species offers sufficiently large sample sizes to allow statistical tests of long-term growth rate and fractional deviations in the age structure of coexisting congeners. In all pairwise comparisons between most-closely related congeners, the less abundant of these interdigitated species pairs showed greater fractional variation in age structure, and greater competitiveness than the more common ($p = 0.00096$, Figure 3.1; Kelly and Bowler 2002).

Of existing models postulating coexistence of competitors, only temporal niche dynamics (storage theory) explains these patterns. Models predicating coexistence through resource 'patchiness' and effective escape from competition (Nee and May 1992, Pacala and Tilman 1994) offer no expectations of any regular difference in the community in age profiles for more and less abundant competitors, nor does the assumption of coexistence through chance immigration (Hubbell 2001). As a third possibility, disturbance theory addresses population fluctuations in competing species by raising the expectation of in-phase pulses, but offers no predictions of relative differences

Caption for figure 3.1 (cont.) taxa representing pairs of taxa more closely related to one another than to any other taxon in the figure (i.e. forming terminal dichotomies in the phylogeny of the community). Each bar represents a 30-year interval. Field growth measures of *Bursera instabilis* indicate that individuals of less than 3 cm dbh, the assumed minimum dbh for canopy individuals in this study, could be greater than 30 years old, necessitating the use of the 60–90-year class as the smallest age class for the two contrasts using this genus. For all other genera, age classes begin at 30–60 years. At Chamela Biological Station, *Jatropha sympetala* had been known historically as *J. standleyi*, and has been displayed in previous studies as such (Pirie *et al.* 2000, Kelly *et al.* 2001, Kelly and Bowler 2002, Buckley and Kelly 2003). E. J. Lott has recently corrected this identification.

between species in the degree of those fluctuations (Hutchinson 1959, Connell 1978). However, neither nearest neighbour tests of patch-based recruitment (Veblen 1989, Kelly and Bowler 2005) nor simultaneity of fluctuations reveal significant patterns for any of the paired comparisons. Instead, the pattern of deviations in Figure 3.1 suggests antiphase fluctuations, an observation in line with storage dynamics in the absence of temporal variation in the number of sites.

Focused interactions and relative abundance

The consistent success of the preceding studies in identifying pattern suggests that the focused, paired structure used to compare closely related species may also reflect the character of the actual temporal process. Although diffuse competition is mathematically reasonable, to support the relationships illustrated above, diffuse competition requires that the competitive ability of each of the species involved have a very tight relationship with the long-term average competitive ability of the forest as a whole: biological fine-tuning. If this fine-tuning does take place in nature, to support the observed relative differences in growth rate and degree of fluctuation, the diffuse model developed to examine this question predicts a simple linear correlation between relative abundance and degree of variation in population structure across species, and no such relationship occurs in the data (Kelly and Bowler 2005).

This inference of focused competition invokes a robust body of ecological theory. General resource partitioning theory posits an underlying structure of niches in which particular species excel at exploiting particular resources (MacArthur 1957, Preston 1962a, b, Armstrong and McGehee 1980, Sugihara 1980, Nee *et al.* 1991, Sugihara *et al.* 2003). In such a niche-structured dynamic, abundance of any one species is directly affected primarily by the abundance of only one to a very few highly similar species. In this way, competition affects community composition through regulating population size and persistence of the species within a focused competitive interaction.

We therefore asked whether focused, pairwise interactions of the kind envisaged by Sugihara and indicated in our earlier work are prevalent and forceful enough in forest trees to be of significance at the level of the community. In particular, in the Sugihara model of sequential breakage (Sugihara 1980, Sugihara *et al.* 2003) each bifurcation of an ecological niche dendrogram has a probability distribution for resource allocation, taken to be reflected in the relative numbers of the two taxa involved. This splitting function is the distribution of the fractional abundance in a pair, r, defined as

$$n_{commoner}/(n_{commoner} + n_{rarer}), \tag{3.1}$$

where $n_{commoner}$ is the number of individuals of the more common member of the pair and n_{rarer} of the less common. This splitting function contains the signature of focused competition when the pairs are chosen from ecological bifurcations. In contrast, if there is no niche structure (i.e. species are interchangeable or 'neutral', as proposed in Hubbell (2001)) then composing pairs according to any meaningful biological algorithm should yield a distribution of fractional abundance indistinguishable from that of pairs chosen at random from the community.

Tests of these ideas were performed for two biome types: the tropical dry forest of Chamela Biological Station, the site of the previously described studies on temporal niche dynamics; and the tropical moist forest of Barro Colorado Island (BCI), Panama, the original stimulus for the development of the idea of neutrality (Hubbell and Foster 1986). For both sites, pairs were identified through membership in terminal dichotomies in their respective community-level phylogenetic trees, and were segregated into two groups: (i) pairs where both species are members of the same genus (congeneric pairs) and (ii) pairs where species do not have congeners within the data set but whose nearest relatives are members of the same family or order (non-congeneric pairs). The distribution of abundances in congeneric pairs was different from that of pairs selected at random from the community for both Chamela and BCI. For the Chamela assay, the distribution of abundances of non-congeneric pairs was not statistically different from randomly selected pairs (Kelly *et al.* 2008); for the BCI data, the distribution of non-congener fractional abundances was significantly different from random in the fifth of five regular censuses, but not in the first four (Table 3.1; Kelly *et al.* 2010).

These observations are consistent with Sugihara's niche apportionment model, but not with the assumption upon which neutral theory is based, that individuals within a community are effectively equivalent. The same studies examined the possibility that the observed distribution of congeneric pairs might conform either to that expected from species interchangeability (rather than neutrality) within congener pairs, or to competitive exclusion (Figure 3.2; Kelly and Bowler

Table 3.1 ***Fractional abundance distributions at BCI for each census year.*** *Species were paired according to their occurrence as terminal dichotomies in the fully resolved molecular phylogeny of tree species in the 50-ha plot on Barro Colorado Island, Panama; population was determined as the number of individuals greater than the size at first flowering time for that species as determined by Robin Foster (potentially more accurate maximum likelihood estimates have been calculated, but only for a small subset of species and individuals within the larger census area (Wright* et al. *2005); the Foster assessments were made by a single individual for all species, and were thus assumed to have more consistent error over the greater range of species). The figures here show the probabilities that the distribution of BCI congener and non-congener pairs (34 and 21, respectively) was significantly different from a random draw of the full species complement.*

	Year of census				
	1985	1990	1995	2000	2005
Congeneric species pairs	0.01	0.039	0.03	0.01	0.024
Non-congeneric species pairs	> 0.25	> 0.14	> 0.25	> 0.25	> 0.25

2009a, Kelly *et al.* 2010). Neither hypothesis was consistent with the observed distribution of paired congeneric species at either site.

Applying maximum likelihood methods, the observed distributions of congeneric fractional abundances have been fitted by the temporal niche model of differential sensitivity between niche partners (Figure 3.3; Kelly and Bowler 2009a, Kelly *et al.* 2010). For either site, stable coexistence through the focused competition of differential sensitivity dynamics is possible over a wide range of environmental fluctuation frequencies, but still consistent with local and regional climate records (Bullock 1986, Ayala 2007). In these distributions, the single assumption was made that the competitive advantage of one species over another will not be immensely greater than it needs to be to coexist successfully with its competitor. Thus, species are unlikely to be at the upper end of the fractional abundance distribution because it takes only a small competitive advantage for the more sensitive, rarer species to increase its population size rapidly. Because it takes an inordinate competitive advantage for the more sensitive species to maintain a population size close to that of its less sensitive niche partner, species are unlikely to be at the lower end of the distribution. The ranges of 'permissible' environmental frequencies also allow the possibility that the fraction of good times

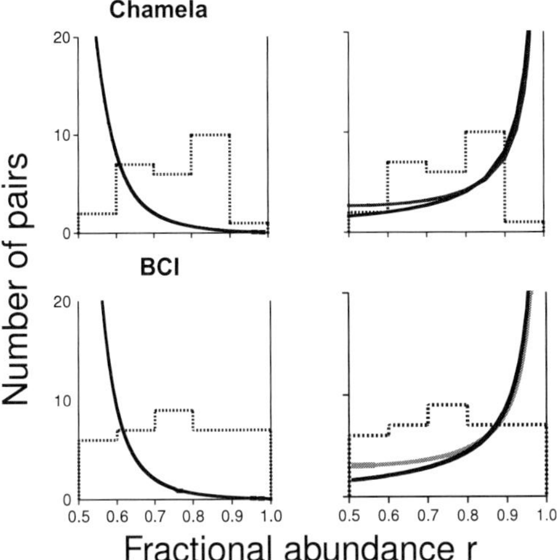

Figure 3.2 Expected fractional abundance distributions under neutrality and competitive exclusion within congeneric pairs at Chamela (top) and BCI (bottom). In all panels, the dashed line represents the distribution of fractional abundances at the site in question. The panels on the left side of the figure show the fractional abundance distribution expected for two species coexisting with equal probability of occupying each of N sites (an approximation of neutrality) rather than interacting with one another. The panels on the right side of the figure show the fractional abundance distributions expected for one species being driven out by an invasive congener (competitive exclusion). Calculations are based on the assumption of pairs being sampled at random times during exclusion. The black line corresponds to a large competitive advantage for the invader, the grey line to a minor advantage. The data for congeneric species (histogram in each part) could not with any significant probability be drawn from the distributions illustrated. Figure after Kelly and Bowler (2009a) and Kelly *et al.* (2010).

may differ among genera: some species may more readily and thus more frequently experience 'bad' conditions than others, and still be able to maintain stable coexistence through a temporal differential sensitivity dynamic.

Conformity of the congener fractional abundance distributions to the temporal niche models for both Chamela and BCI offers evidence that temporal niche dynamics, as a major component of community assembly, are not restricted to a single biome or species type. A more recent simulation study of temporal dynamics at BCI suggests that temporal dynamics may be frequent there in non-congeneric species as well (Usinowicz *et al.* 2012).

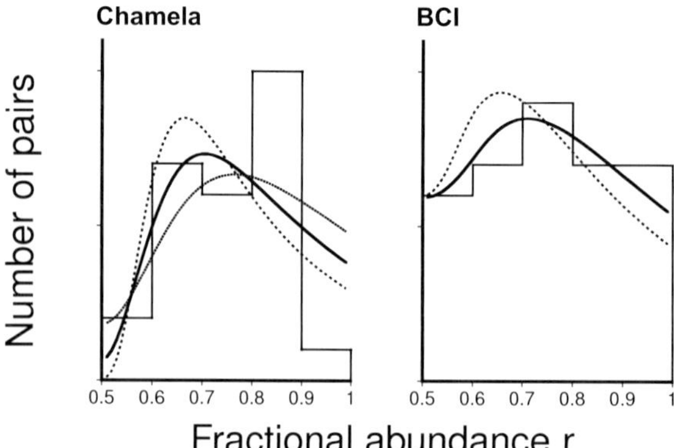

Figure 3.3 Comparison of the data with the model: histogram of fractional abundance data and superimposed [maximum likelihood] fitted distributions. Curves show expected values of r for differing frequencies of environmental fluctuation f overlaid on the histogram representation of the data from Kelly *et al.* (2008). Chamela: The best fit overall is $f = 0.7$ (solid black curve); the cases $f = 0.6$ (dashed black curve), 0.9 (solid grey curve) are essentially one standard deviation removed from the best value. BCI: The best fit overall is $f = 0.9$ (solid black curve). Figure after Kelly and Bowler (2009a) and Kelly *et al* (2010).

The differences in the projected frequency of good years, f, at BCI and Chamela (0.9 at BCI versus 0.7 at Chamela) fall within quite broad standard deviations making any conjecture tentative, but are not inconsistent with documented ecological differences between the two forests. In this regard, the greater water availability in the tropical moist forest at BCI (average 2612 mm rain/year; http://www.sil.si.edu/smithsoniancontributions/EarthSciences) than the tropical dry forest of Chamela (average 731 mm rain/year; http://www.ibiologia.unam.mx/ebchamela, Bullock 1988) and generally less extreme dry season at BCI relative to that of Chamela could have consequences on several levels. It could, for example, directly lower the competitive difference achievable between niche partners and thereby increase the number of high resource years required for persistence (see expression of relationship between a and f in next section). Alternatively, due to the propensity for faster growing plants to be more attractive to herbivores (Kelly and Hanley 2005, Mayrose *et al.* 2011, Hanley this volume), good years may be made less good for the more sensitive species by the higher pest and pathogen loads expected at BCI as a result of the greater water availability relative to Chamela (McCluney *et al.* 2012). Finally, the higher pest and pathogen level at BCI, with its inevitable cycling, could drive the temporal dynamic directly (Chesson and Kuang 2008, Kelly and Bowler 2009b).

The differential sensitivity of temporal niche dynamics provides a mechanism capable of generating the distribution of congeneric fractional abundances seen in Figure 3.2, which is also of the sort described in the hierarchical niche apportionment model of Sugihara (1980, Sugihara *et al.* 2003), for which temporal dynamics may provide a mechanism. Although the hierarchical niche dendrogram of Sugihara *et al.* (2003) is not phylogenetic, the similarity and interactions of congeners may be a signal of its presence. The possibility of interactions nesting within genera in the hierarchical manner described by Sugihara is suggested by available data. In Kelly *et al.* (2001), three rather than two congeners co-occurred as interdigitated populations in two genera (*Bursera* and *Caesalpinia*), and data were collected on all three. In both genera, the less common the species, the greater the level of temporal deviations from a smooth curve.

There are two major inferences to be drawn from the information contained in fractional abundances. First, interaction between species as a function of relatedness is strongest at the congeneric level (see also Jablonski and Finarelli 2009). There is no evidence for any relationship between relatedness and interaction above that level of similarity in the Chamela data with non-congeneric comparisons all at the family level; for the BCI data, the mix of family and order level pairs could not be discriminated from a random distribution. More recently, a similar finding of significant interaction signal restricted to the level of congeners was reported for a group of coastal scrub species (Burns and Strauss 2011). These patterns are not particularly surprising: competitive interaction depends on similarity, as does the taxonomic categorisation of species within genera. However, one implication of this result is that inference of competitive interaction from pattern in phylogenetic structure is not reliable above the level of genus. This does not mean that competitive interaction cannot occur between species more distantly related than congeners (e.g. Brooks 2000, Mayfield and Levine 2010); rather, above the level of congeners, inference of interaction rather than niche differentiation is not reliable. The possibility of nested interactions above the level of most similar pairs is a part of the previously cited work by Sugihara *et al.* (2003), and some of the data presented in section 3.5 and in Kelly and Bowler (2005) are consistent with such a nested structure, although still restricted to congeneric taxa.

A second insight of fundamental ecological process rests upon the distinctive failure of neutral theory to account for the observed patterns of structure or fractional abundance at any point. It has been suggested that coexisting congeners are a grouping most likely to support neutrality (Leibold and McPeek 2006), but the fractional abundance comparison described here is antithetical to neutral dynamics, as are the multiple lines of evidence for focused competition. Furthermore, it is becoming increasingly evident that highly deterministic ecological processes are equally able to produce the previously supposed signature of neutrality (Hubbell 2001), the lognormal species abundance distribution (Pueyo 2006, Pueyo *et al.* 2007, Kelly *et al.* 2011, Bowler and Kelly 2012).

3.3 RECRUITMENT, COMPETITION AND THE ENVIRONMENT

When do temporal storage dynamics occur?

Storage dynamics can occur when there is covariance between competition and the environment (Warner and Chesson 1985). A clear statement in itself, its translation into patterns in nature or experimental results may not always be transparent, in particular with regard to assessing a competitive relationship between species. In essence, if environmental fluctuations can promote coexistence of species that would otherwise suffer competitive exclusion, then there must be covariance between competition and the environment. If competition were indifferent to environmental fluctuations, then the fluctuations could not promote coexistence.

However, there is some advantage in defining competition anyway, in the context of the lottery model for two competing species. The competition is for sites – places with resources for one adult. Let there be x_1 adults (reproductives) of species 1 and x_2 of species 2. These individuals take a fixed number of sites and the problem is the division between them; we might as well define x_1 and x_2 as fractions. Individuals of species i die at a rate δx_i; let both species have a common per capita death rate δ. New sites become available at a rate $\delta(x_1 + x_2)$ and these are up for grabs. Suppose the probability of species 1 successfully germinating a seed (from a seed rain proportional to the number of adults present), in the absence of either intraspecific or interspecific competition, is given by $a_1 x_1$. Then our first guess at writing equations for the coupled population dynamics of the two species would be

$$\frac{dx_1}{dt} = -\delta x_1 + \delta(x_1 + x_2)a_1 x_1 \qquad \frac{dx_2}{dt} = -\delta x_2 + \delta(x_1 + x_2)a_2 x_2. \quad (3.2)$$

The nature of the problem is that populations cannot grow without limit and that to the extent that any dead adult is instantly replaced by an adult of the same or the other species, the sum of the two populations is a constant. This is achieved by multiplying the recruitment term in each equation by the competitive factor

$$\frac{1}{a_1x_1 + a_2x_2} \tag{3.3}$$

so as to obtain the equations

$$\frac{dx_1}{dt} = -\delta x_1 + \frac{\delta(x_1 + x_2)a_1x_1}{a_1x_1 + a_2x_2} \qquad \frac{dx_2}{dt} = -\delta x_2 + \frac{\delta(x_1 + x_2)a_2x_2}{a_1x_1 + a_2x_2}. \tag{3.4}$$

The recruitment terms now contain the effects of both intraspecific and interspecific competition, through the same factor. It is of course the values of the a parameters that fluctuate with the environment, and they are per capita recruitment rates per site in the absence of competition. In principle, one could measure them over the decade(s) it takes to mature an adult; certainly, it seems sensible to measure something like this for a year or two with test seedlings a healthy distance from seedlings of the same age as their own or the other species. Most seedlings die and with maturity taking a decade or more, the important thing will be when a sapling of one or the other species becomes sufficiently demanding on resources that no new seedling can do other than die in the larger individual's presence – seedling survival probability goes from vanishingly small to zero.

It is well accepted (or 'easily shown') that for species 1 to grow from a small population then $a_1 > a_2$ and conversely that for species 2 to grow from a small population then $a_2 > a_1$ (Pimentel *et al.* 1965, Pielou 1977, Turelli 1981). For coexistence to be possible, the ratio a_1/a_2 must fluctuate in response to environmental changes such as humidity level by sufficient amounts over suitable periods. The ratio goes up and down as humidity goes up and down – the interspecific competitive ratio covaries with the level of environmental humidity. In this context, we have essentially a precise definition of the meaning of covariance between competition and the environment.

Things may be less clear-cut in a seedbank model for densely packed annuals. In the seedbank model worked out in Kelly and Bowler (2009b), the condition was imposed that the sum of the populations of adults remained constant from year to year. The recruitment term for adults y was written as

$$y_i \propto \alpha_i E_i x_i, \tag{3.5}$$

where x_i is the number of seeds of species i in the seedbank, E_i is their germination fraction and α_i is a probability that a germinating seed produces an adult plant in the absence of competition. In the herbivore model, we envisaged herbivores attacking the seedlings but there are many possible variants. The coefficient of x_i was factored so that different aspects could be allocated to environmental fluctuations.

With annuals, it may be that competition is not adequately represented in terms of the y_i above. Therefore in the seedbank model of Kelly *et al.* (2005, 2006), the adults y_i (for fractional populations) produced in one season from x_i seeds is written

$$y_i = \frac{\alpha_i \beta_i E_i x_i}{\sum_j \alpha_j \beta_j E_j x_j}. \quad (3.6)$$

The idea is that the *a* parameters represent the survival probability that is affected by environmental fluctuations and were originally defined as measurable in the absence of competition. However, within a season all seedlings that are going to become adults, and the majority that are not, are all growing together and one species might indeed overshadow the other and so modify the survival probabilities, although it need not. It seemed convenient to separate out the isolated survival probability from the survival probability in the hurly-burly, and to this end, the β parameters were introduced. It would also be possible to use these parameters to contain explicitly the cost to a plant of manufacturing herbicide for models of genetically modified herbivore resistance (Bowler *et al.* this volume), but the separation alphabetically then begins to become a bit arbitrary.

For seedbank models, again a competitive ratio can be defined. It is the thing we have called an a^* parameter (Kelly *et al.* 2005, 2006, Kelly and Bowler 2009a, b), and it is this that must covary with the environmental conditions (like herbivore density), so that more of one species survives at high herbivore densities and more of the other at low herbivore densities.

Seeds, seedlings and competition

In the first half of this section, *a* is described as the *population*-level advantage of species 1 over species 2, calculated as the number of individuals of species 1 that make it to adulthood versus the number that do so of species 2 in the absence of competition. For annual plants, this relationship between species can be demonstrated observationally over time as relative abundance in a given year type, or experimentally as a treatment- or condition-dependent population response (e.g. Rice and Menke 1985, Pake and Venable 1995, 1996, Angert *et al.* 2009, Venable and Kimball this volume). Direct observation of covariance between competition and the environment is more problematic in perennial species, where population turnover is in the accretion of individual successes over long periods. While initial conditions will inevitably be important to seedling survival, it is to be expected that over the decade(s) long recruitment period, the particular sequence of year types will on occasion be an important factor in successful recruitment, and thereby able to obscure the critical character of any single year type.

In the original model of Kelly and Bowler (2002) it was assumed that in a fraction f of years, conditions are good and that in the complementary fraction $1-f$ environmental fluctuations prevent any recruitment on the part of species 1 (the more sensitive) but that in good years species 1 has a recruitment advantage a over species 2. Implemented in the lottery-model equations the fraction x_1 of species 1 is

$$x_1 = \frac{f - 1/a}{1 - 1/a}. \tag{3.7}$$

Within the model, the maximum possible value of x_1 is the fraction of good years f. Working from the fractional abundances reported in Kelly *et al.* (2008), Kelly and Bowler (2009a) found that in the relatively harsh conditions of Chamela, f varies around 0.7 and a varies around 2.0 (Figure 3.3), and that species 1 is the rarer niche partner, with x_1 varying around 0.4. From this, we have built an expectation of the relative fraction of the shared spatial niche occupied by species 1, against which can be compared the behavioural (competitive) differences that affect coexistence.

As a measure of population-level differences, a will have some relationship with individual growth rate, which is, after all, a signal of resource acquisition capacity and can be the advantage itself (Tilman 1988). A specific example of the latter is that of more rapidly growing through the period of seedling vulnerability to pathogens (Augspurger and Kelly 1984), but there are a number of threats to seedlings for which greater size is an advantage (e.g. Puertolas *et al.* 2003, Villar-Salvador *et al.* 2012). To determine the condition-dependent competitive effects necessary to create covariance between competition and the environment in this context, experiments, manipulations and observations need to determine the *relative* individual response of niche partners under differing conditions. This has been done for six species (three pairs) in the tree genus *Bursera*, a prominent taxon of the tropical deciduous forest of coastal México (Becerra 2005) that was previously implicated in differential sensitivity temporal dynamics (Kelly *et al.* 2001, Kelly and Bowler 2002, 2005, 2009a).

The genus *Bursera* is thought to have originated in México (Rzedowski 1978), paralleling the development of tropical dry forest (Becerra 2005). *Bursera* has over 80 species in the tropical dry forests of México, a number of which are of economic or cultural importance (Hernández-Apolinar *et al.* 2006, Mabberley 2008). Significant populations of six *Bursera* species, including one documented only recently (*B.* sp. *nova*) are found at Chamela. At Chamela, these six species form three pairs of most-closely related congeners as determined by the molecular phylogenies of Becerra (2003) and De-Nova *et al.* (2012), and distinctive floral and morphological characters of *B.* sp. *nova* according to Toledo (1982).

The six species occur in intermingled populations, but with each pair having its own distinctive pattern within the system of ridge tops, canyon shoulders and slopes that characterises the topology of the region. In the tropical dry forest, interannual (or longer term) variation in water availability is a major factor in plant response (Bullock and Solís-Magallanes 1990, Bullock *et al.* 1995) and the gradient of *Bursera* distribution follows a gradient of water availability and humidity. Conditions grow consistently drier with increasing elevation on the local scale (Segura *et al.* 2003), and decreasing species diversity and canopy density reflect this change (Kelly, unpublished data). Figure 3.4 shows proportional differences in distribution of species pairs among the three distinct components of the local canyon system; this differential distribution of pairs results in each pair being most numerous in a specific section of the habitat: *B. instabilis*/*B. arborea* in the more humid slope areas, *B. excelsa*/*B. heteresthes* along the more-exposed shoulders of the canyons and *B. fagaroides*/*B* sp. *nova* on the flat, dry ridge tops.

In early 2007, seeds of all six species were collected from trees naturally occurring within the limits of the 1600-ha Chamela reserve, and planted individually into 0.5 litre plastic grow-bags containing well-mixed local soil. The seeded bags were arranged within a growing shed open on two sides, with screens on the remaining two sides and roof. Beginning at the inception of the rainy season in mid-July of that year, the bags were watered with non-chlorinated water and from that point forward the soil in the bags was not allowed to dry out. Germination was recorded daily, and in early October, aboveground growth of all resulting seedlings was measured. Concurrent with the measurement of pot-grown seedlings, 30 seedlings of each of the six target species were collected from the field and similarly measured. Supplementary watering of the potted seedlings continued until mid-November of that year. In 2008, the supplementary watering regime

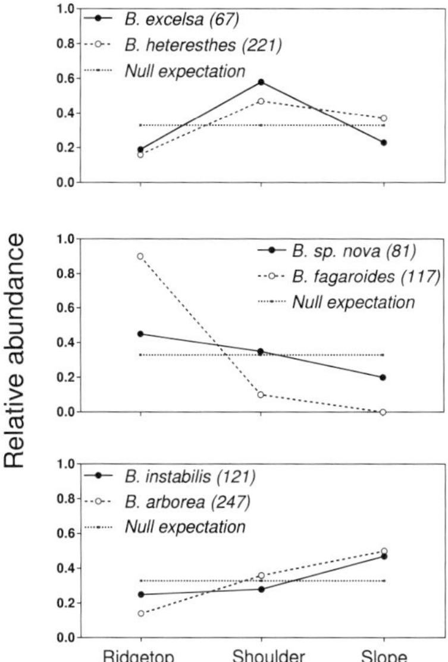

Figure 3.4 Relative distributions of species pairs within the 10-ha study area. In June 2008, all six target species were exhaustively censused over the study area; reproductive status and diameter at base were recorded for all individuals. Shown here are data including all individuals larger than the minimum size of reproduction occurring in the three principal habitat types within the sampled area.

was reinstituted in May, with consequent vegetative budburst and leaf development assessed on a regular basis.

The primary result of this experiment was that degree of difference in growth for niche partners in the field is significantly smaller than that under the well-watered regime in the growing shed (Table 3.2), signifying covariance between competition and the environment (*sensu* Chesson and Huntly 1997). This difference in relative growth rate is an indicator of competitive advantage, much like the parameter a defined earlier; a faster growing seedling is vulnerable to the habitat for a shorter period and more likely to reach a stage at which it can pre-empt a site. Under well-watered greenhouse conditions, the rarer species showed two to three times the aboveground final volume of the common species. It could be useful to determine how this difference relates to the maximum competitive advantage (probability of establishment) of 2–3 assigned to the more sensitive species in the model of Kelly and Bowler (2009a).

Comparing the six *Bursera* species as three pairs of most-closely related congeners, it is evident that an important component of the

Table 3.2 ***Comparison of species traits***. *The rarer species, i.e. that predicted to be the more competitive in a pair, is highlighted in bold for all three pairs. Germination and budburst responses were modelled with accelerated failure time analyses, with Akaike Information Criterion (AIC) showing a Weibull distribution as the best model for germination and a logistic distribution for budburst. Coefficients describe the time-to-event (hence acceleration) for each species, relative to a baseline (here taken to be* Bursera arborea). *More negative coefficients mean quicker responses. The phylogenetic tree of the species group follows the molecular phylogenies of (Becerra 2003, De-Nova* et al. *2012) and is shown in the left-most column. Branch lengths in the figure do not represent branch lengths calculated from base-pair change of neutral gene sequences (see Williams and Kelly 2013 for a discussion on the use of such branch lengths in ecological inference).* B. *sp.* nova *is designated within the* fagaroides *clade based principally on floral morphology but fruit, seed and cotyledon morphology are all consistent with this designation.*

Phylogeny	Species	Abundance (mature trees)	Germination response coefficient	Leaf budburst response coefficient	Growth ($mm^3\ d^{-1}$) field/ +water
	B. excelsa	67	−0.046	−0.127	5.9 / 12.8
	B. heteresthes	225	0.148	0.384	3.1 / 4.2
	B. instabilis	164	−0.141	−1.658	3.2 / 21.1
	B. arborea	287	0	0	3.0 / 9.1
	B. sp. *nova*	81	0.030	−1.658	2.2 / 9.5
	B. fagaroides	111	0.366	0.020	1.5 / 4.8

growth/survival tradeoff is timing. In all pairs the rarer species more quickly released leaf buds from dormancy in the second year of growth and germinated sooner than did the more common species (Figure 3.5; Table 3.2). Both of these responses have the capacity to contribute to greater early growth in their respective seasons (Augspurger 2008, Firn *et al.* 2010), and even calculating growth as a per day response, there is a correspondence between germination date and first season growth. However, the looseness of the relationship between germination date and growth suggests the possibility of innate growth rate differences independent of the advantages of early germination and budburst. Overall, the characterisation of a growth/tolerance tradeoff is robust: the chance that the direction of difference should be as predicted for all three pairs is significant for either two or three

(a)

(c)

Figure 3.5 Differential budburst in congeneric pairs of *Bursera*. All plants were grown in well-mixed and sieved local soil in 0.5 litre bags. The photographs illustrate the difference between sensitive/resistant

ecologically independent responses ($p = 0.016$ for two traits and 0.002 for three; Siegal and Castellan 1988). None of the responses revealed significant correlation with relative abundance when species were treated independently (i.e. without accounting for their phylogenetic relationships; p-values from 0.46 to 0.89; regarding competitive ability and community structure, see also Fraser and Keddy (2005)).

Finally, the observed responses characterise the tradeoff as beyond a simple greater and lesser growth capacity; early germination, early bud-break and faster growth all involve or are correlated with risk. Both early germination and bud-break entail greater exposure of small and vulnerable life stages to the less reliable rains of the early wet season in this habitat, and the greater size that early response can achieve may leave a seedling more visible and less protected from macro-herbivores (Rafferty *et al.* 2010). In effect, the conditions describe disruptive selection ecologically dividing early- from late-responding species.

3.4 SPECIATION

Sympatric speciation via temporal dynamics?

It has been suggested that temporal dynamics, because of its capacity to support population growth from low levels, may also support sympatric speciation (Chesson and Huntly 1997, Kelly and Bowler 2002). However, effective ecological speciation also requires reinforcement, a connection between the ecological factor and reproductive isolation (Dieckmann and Doebeli 1999, Via 2002, Bolnick and Fitzpatrick 2007, Givnish 2010, Hopkins and Rausher 2012). In this section, we will first examine whether the alternating recruitment of temporal dynamics might achieve that end. Second, we present evidence that the differential sensitivity that allows the stable coexistence of temporal dynamics also affects the divergence of

Caption for figure 3.5 (cont.) congeneric pairs in budburst and subsequent early season growth; sample-level differences are in Table 3.2. **(a)** The two plants on the left are *B. excelsa* (sensitive); the two on the right are *B. heteresthes* (resistant). **(b)** The two plants on the left are *B. instabilis* (sensitive); the two on the right are *B. arborea* (resistant). **(c)** The six plants on the left are *B.* sp. *nova* (sensitive); the six plants on the right are *B. fagaroides* (resistant). Differences in budburst were such that for all three pairs, budburst had occurred in all sensitive individuals of the paired species before leaf buds on any individuals of the resistant species began to reveal developing green tissue.

floral phenology in temporal niche partners, potentially fulfilling the requirement of sympatric speciation for reinforcement.

What is going on in temporal dynamics can be explained accurately without going into mathematical detail. In the Kelly and Bowler (2002) lottery model for closely related forest trees, two competing species can coexist in dynamical equilibrium over a significant range of parameters because of differing responses to environmental fluctuations. In that model, species 2 is not affected by environmental fluctuations as badly as species 1; during good conditions species 1 has a competitive advantage over species 2 denoted by a parameter a^* greater than 1 (typically maybe 2 or 3; compare with growth rates in Table 3.1). In bad times (perhaps a third of seasons) species 1 cannot recruit at all and so species 2 always wins the lottery. Either of the species can grow from a very low level, and so if one were an emergent new variety, speciation could apparently take place. Hence, such a temporal dynamic seems a likely way of getting sympatric speciation.

We modified the equations for the two species to include a hybrid with diploid rules for making seeds (see discussion of the diploid model by Bowler *et al.* (this volume)). The most important feature is that species 1 can pollinate species 2 and species 2 can pollinate species 1. A stamen can only be pollinated once. This has two effects: (1) if species 1 and 2 are in a stable dynamical equilibrium, there is a steady input of hybrid (species 3) seeds. If some proportion of these grows to mature adults, there will always be hybrids present in the community. (2) If pollen from species 2 floods the stamens of species 1, the terms of the lottery competition are changed and the conditions for dynamical equilibrium are altered. This makes it difficult or impossible for the population of species 1 in particular to grow from an arbitrary low level. We return to this after setting out the model in more detail.

Suppose we have a dynamic equilibrium. The relative rate at which species 1 fertilises species 1 will be x_1x_1 and the rate at which 2 fertilises 2 will be x_2x_2. The rate at which hybrid seeds are formed is on the same scale, $e_{12}x_1x_2 + e_{21}x_2x_1$, where the coefficients e_{12}, e_{21} are not necessarily the same. If these coefficients are ~ 0.1, then very roughly about 10% of the seeds produced each season will be hybrid. The smaller the hybridisation coefficients are, the smaller the proportion of hybrid seeds produced.

There is another factor important for determining the proportion of adult hybrids and that is the probability of a hybrid seed being successful in producing an adult. As with species 1 we scale to the success of species 2; for species 1 we have the competitive advantage

Table 3.3 ***Equilibrium population fractions.*** *Here we have assumed that conditions were good 66% of the time, that species 1 had an advantage of 3 (over species 2) during good times and that both crossbreeding coefficients for 1 with 2 were 0.05. The 1–3 and 2–3 crossbreeding coefficients were both taken as 0.1 but are less important, it seems. Under good conditions the hybrid, species 3, has an advantage over species 2 of* a_3^* (good) *and under bad conditions of* a_3^* (bad). *Advantages less than 1 are disadvantageous.* x_i *is the fraction of species* i *in the population and shown here is the expected fraction of each type under the specified relationships between species 2 and the hybrid species 3.*

a_3^* (good)	a_3^* (bad)	x_1	x_2	x_3
1.5	0.5	~ 0.45	~ 0.45	~ 0.08
1.0	0.0	~ 0.49	~ 0.49	~ 0.02

of species 1 over species 2, a_1^*, with a value of 2 or 3 in good times and 0 in bad times. For species 3 we introduce a_3^*, which will also have different values in good and bad times; the smaller the average value, the smaller the fraction of hybrids in equilibrium. Table 3.3 gives the results of two calculations made by iterating the full set of equations.

There are many parameters, but it can be concluded at this point that it is possible to have a stable equilibrium and drive the hybrid population as low as wanted by reducing ever further the hybridisation coefficients. Values as low as 0.05 might just be reached by the species 1 and 2 flowering at different times, so that there is not much 1 pollen around when 2 can be fertilised (and vice versa). As an example, distributions of flowers over time were assumed to be normal and pollination was assumed to take place very quickly on the flowering timescale. From this, hybridisation at the level of 0.05 would result if the peaks of the two normal distributions were separated by 3.4 standard deviations. The other possibility is that the pollen of species 2 is in some way incompatible with the stamens of species 1, but we do not consider that here.

We now return to the question of competitive pollination. At least within this very simple model, competitive pollination has the effect of making it impossible for species 1 (the one that does well in good times and cannot recruit in bad times) to grow from an arbitrarily low level. The level above which the population of species 1 will grow to the stable equilibrium depends on the value of the coefficient for pollination of 1 by 2; very roughly, a coefficient of 0.05 implies that species 1 will not grow if started below 0.05 of the total population. The same is not true for species 2 being swamped by species 1. The reason is that in the original simple model there are periods (bad times) during which species 1 cannot recruit at all, and species 2 has no competition. However, if recruitment for species 1 is not completely wiped out in bad times, there is competition. If speciation proceeds by the local appearance of a mutation, then a mutation in species 1 that generates species 2 would probably suffer less from pollen swamping than a mutation of species 2 that generates species 1. Even so, a local mutation is likely to be suppressed by pollination competition, even if the problem of starting with a male tree and a female tree carrying the same mutation could be overcome.

If the closely related species do not have any serious level of hybrids, that could be due to assortative mating resulting from differing flowering times (Fox 2003, Weis 2005, Devaux and Lande 2009). Such assortative mating can lead to a broad distribution of genotypes splitting into two major types as a result of the extreme ends of the distribution responding differently to environmental conditions, with some difference in fitness linked to time of flowering. The nearest to it within the confines of our simple model is to suppose that initially there is a population whose broad distribution of genotypes can be represented by fairly equal populations of species 1, species 2 and the hybrid. Initially they happily cross-pollinate and no variety has any fitness advantage.

The dynamical equations then simply leave the proportions of each where they started out (perhaps 0.25, 0.25, 0.5). Strictly speaking this is a three-variety neutrality and for strictly separate varieties stochastic drift would eventually lead to dominance either of one or of two – it seems unlikely that this proxy for a broad distribution of genotypes can be anywhere near a good representation.

Now change the environment. Changes might be spatial only or temporal only or of course both. Spatially, if species 2 is disadvantaged relative to 1 all the time, then species 1 will take over, and the original broad distribution of genotypes will be replaced by a subset of those genotypes from one end or the other of the spectrum of trait expression. Similarly, if species 1 is disadvantaged relative to species 2, then species 2 will take over. Either thing could happen because of geological or climatic change. In that case, spatial separation of the two varieties 1 and 2 would occur and if the robustness trait were linked to flowering phenology, then two new species could emerge from the original, distinguished by elevation perhaps.

If the environment changed on a purely temporal basis, with species 1 recruiting much better than species 2 in good years but with higher mortality relative to the more robust species 2 in bad years then, again provided the flowering phenology trait were linked to the robustness characteristic, the hybrid would be depressed and a new equilibrium reached on a timescale of perhaps 10 mean tree lifetimes (calculated by starting from the broad band end and suddenly switching on the fluctuating climate pattern.) The hybrid level would be determined by just how low the hybridisation coefficient became. This would seem to fill the definition of sympatric speciation.

In conclusion, we have cast a harsh light on the idea of sympatric speciation being possible simply through either of two species being able to grow from a low level. On the other hand, a broad distribution of genotypes can split because of climate change and this response might be because of spatial variation or temporal variation. In either case, a link between robustness and flowering phenology is required.

Differential sensitivity and floral phenology

We propose that this necessary link between robustness and floral phenology is innate to the differential sensitivity of storage dynamics observed between congeneric species, either through the common control of timing among seasonal plant behaviours (Brachi *et al.* 2010, Yanez-Espinosa *et al.* 2010, Huff and Zilberman 2012), or working through the greater resource availability that sensitivity is expected to impart. Figure 3.6 shows that, as with germination and vegetative budburst, the more competitive species 1 is also the first to flower in all three congeneric pairs. Change in multiple responses because of a single, controlling change in a regulatory function has the potential to affect response to seasonality while at the same time supporting the observed tolerance–sensitivity tradeoff. For example, selection for the enhanced growth that can be gained with early leaf-out in deciduous forests (Augspurger 2008, Lopez *et al.* 2008) would also bring more rapid germination and early flowering as part of a package. At the same time, early germination exposes vulnerable, newly emerging seedlings to the riskier conditions of season changeover. Together, this common change in the character of scheduling would produce a bimodal selective regime of higher risk and greater payoff versus lower payoff but less risk in the context of interannual variation in rainfall in the tropical deciduous forest or late freezes in temperate forests.

The differences in the timing of flowering in *Bursera* are such that there exists a period during which the bulk of the floral crop of one niche partner is unavailable to the pollinators of the other.

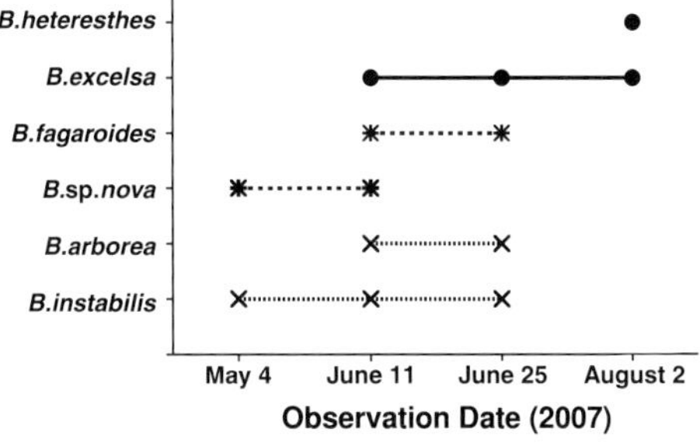

Figure 3.6 *Bursera* flowering phenology. Within all three pairs, the species with the more sensitive seedlings (greater speed of germination, higher growth rate and earlier second year budburst; see Table 3.2) is also the first species to begin flowering. *Bursera* flowering phenology follows a general schedule of an initial large burst of simultaneously open flowers lasting ~1–2 weeks, followed immediately by fruit expansion to approximately final size. Filling of the previously enlarged fruit is initiated in late September and early October, with staggered fruit maturity beginning at some point after the end of the rainy season in mid to late December.

Bursera species seem well able to hybridise (Weeks and Simpson 2004), and *Bursera* at Chamela have a number of common pollinators (P. Lopera Blair, unpublished data). In consequence, the inferred functionality of the observed floral asynchrony existing at Chamela is supported by the general lack of evident hybrids within the exhaustively censused study area.

While we believe this to be a viable and possibly general route for *in-situ* (sympatric) ecological speciation, a general link between flowering phenology and the timing of other vital rates such as germination and budburst would also facilitate the effectiveness of allopatric speciation. A link between ecological and phenological change would maintain reproductive isolation between populations when community assemblages and physical isolation are interrupted by Milankovitch-scale environmental fluctuations (Davis 1986, Schluter 2001).

One proposed criterion of sympatric speciation is that of co-occurrence of sister species (Coyne and Orr 2004). In an early molecular phylogeny of *Bursera*, only one of the Chamela pairs (*B. instabilis*/*B. arborea*) sorted out as sister species (Becerra 2003). However, Bolnick and Fitzpatrick (2007) point out that current distributions of species can give either false negatives or false positives regarding mode of speciation because of subsequent speciation and the plastic character of distributions.

In the previous part of this section, we concluded that the direction of speciation from a temporal dynamic is more likely to generate a more tolerant rather than more sensitive form relative to the parent species. In the following section, examination of changing community structure along a landscape-level gradient of temperature and water reveals distributions of congeneric groups across a number of lineages consistent with this conclusion.

3.5 LANDSCAPE-LEVEL PATTERNS

Fractional abundance and coexistence across an elevation gradient

Among pairs of closely related congeneric species that coexist as intermingled adults at Chamela, recruitment appears to be more strongly affected by environmental fluctuations in the rarer of the two species than in the more common. In every case, the rarer species shows evidence of compensating for this by recruitment advantage when conditions are favourable (Kelly and Bowler 2002, 2005). The lowland coastal site of the original community-based studies, Chamela Biological Station, has the lowest rainfall and highest temperature in the deciduous dry forest that stretches down the Pacific Coast of México from Jalisco to Oaxaca; the major geological uplift from the coastal areas of this region is also accompanied by lower temperatures and greater rainfall (Williams *et al.* 2010, Williams and Kelly 2013). If the lottery model of temporal dynamics is applicable at all elevations in this extensive forest, the more sensitive species of a pair must become more dominant as elevation increases from the coastal lowlands. This can be accommodated within the lottery model, as follows.

In the original model (Section 3.3) the fraction x_1 of species 1 is

$$x_1 = \frac{f - 1/a}{1 - 1/a}. \tag{3.8}$$

Under the harsh conditions of Chamela, Kelly and Bowler (2009a) found that $f \approx 0.7$ and $a \approx 2.0$. Under these conditions, species 1 is the rarer, with $x_1 \approx 0.4$. At higher elevations, species 1 would become commoner if either the fraction of good years f grows or the competitive advantage of species 1 over species 2 grows. Note, however, that the maximum possible value of x_1 is the fraction of good years f.

Suppose we assume that f is in fact high, and we vary the competitive advantage of species 1, so that a decreases as we go to lower elevations. Figure 3.7 shows that if f is sufficiently high to have species 1 overwhelmingly dominant at the most favourable elevations, it remains the commoner until its competitive advantage in good times is $a \approx 1.1$ and it

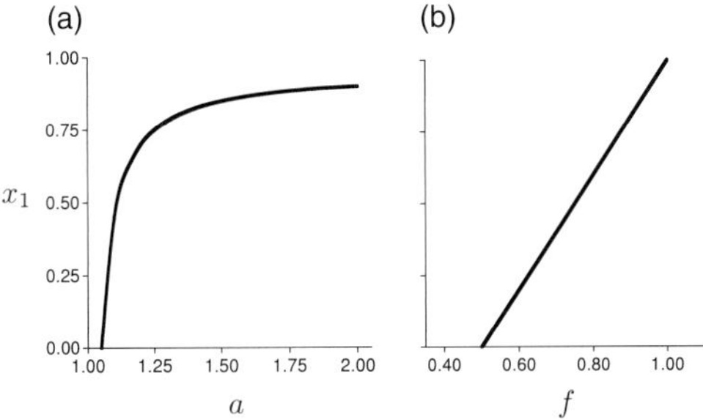

Figure 3.7 Factors affecting the proportion of the sensitive species, x_1, in the original temporal storage model of Kelly and Bowler (2002). **(a)** The effect of changing relative competitive ability of the sensitive species during high resource years, a^+; the frequency of good years, f, is held constant at 0.095. **(b)** The relative competitive ability of the more sensitive species (a^+) is held constant at 2.0 while f, the relative frequency of high resource years, is allowed to vary.

goes extinct for only a marginally smaller advantage. Roughly speaking, it is either dominant or it is not present. This is not impossible but it is implausible, being inconsistent with the pattern at Chamela of the faster growing species being associated with the lower population size.

There is another way to have species 1 dominant, even overwhelmingly so, in the balmy uplands, while being the rarer species close to sea level, as seen by Williams and Kelly (2013). Take the advantage parameter, a, as fixed and vary the fraction of good years, f. At higher elevations, f can be equal to 1 and decrease as we go downhill. This relationship is shown in Figure 3.7.

This is reasonable enough, in that species 1 is as dominant as desired at high elevations and has become the rarer as the fraction of good years f drops with descending altitude to a value of about 0.7. We would suppose that is about Chamela altitude. The uncomfortable thing about it is that the fraction of good years decreases at lower elevations (in Figure 3.7 the advantage factor is independent of elevation at a value of 2). If the advantage factor a grew with increasing harshness, then the fraction x_1 would drop less rapidly with f.

There is in fact a way of varying x_1 over the full range from 1 at high elevations to values typical of Chamela without varying either the fraction of good years f or the advantage of species 1 over species 2 during good years. In the original model, it was supposed that during bad years species 1 could not recruit at all, so that the advantage parameter in bad years is zero. If this assumption is relaxed so that in bad years species 1 is

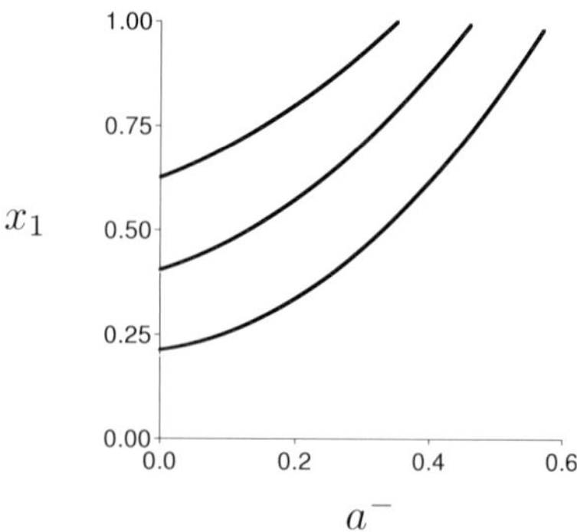

Figure 3.8 Changing relative abundance in the complex lottery model. The complex lottery model is applied to examine variation in proportional abundance of the more sensitive species 1 (x_1) as a response to variation in the negative impact of bad years on the more sensitive species (a^-). This response is placed in the context of specified frequency of good years, f, and the relative competitive advantage of species 1 over the tolerant species 2 during good years, a^+; top curve: $f = 0.7$, $a^+ = 5$; centre curve: $f = 0.7$, $a^+ = 2$; bottom curve: $f = 0.3$, $a^+ = 5$.

at a competitive disadvantage (that is, the parameter $a < 1$) then the full range is covered for f considerably less than 1. In this development of the original model, the analytic solution is more complicated than the solution for the original lottery model solved in Kelly and Bowler (2005).

The extended two-component lottery model has an analytic solution valid under the same conditions as the original. The model differs from the original only in that the more sensitive species is able to recruit during bad times. The fraction of good times is f and during these times, the more sensitive species has the advantage a^+ over the slow but steady species. In bad times, the advantage factor is a^- (less than 1). Then the population fraction x_1 of the sensitive species is

$$x_1 = \frac{(1 - a^-/a^+)f - (1/a^+ - a^-/a^+)}{(a^+ - 1)\{1/a^+ - a^-/a^+\}}. \tag{3.9}$$

Figure 3.8 serves as an illustration of this more complex model. These results show that it is perfectly possible to have the same time structure in environmental conditions at all elevations, maintain the Chamela results at Chamela elevations, and yet have the more sensitive species as dominant as desired at the highest (most benign) elevations, even though in hard times it is at a significant disadvantage relative to the more tolerant species.

Some mixture of these effects here treated individually could be constructed to match exactly output with observation, but the goal here is general agreement with expectations, with further development of the theory in a more appropriate treatment dedicated to it specifically. The essential point is that the time-structured lottery model can account for the more competitive of a pair of congeneric species being very dominant high up where the environment is less water stressed, and yet the rarer at sea level, the hottest and driest environment on a larger scale.

Observations

The first half of this section has described changes in the temporal niche dynamic to be expected from changes in the frequency f of good years. As f increases, the proportion of the more competitive (more sensitive, faster growing) niche partner should increase, eventually to the point that the more tolerant species is unable to sustain a population. One outcome of this could be a decreased number of interdigitated congeneric species as the environment becomes increasingly benign. This is a pattern that occurs within the local canyon system at Chamela: in the shadier, more humid lower reaches of canyons within the reserve, *Bursera* is represented principally by *B. instabilis* and *B. arborea*; in the more open conditions of the canyon shoulders, these two species are joined in force by *B. excelsa* and *B. heteresthes*, and on the exposed, low-statured forest of the hot, dry ridge tops (harsher on a local scale), *B. fagaroides* and *B.* sp. *nova* bring the numbers of *Bursera* species to six (Figure 3.4).

The above expectation from temporal niche dynamics of condition-dependent changes in the representation of species within genera is also manifest on the landscape scale. Here we draw on an existing study of tree community composition along a 600 m elevation gradient in the tropical deciduous forest of the Mexican state of Oaxaca (Williams *et al.* 2010). The study assayed 4300 woody individuals in 43 transects (100 individuals/transect) scattered over an area of approximately 1800 km^2. Transect sites were selected using criteria of similar aspect and stand type (undisturbed mature forest). Travelling up this elevation span, temperature decreases by 3°C and rainfall increases from ~1000 mm y^{-1} to ~3000 mm y^{-1} (Figure 3.9), significantly decreasing the level of evapotranspirational demand in this typically hot, dry biome type (Olson *et al.* 2001). This degree of change is also approximately that projected by climate change models for the region (Anonymous 2011).

Analyses show that the total number of species, genera and families did not change significantly across the elevation/environment

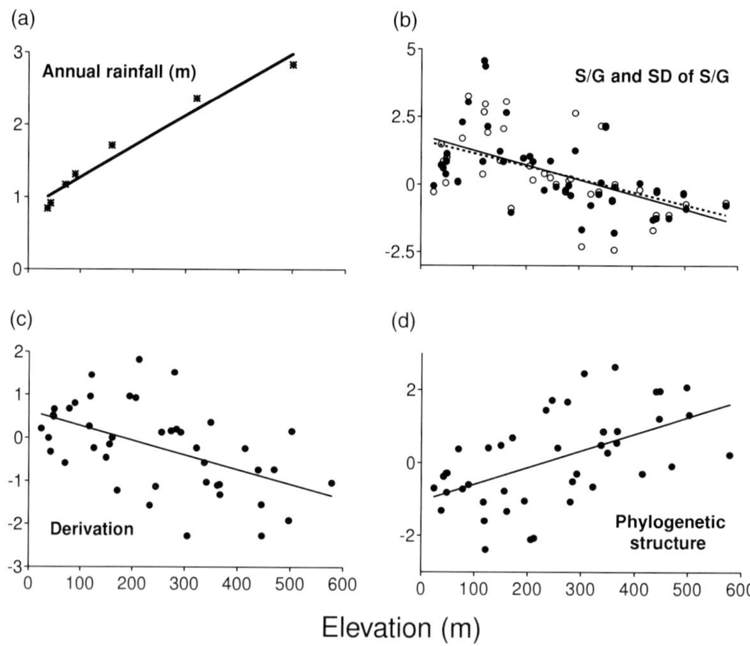

Figure 3.9 Oaxaca gradient analyses. Average annual rainfall relative to the elevation gradient **(a)** was calculated from raw data of seven weather stations occurring within the 1800-ha study area, and made available by WorldClim (Hijmans *et al.* 2005). Species per genus (S/G) and standard deviation (SD) of S/G (**b**; S/G = filled circles, SD of S/G = open circles) both showed significant negative relationships with elevation when tested independently (S/G: $p < 0.0001$, $r^2 = 0.33$; SD of S/G: $p = 0.0005$, $r^2 = 0.27$; S/G values shown are corrected for possible effect of sample size (Gotelli and Colwell 2001). When entered into a general linear model, S/G and SD of S/G together are able to account for all of the significant variation in the relationship between phylogenetic structure **(d)** and elevation (independent regression analysis shows elevation accounting for 30% of the variation in the commonly used measure of phylogenetic structure applied here (see text). Elevation accounts for 25% of the variation in the degree of species derivation among samples (**c**; derivation = species distance from community phylogeny base, summed across species in sample), but S/G accounts for all of that relationship, plus an additional independently significant 20% of variation among samples in species derivation ($p = 0.0011$).

gradient. However, the phylogenetic structure (NRI; Webb 2000) of the community did, in terms of the distribution of species among genera (S/G; Williams and Kelly 2013). In the relatively benign conditions at high elevation, species were quite evenly distributed among genera, with only one or occasionally two species per genus. With the increasingly harsh environment at lower elevations, species became significantly more unequally distributed among genera. In these locations, most genera have only one species, with a few reaching four to five or as many as six congeners in a single transect. Furthermore, the changes in S/G and the standard deviation of S/G account for most or all of the response of phylogenetic structure to the elevation/environment gradient (Figure 3.9).

The documented changes are also consistent with the potential capacity of temporal dynamics to support speciation that is outlined in the previous section of this review. Lower elevation transects comprise significantly more derived species as, within clades, the average distance of species from the root of the phylogenetic tree becomes greater with decreasing elevation (Figure 3.9; Williams and Kelly unpublished manuscript). This is consistent with the classic ecological expectation that more benign habitats allow more inclusive species rosters and that species require more specialised functions to cope with harsher conditions (e.g. Woodward and Diament 1991, Kleidon and Mooney 2000). In the analyses of Williams and Kelly, the increase in degree of derivation is fully accounted for statistically by the changing distribution of species among the major genera (genera with $\geq$4 species) with elevation. Thus, the prediction of greater specialisation with greater environmental harshness is framed by a system of adaptive radiations within a subset of lineages (Losos 2010). The observed changes indicate multiple adaptive radiations from mesic to dry adapted species, a pattern consistent with the uplift event 5 million years ago that led to the transformation of a previously mesic habitat into the Mexican dry forest of today (Becerra 2005).

In conclusion, if the condition-dependent competition of temporal dynamics is responsible for the restriction of low elevation species to lower elevations, then these low elevation forests may serve as reservoirs of species redundancy. With the expected climate change effect of hotter, dryer conditions in this area (Anonymous 2011), species that do not currently exist at higher elevation may migrate upward to fill the niches vacated by less drought-adapted species as climate changes (Alexander *et al.* 2011, Schwinning *et al.* this volume). Therefore, maintaining dry forest diversity at low elevations may be especially critical in the face of climate change. This is especially important since these are also the areas currently most affected by commercial development. Evolutionarily, if low elevation forests are harbours for the multiple adaptive radiations that our evidence thus far suggests, then they will also be reservoirs for unique genes and gene combinations, a second, but not secondary, reason for urgency in understanding and conserving these forests (Vane-Wright *et al.* 1991, Faith 1992, Magnuson-Ford *et al.* 2010).

3.6 LARGER CONTEXT AND DISCUSSION

Our attention so far has been on the action of temporal niche dynamics with regard to congeneric species, but temporal niche dynamics in perennials may also occur between unrelated species (see Adler this volume, Chesson *et al.* this volume). As a specific example for trees, working with coexisting populations of native *Fraxinus excelsior* (ash) and non-native *Acer pseudoplatanus* (sycamore) in the UK, Ellis *et al.* (unpublished) were able to use seedling age distributions combined with well-kept local weather records to link recruitment of each species to environmental factors (see also Mason 2005, Ellis 2007). Ellis *et al.* found that recent ash recruitment correlates with the sunny summers needed for seed production, and that best sycamore recruitment occurs in years where early spring rainfall supports seedling survival. The functionality of these correlations was shown by earlier experimental studies on seed and seedling ecology of the two species (Helliwell 1965, Dethioux 1970, Waters and Savill 1992, Tapper 1996, Binggeli and Blackstock 1999). The relationship between climate and first stage recruitment was then used to assess potential coexistence response to locally projected climate change; the analysis predicted that in future the dynamic might be expected to reverse, with the native ash favoured over the invading sycamore.

Each of these species is a large component of British woodland, and the specific findings show how an understanding of temporal dynamics might be useful for management purposes. However, this study illustrates a difference in the character of temporal dynamics in congeneric versus non-congeneric niche partners. The evolutionary source of temporal dynamics between congeneric species dictates that niche partners usually differentiate temporally along a single axis, with anticorrelation of effect: favouring one partner disfavours the other. Niche partners are more and less responsive to the stimulus, segregating out vital functions (germination, budburst) that expose the individual to disruptive risk/growth tradeoffs. In contrast, the ecological accident that produces non-congeneric niche partners is more likely to result in an uncorrelated, multi-axis regulation of the interaction, affecting different requirements or parts of the life cycles of each of the species involved.

Interannual variation in the timing and amount of rainfall is an important factor in temporal dynamics and tree coexistence, but certainly not the only one in either the tropical or the temperate zones (*contra* Penuelas *et al.* 2011). Temperature variation can affect successful recruitment through interaction with water availability, or through the direct action of cooling or freezing damage to seedlings or small plants. At any scale, warmer or cooler years will produce a different evapotranspirational environment even with the same schedule of water availability, thereby affecting the growth responses and mortality that support stable coexistence (Williams and Kelly 2013). As a primary rather than a companion factor, interannual variation in temperate zone spring warming schedules is also implicated in a risk/growth tradeoff, with seedling frost damage producing disruptive selection parallel to that of drought in more water-restricted environments (Kelly *et al.* 2003). Further evidence of temperature-based disruptive selection in trees can be found in Jump *et al.* (2006) and Augspurger (2008, 2009).

Pests and pathogens may also support temporal dynamics in trees. Damping off of tree seedlings is fairly ubiquitous and variable in space and time (Augspurger 1984, Augspurger and Kelly 1984, Packer and Clay 2000, 2003, Reinhart *et al.* 2003), with differential pathogenicity and susceptibility sufficient for the necessary resistant/non-resistant dynamic between competing tree species (Augspurger and Wilkinson 2007). Herbivores can support temporal dynamics in herbaceous plants (Hanley *et al.* 1995, 1996, Hanley and Sykes this volume), but folivory may be a less common factor in temporal dynamics between tree species. In the seedlings of many tree species, herbivore damage is very low (Eichhorn *et al.* 2006); sparsely dispersed tree seedlings, as they are for many tree species, will not represent an effective resource to the herbivore.

Nonetheless, herbivory on mature/large trees has notable interannual variation (Filip *et al.* 1995), and an asynchronous schedule of attack on congeners might be able to support a temporal dynamic through differentially disrupting seed set. In the tropical dry forest, folivores are sometimes able to fully denude mature trees. Although this is restricted to a subset of species (Kelly, pers. obs.), intense herbivory of this sort may be able to affect relative recruitment through limiting viable seed production. A pattern consistent with this may be found in Becerra (2007), in which at the very local scale congeners of *Bursera* are both more closely related and more divergent in their defence chemistry, than expected by chance. On the other hand, it may be that a temporal dynamic resulting from interannual variation in rain or temperature has allowed the development of divergent defensive chemicals in co-occurring congeners. However, there is no reason that herbivory and variable climate cannot both play a role in ongoing species coexistence, and it would be surprising if they did not. A factor to consider in the development of such a dynamic is the reproductive isolation innate to the resistant/responsive dichotomy of tree species response to ambient conditions, a factor that would facilitate either allopatric or sympatric speciation. In so far as is currently known, chemical defence strategies of trees do not have any such innate capacity to simultaneously foster reproductive isolation (but see Agrawal *et al.* 2012 for a response to experimental escape from herbivory in an herbaceous species).

Finally, the action of temporal niche processes in species coexistence has general implications for assessing competition in trees and other long-lived species. Fundamental to this is the temporal process as it occurs in nature. Year-type specialisation at the seedling level means that interspecific competition for a site will not be between same age seedlings. The degree of asymmetry *sensu* Schwinning and Fox (1995) and Schwinning and Weiner (1998) and potential pre-emption of one seedling by the other will initially depend on the year type of the primary coloniser, and the sequence of subsequent years before a second, non-conspecific arrives; alternatively, the sequence of year types may switch the asymmetry between the competitors. Either way, the advantage of one species over the other is *condition dependent*, and the crucial measure to establish this is the comparison between the two species of the relative response of each to resource enhancement (or deprivation; section 3.3). Information of this sort can be used in simulation and vital rate models to project population-level change or stability relative to climate variables, as illustrated in several chapters in this volume).

The idea of condition-dependent advantage has a long history in ecology with regard to the divergent spatial distribution of species; the idea that it might also contribute to species coexistence through temporal dynamics is less well integrated into the ecological workbook. A consequence of this is that the components of such temporal dynamics – condition-dependent advantage, asymmetric competition, seedling dynamics and pre-emption – have not been incorporated effectively into many commonly used experimental techniques or statistical analyses aimed at understanding the determinants of coexistence or the lack of it (e.g. Armas *et al.* 2004, Cahill *et al.* 2008; for a counter-example see Daehler 2003). The central importance of this is that if temporal processes are not appropriately accounted for they are quite likely to be misidentified even when a major factor. Moreover, the presence of unidentified temporal processes may confound effects of non-temporal processes. Neither outcome is desirable – even if the world were not currently confronting rapid environmental change.

ACKNOWLEDGEMENTS

CKK thanks Chamela Biological Station and Merton College each for many years of hospitality and support. A large portion of the results described in this paper were obtained through grants from the US National Science Foundation (DEB-0614468 to GAF and CKK, and DEB-1120330 to GAF) and the National Geographic Society (CKK, GAF, ASM).

REFERENCES

Agrawal, A. A., Hastings, A. P., Johnson, M. T. J., Maron, J. L. and Salminen, J. P. (2012). Insect herbivores drive real-time ecological and evolutionary change in plant populations. *Science* **338**, 113–116.

Alexander, J. M., Kueffer, C., Daehler, C. C. *et al.* (2011). Assembly of nonnative floras along elevational gradients explained by directional ecological filtering. *Proceedings of the National Academy of Sciences, USA* **108**, 656–661.

Angert, A. L., Huxman, T. E., Chesson, P. and Venable, D. L. (2009). Functional tradeoffs determine species coexistence via the storage effect. *Proceedings of the National Academy of Sciences, USA* **106**, 11641–11645.

Anonymous (2011). Climate: observations, projections and impacts. Mexico. Exeter, UK: UK Met Office Hadley Centre.

Armas, C., Ordiales, R. and Pugnaire, F. I. (2004). Measuring plant interactions: a new comparative index. *Ecology* **85**, 2682–2686.

Armstrong, R. A. and McGehee, R. (1980). Competitive exclusion. *American Naturalist* **115**, 151–170.

Augspurger, C. K. (1984). Seedling survival of tropical tree species: interactions of dispersal distance, light-gaps, and pathogens. *Ecology* **65**, 1705–1712.

Augspurger, C. K. (2008). Early spring leaf out enhances growth and survival of saplings in a temperate deciduous forest. *Oecologia* **156**, 281–286.

Augspurger, C. K. (2009). Spring 2007 warmth and frost: phenology, damage and refoliation in a temperate deciduous forest. *Functional Ecology* **23**, 1031–1039.

Augspurger, C. K. and Kelly, C. K. (1984). Pathogen mortality of tropical tree seedlings: experimental studies of the effects of dispersal distance, seedling density, and light conditions. *Oecologia* **61**, 211–217.

Augspurger, C. K. and Wilkinson, H. T. (2007). Host specificity of pathogenic *Pythium* species: Implications for tree species diversity. *Biotropica* **39**, 702–708.

Ayala, R. (2007) Datos climaticos estación meteorológica Chamela, Available at http://www.ibiologia.unam.mx/ebchamela/www/clima.html.

Becerra, J. X. (2003). Evolution of the Mexican *Bursera* (Burseraceae) inferred from ITS, ETS and 5S ribosomal DNA sequences. *Molecular Phylogenetics and Evolution* **26**, 300–309.

Becerra, J. X. (2005). Timing the origin and expansion of the Mexican tropical dry forest. *Proceedings of the National Academy of Sciences, USA* **102**, 10919–10923.

Becerra, J. X. (2007). The impact of herbivore–plant coevolution on plant community structure. *Proceedings of the National Academy of Sciences, USA* **104**, 7483–7488.
Binggeli, P. and Blackstock, P. (1999). Impact of exposure on extension growth and stem quality of young ash and sycamore in farm forestry plantations. *Agroforestry Forum* **9**, 19–23.
Bolnick, D. I. and Fitzpatrick, B. M. (2007). Sympatric speciation: models and empirical evidence. *Annual Review of Ecology, Evolution and Systematics* **38**, 459–487.
Bowler, M. G. and Kelly, C. K. (2012). On the statistical mechanics of species abundance distributions. *Theoretical Population Biology* **82**, 85–91.
Brachi, B., Faure, N., Horton, M. *et al.* (2010). Linkage and association mapping of *Arabidopsis thaliana* flowering time in nature. *PLoSGenetics* **6**, e1000940. doi:10.1371/journal.pgen.1000940.
Brooks, M. L. (2000). Competition between alien annual grasses and native annual plants in the Mojave Desert. *American Midland Naturalist* **144**, 92–108.
Buckley, Y. M. and Kelly, C. K. (2003). Comparison of population structures and ecology of a congeneric pair of common and rare neotropical tree species. *Plant Ecology* **167**, 45–56.
Bullock, S. H. (1986). Climate of Chamela, Jalisco, and trends in the south coastal region of México. *Archives of Meteorology, Geophysics and Bioclimatology, Series B.* **36**, 297–316.
Bullock, S. H. (1988). Rascos del ambiente fisico y biológico de Chamela, Jalisco, México. *Folio Entomologica Mexicana* **77**, 5–17.
Bullock, S. H. and Solís-Magallanes, J. A. (1990). Phenology of canopy trees of a tropical deciduous forest in México. *Biotropica* **22**, 22–35.
Bullock, S. H., Mooney, H. A. and Medina, E. (1995). *Seasonally Dry Tropical Forests.* New York: Cambridge University Press.
Burns, J. H. and Strauss, S. (2011). More closely related species are more ecologically similar in an experimental test. *Proceedings of the National Academy of Sciences, USA* **108**, 5302–5307.
Cahill, J. F., Kembel, S. W., Lamb, E. G. and Keddy, P. A. (2008). Does phylogenetic relatedness influence the strength of competition among vascular plants? *Perspectives in Plant Ecology, Evolution and Systematics* **10**, 41–50.
Chesson, P. and Huntly, N. (1997). The roles of harsh and fluctuating conditions in the dynamics of ecological communities. *American Naturalist* **150**, 519–553.
Chesson, P. and Kuang, J. J. (2008). The interaction between predation and competition. *Nature* **456**, 235–238.
Cole, K., Ironside, K., Eischeid, J. *et al.* (2011). Past and ongoing shifts in Joshua tree support future modeled range contraction. *Ecological Applications* **21**, 137–149.
Connell, J. H. (1978). Diversity in tropical rainforests and coral reefs. *Science* **199**, 1302–1310.
Connell, J. H. and Green, P. T. (2000). Seedling dynamics over thirty-two years in a tropical rain forest tree. *Ecology* **81**, 568–584.
Coyne, J. A. and Orr, H. A. (2004). *Speciation.* Sunderland, MA: Sinauer.
Daehler, C. C. (2003). Performance comparisons of co-occurring native and alien invasive plants: implications for conservation and restoration. *Annual Review of Ecology, Evolution, and Systematics* **34**, 183–211.
Davis, M. B. (1986). Climatic instability, time lags, and community disequilibrium. In J. Diamond and T. J. Case (eds), *Community Ecology.* New York: Harper and Row, pp. 269–284.
De-Nova, J. A., Medina, R., Montero, J. C. *et al.* (2012). Insights into the historical construction of species-rich Mesoamerican seasonally dry tropical forests: the diversification of *Bursera* (Burseraceae, Sapindales). *New Phytologist* **193**, 276–287.

Dethioux, M. (1970). The sycamore seedling. *Notes techniques du Centre d'écologie forestière et rurale, Belgium No.* 10.
Devaux, C. and Lande, R. (2009). Displacement of flowering phenologies among plant species by competition for generalist pollinators. *Journal of Evolutionary Biology* **22**, 1460–1470.
Dieckmann, U. and Doebeli, M. (1999). On the origin of species by sympatric speciation. *Nature* **400**, 354–357.
Eichhorn, M. P., Compton, S. G. and Hartley, S. E. (2006). Seedling species determines rates of leaf herbivory in a Malaysian rain forest. *Journal of Tropical Ecology* **22**, 513–519.
Ellis, C. (2007). Temporal niche axes in *Acer pseudoplatanus* L. and *Fraxinus excelsior* L: a climate change perspective. Final Honours Thesis, University of Oxford, Oxford.
Faith, D. P. (1992). Conservation evaluation and phylogenetic diversity. *Biological Conservation* **61**, 1–10.
Felsenstein, J. (1985). Phylogenies and the comparative method. *American Naturalist* **125**, 1–15.
Filip, V., Dirzo, R., Maass, J. M. and Sarukhán, J. (1995). Within- and among-year variation in the levels of herbivory on the foliage of trees from a Mexican tropical deciduous forest. *Biotropica* **27**, 78–86.
Firn, J., MacDougall, A., Schmidt, S. and Buckley, Y. M. (2010). Early emergence and resource availability can competitively favour natives over a functionally similar invader. *Oecologia* **163**, 775–784.
Fox, G. A. (2003). Assortative mating and plant phenology: evolutionary and practical consequences. *Evolutionary Ecology Research* **5**, 1–18.
Fraser, L. H. and Keddy, P. A. (2005). Can competitive ability predict structure in experimental plant communities? *Journal of Vegetation Science* **16**, 571–578.
Gentry, A. H. (1991). Alwyn H. Gentry forest transect data set. Available at: http://www.wlbcenter.org/gentry_data.htm.
Givnish, T. J. (2010). Ecology of plant speciation. *Taxon* **59**, 1326–1366.
Gotelli, N. J. and Colwell, R. K. (2001). Quantifying biodiversity: procedures and pitfalls in the measurement and comparison of species richness. *Ecology Letters* **4**, 379–391.
Hanley, M. E., Fenner, M. and Edwards, P. J. (1995). An experimental field study of the effects of mollusc grazing on seedling recruitment and survival in grassland. *Journal of Ecology* **83**, 621–627.
Hanley, M. E., Fenner, M. and Edwards, P. J. (1996). The effect of mollusc grazing on seedling recruitment in artificially created grassland gaps. *Oecologia* **106**, 240–246.
Hara, T., Kimura, M. and Kikuzawa, K. (1991). Growth patterns of tree height and stem diameter in populations of *Abies veitchii, A. mariesii* and *Betula ermanii*. *Journal of Ecology* **79**, 1085–1098.
Harvey, P. H. (1996). Phylogenies for ecologists. *Journal of Animal Ecology* **65**, 255–263.
Harvey, P. H. and Pagel, M. D. (1991). *The Comparative Method in Evolutionary Biology*. Oxford: Oxford University Press.
Helliwell, D. R. (1965). Factors influencing the growth of seedlings of sycamore and Norway maple. *Quarterly Journal of Forestry* **59**, 327–337.
Hernández-Apolinar, M., Valverde, T. and Purata, S. (2006). Demography of *Bursera glabrifolia*, a tropical tree used for folk woodcrafting in southern México: an evaluation of its management plan. *Forest Ecology and Management* **223**, 139–151.

Hijmans, R. J., Cameron, S. E., Parra, J. L., Jones, P. G. and Jarvis, A. (2005). Very high resolution interpolated climate surfaces for global land areas. *International Journal of Climatology* **25**, 1965–1978.

Hopkins, R. and Rausher, M. D. (2012). Pollinator-mediated selection on flower color allele drives reinforcement. *Science* **335**, 1090–1092.

Hubbell, S. P. (2001). *The Unified Neutral Theory of Biodiversity and Biogeography*. Princeton, NJ: Princeton University Press.

Hubbell, S. P. and Foster, R. B. (1986). Biology, chance, and history and the structure of tropical rain forest tree communities. In J. Diamond and T. J. Case (eds), *Community Ecology*. New York: Harper & Row, pp. 314–329.

Huff, J. T. and Zilberman, D. (2012). Regulation of biological accuracy, precision, and memory by plant chromatin organization. *Current Opinion in Genetics & Development* **22**, 132–138.

Hutchinson, G. F. (1959). Homage to Santa Rosalia, or why are there so many kinds of animals? *American Naturalist* **93**, 145–159.

Jablonski, D. and Finarelli, J. A. (2009). Congruence of morphologically defined genera with molecular phylogenies. *Proceedings of the National Academy of Sciences, USA* **106**, 8262–8266.

Johnson, E. A. and Frier, G. I. (1989). Population dynamics in lodgepole pine-Engleman spruce forests. *Ecology* **70**, 1335–1345.

Jump, A. S., Hunt, J. M., Martinez-Izquierdo, J. A. and Penuelas, J. (2006). Natural selection and climate change: temperature-linked spatial and temporal trends in gene frequency. *Molecular Ecology* **15**, 3469–3480.

Kelly, C. K. and Bowler, M. G. (2002). Coexistence and relative abundance in forest tree species. *Nature* **417**, 437–440.

Kelly, C. K. and Bowler, M. G. (2005). A new application of storage dynamics: differential sensitivity, diffuse competition and temporal niches. *Ecology* **86**, 1012–1022.

Kelly, C. K. and Bowler, M. G. (2009a). Temporal niche dynamics, relative abundance and phylogenetic signal in coexisting species. *Theoretical Ecology* **2**, 161–169.

Kelly, C. K. and Bowler, M. G. (2009b). Investigating the role of enemies in temporal niche dynamics: differential sensitivity, competition, and stable coexistence. *Theoretical Population Biology* **76**, 278–284.

Kelly, C. K. and Hanley, M. E. (2005). Juvenile growth and palatability in congeneric British herbs. *American Journal of Botany* **92**, 1586–1589.

Kelly, C. K., Banyard Smith, H., Buckley, Y. M. *et al.* (2001). Investigations in commonness and rarity: a comparative analysis of co-occurring, congeneric Mexican trees. *Ecology Letters* **4**, 618–627.

Kelly, C. K., Blundell, S. J., Bowler, M. G. *et al.* (2011). The statistical mechanics of community assembly and species distribution. *New Phytologist* **191**, 819–827.

Kelly, C. K., Bowler, M. G. and Breden, F. M. (2006). An analytical model assessing the potential threat to natural habitats from insect resistance transgenes: continuous transgene input. *Biology Letters* **2**, 293–297.

Kelly, C. K., Bowler, M. G., Breden, F. M., Fenner, M. and Poppy, G. M. (2005). An analytical model assessing the potential threat to natural habitats from insect resistance transgenes. *Proceedings of the Royal Society of London B* **272**, 1759–1767.

Kelly, C. K., Bowler, M. G., Joy, J. B. and Williams, J. N. (2010). Fractional abundance and the ecology of community structure. *ArXiv*1008.2527v1.

Kelly, C. K., Bowler, M. G., Pybus, O. G. and Harvey, P. H. (2008). Phylogeny, niches and relative abundance in natural communities. *Ecology* **89**, 962–970.

Kelly, C. K., Chase, M., Fay, M. F., de Bruijn, A. and Woodward, F. I. (2003). Temperature-based population segregation in birch. *Ecology Letters* **6**, 87–89.

Kleidon, A. and Mooney, H. A. (2000). A global distribution of biodiversity inferred from climatic constraints: results from a process-based modelling study. *Global Change Biology* **6**, 507–523.

Leibold, M. A. and McPeek, M. A. (2006). Coexistence of the niche and neutral perspectives in community ecology. *Ecology* **87**, 1399–1410.

Lopez, O. R., Farris-Lopez, K., Montgomery, R. A. and Givnish, T. J. (2008). Leaf phenology in relation to canopy closure in southern Appalachian trees. *American Journal of Botany* **95**, 1395–1407.

Losos, J. B. (2010). Adaptive radiation, ecological opportunity and evolutionary determinism. *American Naturalist* **175**, 623–639.

Mabberley, D. J. (2008). *Mabberley's Plant Book*. Cambridge: Cambridge University Press.

MacArthur, R. H. (1957). On the relative abundance of bird species. *Proceedings of the National Academy of Sciences, USA* **43**, 293–295.

Magnuson-Ford, K., Mooers, A. Ø., Paquette, S. R. and Steel, M. (2010). Comparing strategies to preserve evolutionary diversity. *Journal of Theoretical Biology* **266**, 107–116.

Mason, S. (2005). Reproduction and recruitment in native ash and its invasive competitor sycamore. MS Thesis, University of Southampton, Southampton.

Mayfield, M. M. and Levine, J. M. (2010). Opposing effects of competitive exclusion on the phylogenetic structure of communities. *Ecology Letters* **13**, 1085–1093.

Mayrose, M., Kane, N. C., Mayrose, I., Dlugosch, K. M. and Rieseberg, L. H. (2011). Increased growth in sunflower correlates with reduced defences and altered gene expression in response to biotic and abiotic stress. *Molecular Ecology* **20**, 4683–4694.

McCluney, K. E., Belnap, J., Collins, S. L. *et al.* (2012). Shifting species interactions in terrestrial dryland ecosystems under altered water availability and climate change. *Biological Reviews* **87**, 563–582.

Nee, S. and May, R. M. (1992). Dynamics of metapopulations: habitat destruction and competitive coexistence. *Journal of Animal Ecology* **61**, 37–40.

Nee, S., Harvey, P. H. and May, R. M. (1991). Lifting the veil on abundance patterns. *Proceedings of the Royal Society of London B* **243**, 161–163.

Olson, D. M., Dinerstein, E., Wikramanayake, E. D. *et al.* (2001). Terrestrial ecoregions of the world: a new map of life on earth. *Bioscience* **51**, 933–938.

Pacala, S. W. and Tilman, D. (1994). Limiting similarity in mechanistic and spatial models of plant competition in heterogeneous environments. *American Naturalist* **143**, 222–257.

Packer, A. and Clay, K. (2000). Soil pathogens and spatial patterns of seedling mortality in a temperate tree. *Nature* **404**, 278–281.

Packer, A. and Clay, K. (2003). Soil pathogens and *Prunus serotina* seedling and sapling growth near conspecific trees. *Ecology* **84**, 108–119.

Pake, C. E. and Venable, D. L. (1995). Is coexistence of Sonoran Desert annuals mediated by temporal variability in reproductive success? *Ecology* **76**, 246–261.

Pake, C. E. and Venable, D. L. (1996). Seed banks in desert annuals: implications for persistence and coexistence in variable environments. *Ecology* **77**, 1427–1435.

Parker, K. C., Parker, A. J., Beaty, R. M., Fuller, M. M. and Faust, T. D. (1997). Population structure and spatial pattern of two coastal populations of Ocala sand pine (*Pinus clausa* (Chapm. ex Englem.) Vasey ex Sarg. var. *clausa* D. B. Ward). *Journal of the Torrey Botanical Society* **124**, 22–33.

Penuelas, J., Terradas, J. and Lloret, F. (2011). Solving the conundrum of plant species coexistence: water in space and time matters most. *New Phytologist* **189**, 5–8.

Pielou, E. C. (1977). *Mathematical Ecology*. New York: Wiley.

Pimentel, D., Feinberg, E. H., Wood, P. W. and Hayes, J. T. (1965). Selection, spatial distribution, and coexistence of competing fly species. *American Naturalist* **99**, 97–110.

Pirie, C. D., Walmsley, S., Ingle, R. *et al.* (2000). Investigations in commonness and rarity: a comparison of seed dispersal in the common *Jatropha standleyi* and the rare and endangered *J. chamelensis* (Euphorbiaceae). *Biological Journal of the Linnean Society* **71**, 501–512.

Preston, F. W. (1962a). The canonical distribution of commonness and rarity: Part I. *Ecology* **43**, 185–215.

Preston, F. W. (1962b). The canonical distribution of commonness and rarity: Part II. *Ecology* **43**, 410–432.

Puertolas, J., Gil, L. and Pardos, J. A. (2003). Effects of nutritional status and seedling size on field performance of *Pinus halepensis* planted on former arable land in the Mediterranean basin. *Forestry* **76**, 159–168.

Pueyo, S. (2006). Diversity: between neutrality and structure. *Oikos* **112**, 392–405.

Pueyo, S., He, F. and Zillio, T. (2007). The maximum entropy formalism and the idiosyncratic theory of biodiversity. *Ecology Letters* **10**, 1017–1028.

Rafferty, C. M., Lamont, B. B. and Hanley, M. E. (2010). Herbivore feeding preferences in captive and wild populations. *Austral Ecology* **35**, 257–263.

Reinhart, K. O., Packer, A., Van der Putten, W. H. and Clay, K. (2003). Plant–soil biota interactions and spatial distribution of black cherry in its native and invasive ranges. *Ecology Letters* **6**, 1046–1050.

Rice, K. J. and Menke, J. W. (1985). Competitive reversals and environment-dependent resource partitioning in *Erodium*. *Oecologia* **67**, 430–434.

Rzedowski, J. (1978). *Vegetación de México*. México, D. F.: Limusa.

Schluter, D. (2001). Ecology and the origin of species. *Trends in Ecology and Evolution* **16**, 372–380.

Schwinning, S. and Fox, G. A. (1995). Population dynamic consequences of competitive symmetry in annual plants. *Oikos* **72**, 422–432.

Schwinning, S. and Weiner, J. (1998). Mechanisms determining the degree of size asymmetry in competition among plants. *Oecologia* **113**, 447–455.

Segura, G., Balvanera, P., Duran, E. and Perez, A. (2003). Tree community structure and stem mortality along a water availability gradient in a Mexican tropical dry forest. *Plant Ecology* **169**, 259–271.

Siegal, S. and Castellan, N. J. (1988). *Nonparametric Statistics for the Behavioral Sciences*, 2nd edn. New York: McGraw-Hill.

Sugihara, G. (1980). Minimal community structure: an explanation of species abundance patterns. *American Naturalist* **116**, 770–787.

Sugihara, G., Bersier, L.-F., Southwood, T. R. E., Pimm, S. L. and May, R. M. (2003). Predicted correspondence between species abundances and dendrograms of niche similarities. *Proceedings of the National Academy of Sciences, USA* **100**, 5246–5251.

Tapper, P.-G. (1996). Long-term patterns of mast fruiting in *Fraxinus excelsior*. *Ecology* **77**, 2567–2572.

Tilman, D. (1988). *Plant Strategies and the Dynamics and Structure of Plant Communities*. Princeton, NJ: Princeton University Press.

Toledo M. C. A. (1982). El genero *Bursera* (Burseraceae) en el estado de Guerrero (México). Unpublished BS thesis, Facultad de Ciencias, Universidad Nacional Autónoma de México, México, D. F.

Turelli, M. (1981). Niche overlap and invasion of competitors in random environments. 1. Models without demographic stochasticity. *Theoretical Population Biology* **20**, 1–56.

Usinowicz, J., Wright, S. J. and Ives, A. R. (2012). Coexistence in tropical forests through asynchronous variation in annual seed production. *Ecology* **93**, 2073–2084.

Vane-Wright, R. I., Humphries, C. J. and Williams, P. H. (1991). What to protect? Systematics and the agony of choice. *Biological Conservation* **55**, 235–254.

Veblen, T. T. (1989). Tree regeneration responses to gaps along a trans-Andean gradient. *Ecology* **70**, 541–543.

Veblen, T. T. (1992). Regeneration dynamics. In D. C. Glenn-Lewin, R. K. Peet and T.T. Veblen (eds), *Plant Succession: Theory and Prediction*. London, UK: Chapman and Hall, pp. 152–187.

Via, S. (2002). The ecological genetics of speciation. *American Naturalist* **159**, S1–S7.

Villalba, R. and Veblen, T. T. (1998). Influences of large scale climatic variability on episodic tree mortality in northern Patagonia. *Ecology* **79**, 2624–2640.

Villar-Salvador, P., Puertolas, J., Cuesta, B. *et al.* (2012). Increase in size and nitrogen concentration enhances seedling survival in Mediterranean plantations. Insights from an ecophysiological conceptual model of plant survival. *New Forests* **43**, 755–770.

Warner, R. R. and Chesson, P. L. (1985). Coexistence mediated by recruitment fluctuations: a field guide to the storage effect. *American Naturalist* **125**, 769–787.

Waters, T. L. and Savill, P. S. (1992). Ash and sycamore regeneration and the phenomenon of their alternation. *Forestry* **65**, 417–433.

Webb, C. O. (2000). Exploring the phylogenetic structure of ecological communities: an example for rain forest trees. *American Naturalist* **156**, 145–155.

Weeks, A. and Simpson, B. B. (2004). Molecular genetic evidence for interspecific hybridization among endemic Hispaniolan *Bursera* (Burseraceae). *American Journal of Botany* **91**, 976–984.

Weis, A. E. (2005). Direct and indirect assortative mating: a multivariate approach to plant flowering schedules. *Journal of Evolutionary Biology* **18**, 536–546.

Williams, J. N. and Kelly, C. K. (2013). Deconstructing the signal: phylogenetic structure, elevation change and the implications for species coexistence. *Evolutionary Ecology Research* Published online 8 August 2013.

Williams, J. N., Viers, J. H. and Schwartz, M. W. (2010). Tropical dry forest trees and the relationship between local abundance and geographic range. *Journal of Biogeography* **37**, 951–959.

Woodward, F. I. and Diament, A. D. (1991). Functional approaches to predicting the ecological effects of global change. *Functional Ecology* **5**, 202–212.

Wright, S. J., Muller-Landau, H. C., Calderon, O. and Hernandez, A. (2005). Annual and spatial variation in seedfall and seedling recruitment in a neotropical forest. *Ecology* **86**, 848–860.

Yanez-Espinosa, L., Terrazas, T. and Lopez-Mata, L. (2010). Phenology and radial stem growth periodicity in evergreen subtropical rainforest trees. *IAWA Journal* **31**, 293–307.

PETER B. ADLER

4

Testing the storage effect with long-term observational data

4.1 INTRODUCTION

Over 50 years ago, Hutchinson (1941) noted that variation in environmental conditions could alter the outcome of competition. One implication of his observation was that environmental fluctuations could promote coexistence, allowing many species to persist in a habitat where all but one would be excluded under constant conditions. By the end of the 1980s, Chesson and colleagues had clearly described the theoretical requirements for coexistence via the storage effect (Chesson and Warner 1981, Warner and Chesson 1985, Chesson and Huntly 1989). Yet despite the long history of these ideas, relatively few direct empirical tests of the storage effect exist. Studies from a variety of natural ecosystems provide partial evidence for the storage effect (Pake and Venable 1995, 1996, Kelly and Bowler 2002, Descamps-Julien and Gonzalez 2005, Facelli *et al.* 2005, Kelly *et al.* 2008), but tests of all the required conditions or quantification of the strength of the effect are much rarer (Cáceres 1997, Adler *et al.* 2006, 2009, Angert *et al.* 2009).

The lack of rigorous case studies limits our ability to generalise about the role of the temporal storage effect in maintaining diversity. We know that multiple coexistence mechanisms will operate in different communities, but currently we cannot say where the storage effect makes an especially important contribution. This information will be essential for understanding the consequences of expected increases in climate variability (Karl and Trenberth 2003, Jain *et al.* 2005, Salinger 2005, Allan and Soden 2008), which could impact species diversity in

Temporal Dynamics and Ecological Process, ed. C. K. Kelly, M. G. Bowler, G. A. Fox.

systems where the storage effect is important (Adler and Drake 2008). Understanding the influence of the storage effect on coexistence across a variety of ecosystems is therefore a prerequisite for anticipating future changes in species diversity.

Perhaps the most efficient way to gain a general understanding of the importance of the storage effect is by analysing existing long-term, observational data sets. Testing the storage effect requires information about the intrinsic performance and competitive effects of co-occurring species under a wide variety of environmental conditions. Although field experiments can directly test the effects of a few precipitation or temperature treatments, imposing the broad range of environmental conditions found in long-term data would be a logistical nightmare with a high cost. In addition, recent statistical advances have made it easier to link observational data with theoretical models (Clark 2005).

My first objective is to illustrate tests of the storage effect using observational data sets. The basic approach is to use the data to parameterise a model, and then simulate the model to test for the conditions of the storage effect and quantify the overall effect of environmental variation on coexistence (following Adler *et al.* 2006, 2009). While the data sets I analyse are unique because they come from annually mapped quadrats (Albertson and Tomanek 1965, West *et al.* 1979), the same approach could also work for non-spatial data. My second objective is to compare the results of this analysis for two plant communities, a mixed prairie in western Kansas, USA, and a sagebrush steppe in eastern Idaho, USA. My hope is that the comparison of these two communities will sharpen our thinking about how we should expect the strength of the storage effect to vary across ecosystems.

Temporal storage effect theory

The temporal storage effect operates when different competitors experience fitness advantages at different times and can 'store' the gains made during favourable periods. For this to occur, three conditions must be satisfied (Chesson and Warner 1981, Warner and Chesson 1985, Chesson and Huntly 1989, Chesson 1990, 2000). First, organisms must have some mechanism for persisting during unfavourable periods, such as a seedbank, quiescence or diapause. This condition, which gives the storage effect its name, buffers negative population growth; without it, populations would go extinct after a brief unfavourable period and environmental variation could never promote coexistence. Second, species must respond differently to environmental variation, making it possible for one species to experience a relatively good year while its competitor is experiencing a relatively bad year. The third condition, which is more of an outcome of the previous two than an independent criterion, requires covariance between competition and the environment. Specifically, competition must limit growth more in environmentally favourable than unfavourable years. Given condition two, which tends to partition intraspecific and interspecific competition into different years, this environment–competition covariance ensures that intraspecific competition will be stronger than interspecific competition, one basic requirement for stabilised coexistence (Rosenzweig and MacArthur 1963, Chesson 2000).

Testing for the three conditions of the storage effect determines whether or not it is operating, but does not quantify its strength. The overall effect of temporal environmental variability on coexistence can be estimated using an invasibility approach. Stable coexistence occurs when each species can recover from low abundance (its 'invader' state) in the presence of competitors (the 'resident' species) at their stochastic equilibrium abundances (Chesson 2000). If climate variability helps stabilise coexistence, then, on average, species should recover from a fall to low density more quickly in a variable climate than in a constant environment. The difference in the low-density growth rate between the variable and constant scenarios measures the overall effect of climate variability on coexistence (the overall effect includes contributions from both the storage effect and relative nonlinearity (Chesson 2000)).

Comparison of prairie and sagebrush steppe communities

Should we expect to find a stronger storage effect in a prairie grassland or a sagebrush steppe? Three features of the environment could be relevant. First, a lower level of average resource availability might select for life-history traits necessary to the storage effect, such as seed dormancy. Second, a greater magnitude of environmental variation could increase the potential for temporal niche partitioning. After all, the storage effect cannot operate in a constant environment. Third, the predictability of the environmental variation could determine how successful organisms are in taking advantage of favourable conditions while avoiding unfavourable conditions (e.g. predictive germination).

Before evaluating whether the prairie and sagebrush steppe environments differ in any of these three ways, some background information is necessary. The prairie study site is located near Hays, Kansas (38.8°N, 99.3°W), at approximately 630 m above sea level. Mean annual

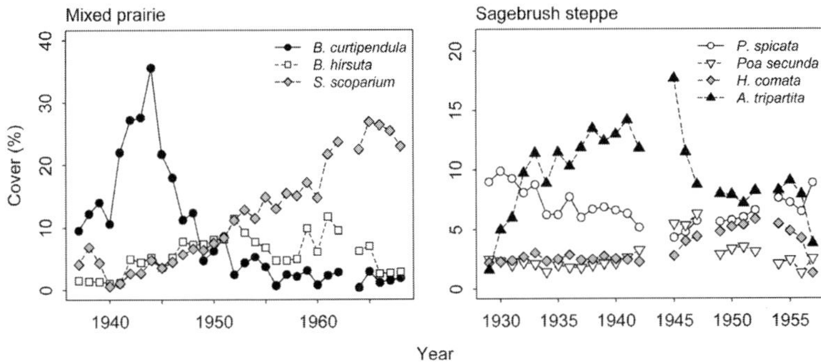

Figure 4.1 Observed abundances of the modelled species in the mixed prairie and sagebrush steppe data sets. Abundance is measured as absolute basal cover for the grass species, and absolute canopy cover for the shrub *Artemisia tripartita*, calculated after transforming the original maps to grids of 2-cm cells. Mean annual abundances were estimated using a linear mixed model with quadrats as random effects and year as a fixed effect.

precipitation is 580 mm, 75% falling during spring and summer, and mean annual temperature is 12°C. I analysed ungrazed quadrats located on shallow limestone soils and dominated by the perennial C4 grasses *Bouteloua curtipendula*, *Bouteloua hirsuta* and *Schizachyrium scoparium*. These three species account for almost 95% of total plant basal cover. Over the period of the data set, the basal cover of the three species fluctuated, mostly in response to the Great Drought of the 1930s (Figure 4.1). The sagebrush steppe data set comes from the US Sheep Experiment Station (USSES) near Dubois, Idaho (44.2°N, 112.1°W). The site is approximately 1500 m above sea level, has a mean annual temperature of 6.2°C, and receives on average 307 mm per year of precipitation, distributed throughout the year but peaking in May and June. All quadrats included in the analysis were located in ungrazed exclosures. The dominant perennial grasses (all C3) are *Pseudoroegneria spicata*, *Poa secunda* and *Hesperostipa comata*, and the dominant shrub is *Artemisia tripartita*. These four species, which account for over 70% of total basal cover and 60% of canopy cover, were the focus of my analysis. Changes in their cover over the period of record appear small in comparison to the mixed prairie community (Figure 4.1).

If low resource availability selects for greater investment in life-history traits such as dormancy or long-lived adults, then we might expect a stronger storage effect in the sagebrush community. Primary productivity, a proxy for resource availability, is up to 50% lower in the sagebrush site than in the prairie (P. B. Adler, unpublished data). On the other hand, if we find long-lived, competition-insensitive life stages in both communities, then this difference in resource availability may not be important for explaining differences in the storage effect.

The magnitude of interannual environmental variability may influence the potential for temporal niche partitioning. However, the coefficient of variation for annual precipitation differs only slightly between the sagebrush site (23.5%) and the prairie site (27%). The difference in the seasonality of precipitation might offer a stronger argument for lower interannual variability of the sagebrush steppe environment. The prairie receives 75% of its precipitation between April and September, when temperatures are high, whereas precipitation occurs mostly in autumn through spring in the sagebrush site. Because of low evaporation and transpiration during the wet sagebrush steppe winters, soil water availability peaks reliably in early spring. Furthermore, interannual variability in water availability may be lower at the sagebrush steppe site if even relatively dry winters can recharge the soil profile. If we are willing to assume slightly higher interannual variability in soil moisture in the prairie than in the sagebrush site, we might expect more potential for temporal niche differentiation and a stronger storage effect in the prairie.

Greater predictability of environmental variation could also favour a strong storage effect. In theoretical studies, temporal autocorrelation is often used as a proxy for 'predictability', based on the notion that a more predictable environment offers more reliable demographic cues. I tested for interannual temporal autocorrelation in seasonal and annual precipitation and temperature and the only significant result was for annual temperature at the prairie site, with positive autocorrelation over 1- and 2-year lags. In the absence of any other information about environmental predictability, I will use this result to predict that plants in the prairie site should more successfully exploit environmental variation than plants at the sagebrush site.

Of the three environmental descriptors I compared between sites, differences in one (resource availability) suggest a stronger storage effect at the sagebrush site than in the prairie, while differences in two (the magnitude and predictability of environmental variation) suggest the opposite. The clearest difference between the sites is in resource availability, with lower productivity in the sagebrush site implying greater investment in life-history traits that would satisfy the first condition of the storage effect. However, this condition may be the easiest to satisfy, meaning that the subtle differences in patterns of environmental variation, relating to the second and third conditions of the storage effect, could be more important.

4.2 STATISTICAL MODELLING METHODS

Data set description

During the early 1900s, rangeland ecologists in western North America established permanent 1-m^2 quadrats and used pantographs to map annually the basal or canopy cover of all the individual plants in each quadrat (Hill 1920). The decades-long extent and demographic resolution of these data sets makes them ideal for testing the storage effect. My analysis focuses on a set of four quadrats censused between 1937 and 1972 in a Kansas mixed prairie (Albertson and Tomanek 1965) and a set of 18 quadrats censused between 1929 and 1957 in an Idaho sagebrush steppe. The original maps were digitised into a geographic information system, with each individual plant represented as a polygon.

Statistical modelling overview

My collaborators and I used statistical and simulation models to link the historical data with storage effect theory. More specifically, we fitted a statistical model to the observational data, and then simulated this model to test the conditions of the storage effect and to estimate the overall effect of climate variability on coexistence. Here I describe the main features of the statistical and simulation approaches (see Adler *et al.* (2006 and 2009) for additional detail).

We took a lattice approach, turning the quadrat maps into grids of 2-cm cells, with each cell occupied by one of the three grass species or bare ground. For the sagebrush steppe data set, the shrub species is included in a second, 'canopy' layer. Two grass species cannot co-occur in one cell, but it is possible for grass and the shrub to overlap: a basal cell can be occupied by a grass species while the canopy cell at the same coordinate is occupied by the shrub. Large plants appear as clusters of cells all in the same state. The goal of the model is to predict how the states of each cell change from one time step to the next (Figure 4.2a).

Statistical modelling of survival and colonisation

We assume that the transition probabilities for each cell are a function of two unobserved processes, survival and colonisation. Consider a cell occupied by *Poa secunda* in two successive time steps, t and t+1. The *Poa*

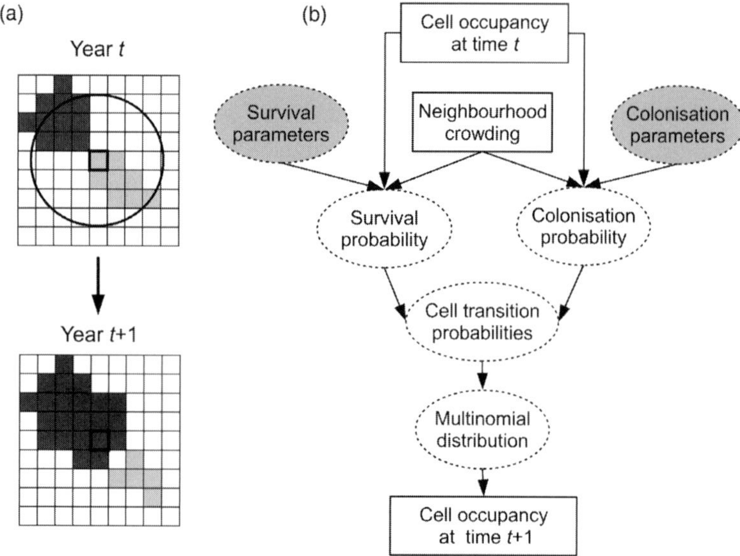

Figure 4.2 Flowchart of modelling approach. **(a)** The goal of the model is to predict the state of cells (occupancy by one species or bare ground) at time $t + 1$ based on their states at time t. In this cartoon example, the focal cell (thick border) makes a transition from occupancy by the light grey species to the dark grey species. The probability of this transition depends on the composition of the local neighbourhood, indicated by the black circle, and conditions in year t. **(b)** Statistical approach: Predictions depend on observed conditions (rectangles with solid borders) and unobserved survival and colonisation processes (dashed ovals). The survival and colonisation probabilities for each species in each cell depend on the observed states of cells at time t, the observed local neighbourhood surrounding each cell at time t, and fitted survival and colonisation parameters which vary among species and among years (shaded ovals). Once the survival and colonisation parameters are calculated, we can calculate the probability that each species will occur in a given cell at time $t + 1$. Those predicted cell probabilities are compared to the observed cell occupancies using a multinomial distribution. To fit the statistical model, I look for the survival and colonisation parameters that best predict the observed cell occupancies. In stochastic simulations of the model, cell occupancies at time $t+ 1$ are random draws from the multinomial distribution for each cell.

tillers in that cell could have survived from t to $t+1$ or, alternatively, the plant could have died but then successfully recolonised the cell. The processes driving other transitions are more clear-cut: mortality is the only explanation for why *Poa* would present at t but not $t+1$, and colonisation is the only explanation for the appearance of a different species in the cell at $t+1$.

The survival and colonisation probabilities depend on three different factors. First, they vary among species. Some species may consistently have higher survival or colonisation rates than others. Second, the probabilities may vary from one year to another. To accommodate this temporal variability, we introduce random effects that allow the survival and colonisation rates to vary around each species' average rate. The random effects do not explain *why* survival may be higher in one year than another (e.g. higher precipitation); they just quantify the magnitude of the interannual variability. Third, survival and colonisation rates in a given cell may be influenced by interactions with other plants in the local neighbourhood around that cell. For example, survival of *Poa* may be lower in a cell experiencing intense crowding by *Hesperostipa*, and colonisation of *Hesperostipa*, via vegetative growth, may be higher if *Hesperostipa* is abundant in the local neighbourhood. We assume that the influence of neighbouring cells decreases as an exponential function of distance from the focal cell. These local interactions mean that a cell in the centre of a large plant can behave quite differently from an isolated cell. The parameters that determine the effect of each species' local density on the survival or colonisation of other species can be thought of as competition coefficients, although facilitative interactions are equally possible.

Figure 4.2b shows how we use the observed data to learn about the survival and colonisation parameters. We start the iterative model-fitting algorithm (Markov Chain Monte Carlo implemented in a Bayesian hierarchical framework) by guessing at values of the survival and colonisation parameters for each species in each year. Based on (1) these parameter values, (2) the observed cell occupancy states at time $t+1$, and (3) the observed local neighbourhood around each cell at time t, we calculate the probability of survival and colonisation for each species in each cell. The survival and colonisation parameters in turn determine the probability that each species will be present in each cell at time $t+1$. We compare these predictions with the observed cell occupancy states at time $t+1$ using a multinomial distribution. Next, we assign new values to the survival and colonisation parameters, and then repeat these steps iteratively until finding the survival and colonisation parameters that best predict the observed cell occupancy. The fitted models do a reasonable job of reproducing the observed year-to-year changes in cover (Adler *et al.* 2006, 2009).

Simulation modelling

Once the species- and year-specific survival and colonisation parameters are fit, we can simulate the model to explore community dynamics under a variety of scenarios. At each time step in the simulation, we perform the calculations illustrated in Figure 4.2, except rather than starting from the observed cell occupancy states, we can start from any arbitrary initial conditions and project the model forward in time. For example, we might ask how quickly each species increases from low density in an otherwise empty landscape, and then repeat this simulation with other species at their (stochastic) equilibrium abundances. Alternatively, we could compare simulations that incorporate temporal variability by drawing randomly from each set of year-specific survival and colonisation parameters at each time step to simulations of a 'constant' environment in which time-averaged survival and colonisation parameters are used at every time step.

4.3 SIMULATIONS AND RESULTS

Condition 1: long-lived life stage

My collaborators and I tested for the first condition of the storage effect, involving mechanisms of persistence, by performing a survival analysis for each species, using spatial locations to track individual genets through time. We followed the methods of Lauenroth and Adler (2008) with one modification: we allowed genets to 'survive' across 2 years of missing data, which reduced the number of records to be censored. Even if juvenile survival is low, high adult survival can buffer species against unfavourable periods, provided that adult survival is relatively insensitive to competition. In fact, storage of reproductive potential in the rhizomes and dormant meristems of long-lived perennials provides a direct analogy to the seed storage that forms the basis of much of the theory (Chesson and Huntly 1989, Chesson *et al.* 2004). Only when these rhizomes or meristems are activated, and vegetative growth initiated, will they become vulnerable to negative effects of competition.

The survival analysis showed strong evidence for the first condition of the storage effect, a long-lived life stage. For all species in both the Kansas and Idaho data sets, survival during the first years of life was low, but once individuals reached 3–4 years of age, annual survival rates often exceeded 90% (Figure 4.3). The maximum observed lifespans, which are constrained by the length of the data set, were greater than 20 years for all four species.

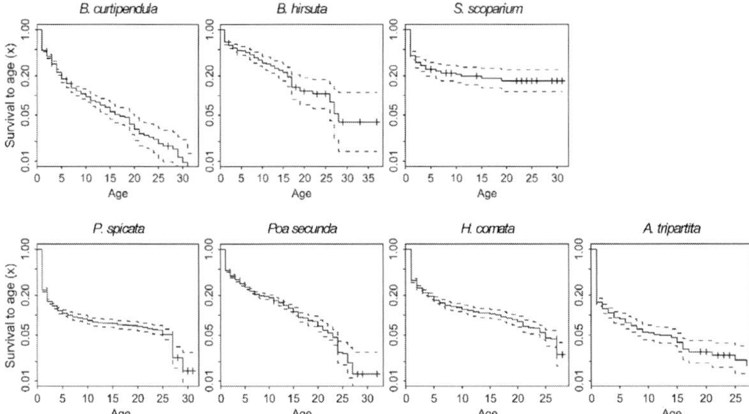

Figure 4.3 Kaplan–Meier survival curves for the prairie (top row) and the sagebrush steppe species (bottom row).

Condition 2: species-specific environmental responses

To test for condition 2, species-specific responses to the environment, we used the lattice model to simulate each species' yearly intrinsic population growth rate. In storage effect theory, a species' response to the environment is measured when it is unconstrained by intra- or interspecific competition (Chesson 2000). For each species in each year, we initialised an empty lattice with one individual (3-cm radius) of the focal species, drew survival and regression parameters for the appropriate year and calculated population growth as

$$r = \log[E(P_1)] - \log(P_0), \tag{4.1}$$

where P_0 is the proportion of the lattice in which the focal species is present at time 0 and P_1 is the proportion expected at time 1. For each set of year-specific parameters we repeated this one time step projection 250 times to account for parameter uncertainty. After estimating the yearly intrinsic population growth rates, we calculated species pairwise correlations. The weaker, or more negative, the pairwise correlations, the greater the potential for species to exploit unique temporal niches.

Pairwise correlations in the yearly intrinsic growth rates among grasses were similar in both the prairie and sagebrush communities, ranging from 0.13 to 0.44 (Table 4.1). These relatively weak positive correlations satisfy condition 2 of the storage effect by showing that species do not respond identically to temporal environmental variation. In the sagebrush community, pairwise correlations between the grasses

Table 4.1 *Correlation matrices for the simulated yearly intrinsic growth rates of the modelled species in the mixed prairie and sagebrush steppe communities.*

Mixed prairie	*B. curtipendula*	*B. hirsuta*	*S. scoparium*	
B. curtipendula	–	0.17	0.17	
B. hirsuta		–	0.44	
S. scoparium			–	
Sagebrush steppe	*P. spicata*	*Poa secunda*	*H. comata*	*A. tripartita*
P. spicata	–	0.28	0.44	−0.22
Poa secunda		–	0.13	−0.05
H. comata			–	−0.03
A. tripartita				–

and the sagebrush species were actually negative, showing even stronger potential for species-specific responses (Table 4.1).

The analysis is not focused on explaining why different species perform better in different years. Much of the variation may be explained by interannual variation in temperature or in the timing and amount of precipitation (Adler *et al.* 2006, 2009). While understanding these mechanisms will be critical for ecological forecasting, it is not essential for testing the role of the storage effect and environmental variation per se in stabilising coexistence.

Condition 3: covariance between environment and competition

We tested for condition 3, environment–competition covariance, by combining the estimated yearly intrinsic population growth rates with results from a second simulation to determine the effect of competition, or crowding, on growth. We initialised a series of grids in which we randomly varied the total abundance and the relative abundances of the three basal grasses and the canopy shrub. To create initial conditions with realistic spatial structure, we used sequential indicator kriging (as in Adler *et al.* 2006), assuming random spatial associations among the four species. For each year, we drew the appropriate survival and colonisation parameters and projected species' population growth rates forward one time step. Again, we repeated each projection many times to account for parameter uncertainty. We then regressed the projected growth rates for each species against square root transformed total cover, which serves as a proxy for intra- plus interspecific crowding. The slope of this relationship represents the effect of that crowding on population growth rates for a particular year. Finally, for each species, we plotted the slope of this

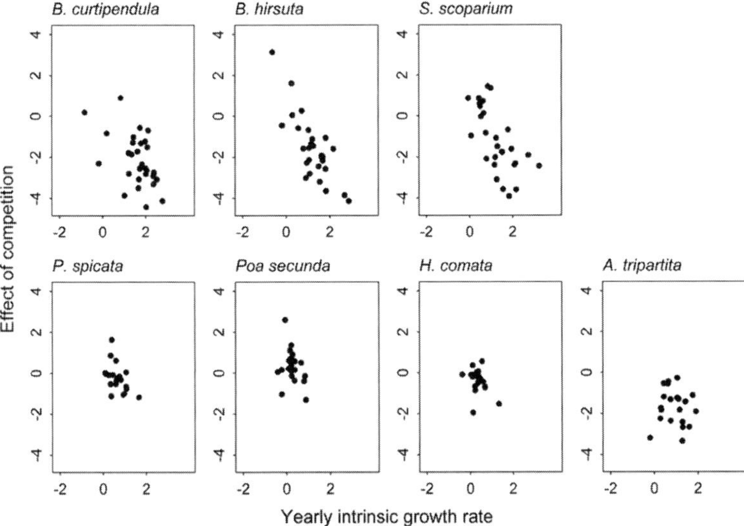

Figure 4.4 Covariance between the environment and the effect of competition on growth for the prairie (top row) and sagebrush steppe (bottom row) species. The favourability of the environment is measured by each species' simulated yearly intrinsic growth rate. The effect of competition in each year is measured by the slope relating total cover to the growth rate: negative values indicate that an increase in total cover decreases the simulated growth rate.

competition relationship for each year against the corresponding yearly intrinsic growth rate. Negative covariance, meaning stronger negative effects of crowding in more favourable years (higher intrinsic growth rates), is required for fluctuations to stabilise coexistence.

In the prairie community, we found strong negative relationships between the yearly intrinsic population growth rates and the per capita effect of crowding on growth: crowding had weak or even facilitative effects in unfavourable years (low intrinsic growth) but strong negative effects in favourable years (Figure 4.4, top row). For these three species, the covariances ranged from −0.50 to −1.12. These results provide strong evidence for condition 3 of the storage effect.

In the sagebrush steppe community, the relationships between the yearly intrinsic growth rates and the effect of competition on growth were negative, as for the prairie species, but were quite weak (Figure 4.4, bottom row). The covariances were up to an order of magnitude weaker, ranging from −0.02 to −0.13. Based on these results, it appears that condition 3 of the storage effect is not satisfied in the sagebrush steppe community.

Quantifying the strength of fluctuation-mediated coexistence

If all three conditions of the storage effect are satisfied, as they are for the prairie community if not the sagebrush steppe community, we can conclude that temporal environmental fluctuations should have a stabilising effect on coexistence. The next step is to determine the strength of this effect. To quantify the overall effect of interannual variability on coexistence (the combined effects of the storage effect and relative nonlinearity), we need to compare species' long-term average low-density growth rates in constant and variable environments.

We simulated each species' long-term low-density population growth rate by initialising 500 grids with the resident species at their stochastic equilibrium abundances (using sequential indicator kriging to create realistic spatial structure). We then introduced one individual (3-cm radius) of the focal or invader species at a random coordinate. To simulate a variable environment, we randomly selected one set of year-specific parameters and projected the focal species' growth, r, over one time step. We repeated these steps for each of the 500 grids. To simulate a constant environment, we used the mean parameters to project population growth from the same initial grids. For both the constant and variable environments, the long-term low-density growth rate is the mean of the respective r values over all replicate grids (the geometric mean of the annual population growth rates).

The crux of these simulations is defining the stochastic equilibrium of the resident species. This cannot be observed directly in our data, because we never observe our focal species invading a community in which they are not present. In addition, we want to define the stochastic equilibrium of the resident species for both variable and constant environments. We estimated the stochastic equilibria of the resident species with simulations of the lattice model: we initialised a grid with the residents at low abundance, but with the abundance of the focal species set to zero, and projected the system forward until some stochastic equilibrium was reached, using parameters representing either a constant or variable environment for one entire run. We repeated this projection many times for each focal species, saving values from the last half of each simulation. In the Discussion, I explore how these assumptions about the structure of the resident community influence the results.

In the prairie community, all three grass species had much higher long-term average population growth rates when rare under the variable environment than under constant mean conditions (Figure 4.5, top row). *B. curtipendula's* growth rate was almost twice as high in the variable

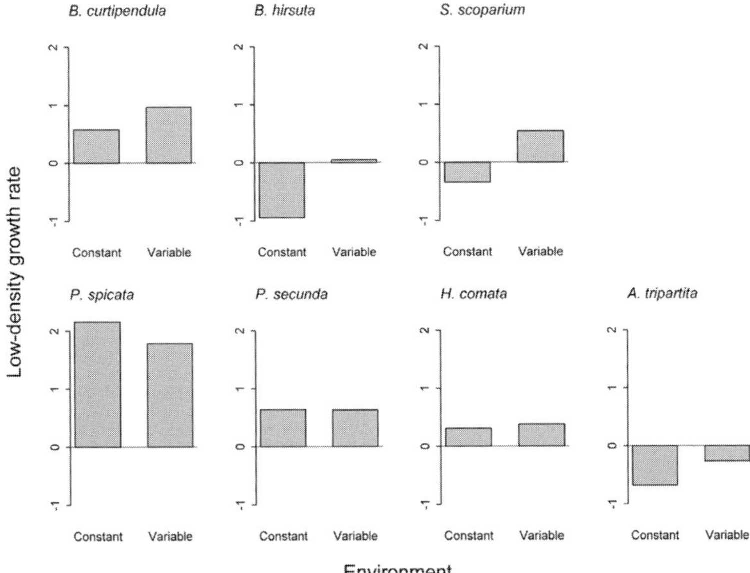

Figure 4.5 The overall effect of climate variability on coexistence is measured as the difference in low-density growth rates projected for constant and variable environments. Results for the prairie species are shown on the top row and sagebrush steppe species on the bottom row.

environment. Differences in growth rates were even larger for *B. hirsuta* and *S. scoparium*, which both had negative growth rates in a constant environment but positive growth rates in a variable one. The switch from negative to positive low-density growth implies that temporal variation may be essential for the persistence of these two species.

The results were much different for the sagebrush steppe community. For the three grasses, the simulated long-term, low-density population growth rates were either higher in the constant environment (*P. spicata*), or virtually equal in the constant and variable environments (*Poa secunda* and *H. comata*) (Figure 4.5, bottom row). *A. tripartita*'s long-term, low-density growth rate was slightly higher in the variable environment. The weak effect of environmental variation on coexistence in this community is consistent with the very weak environment–competition covariance.

4.4 DISCUSSION

The results show a dramatic contrast between a strong storage effect in the prairie grassland and a very weak storage effect in the sagebrush steppe. While we found evidence for the first two conditions of the storage effect in both communities, it was the third condition, environment–competition covariance, which made the difference. The covariances for the prairie species were up to an order of magnitude higher than those for the sagebrush steppe species. In the next sections, I first discuss how environmental differences between the sites may or may not help explain these contrasting results. Second, I discuss the influence of modelling assumptions on the results. Third, I address future directions in testing fluctuation-mediated coexistence.

Comparison of mixed prairie and sagebrush steppe

My comparison of environmental factors at the two study sites suggested that the lower resource availability in the sagebrush community might select for greater investment in long-lived life stages and thus a stronger storage effect. However, species from both sites were capable of living for decades, implying an adult stage insensitive to competition. Instead, subtle differences in the magnitude and predictability of environmental variation may provide a better explanation for the contrasting results. In the prairie system, the higher coefficient of variation for precipitation, uncertain summer rather than winter precipitation and a hint of autocorrelation in temperature might increase the potential for temporal niches and a strong storage effect.

Consistent with this hypothesis, the prairie community appeared more demographically dynamic than the sagebrush steppe community. The stronger environment–competition covariances of the prairie species reflect relatively high interannual variability in both intrinsic population growth rates (which represents species' response to the environment) and in competition. Yearly intrinsic growth rates had variances ranging from 0.58 to 1.07 for the prairie species, compared to a range of 0.11 to 0.30 for the sagebrush steppe species. For the prairie species, variances of the effect of competition on population growth ranged from 1.6 to 2.6, compared to a range of 0.33 to 0.78 for the sagebrush steppe species. These differences in demographic variability seem quite large relative to the subtle differences in the magnitude and predictability of environmental variation. My inability to explain such striking differences in demographic variability between sites shows how far we are from understanding cross-system variation in the importance of the storage effect.

The sagebrush steppe results also provide an important reminder that species-specific responses to the environment do not guarantee that the storage effect is operating. Without the third condition, environment–competition covariance, the storage effect cannot stabilise coexistence. Unfortunately, environment–competition covariance is the hardest condition of the storage effect to quantify, and previous studies of desert annuals (Pake and Venable 1996, Adondakis and Venable 2004, Facelli *et al.* 2005) have inferred an important role for the storage effect in coexistence based only on evidence for the first two conditions. My analysis highlights the danger of drawing such conclusions without information on the strength of environment–competition covariance.

Modelling assumptions

Models play an essential role in coexistence research by extending inference from observational studies and also from field experiments (e.g. Levine and HilleRisLambers 2009). In many cases, a model may offer the only feasible way to 'remove' stabilising mechanisms in order to quantify their role in maintaining diversity, as in my simulations of a constant environment. The critical role that models play in testing coexistence demands a careful evaluation of model assumptions. My collaborators and I have directly tested two aspects of our modelling approach, the first involving the simulation of resident communities, the second involving model structure.

For the analyses presented here, we used the empirical model to simulate the cover of resident species after removing one species from the community. An alternative approach is to base the projected resident community on historical abundances. Neither approach is perfect: the simulations assume that the model can project realistic patterns of abundance, while the historical approach extrapolates from observations based on arbitrary decisions. Adler *et al.* (2009) simulated low-density population growth rates in variable and constant environments using both methods for defining the resident communities. The quantitative results differed, with higher growth rates observed for the historically based resident communities than the simulated resident communities. For example, *A. tripartita* had negative low-density growth rates in constant and variable environments using simulated resident abundances (as in Figure 4.5), implying local extinction, but positive low-density rates using historical extrapolations for resident abundances. This absolute difference in growth rates reflects higher total resident cover in the simulated resident communities. However, the qualitative conclusion, that fluctuations have little effect on coexistence in the sagebrush steppe, did not change. I am confident that dramatic differences in the strength of the storage effect for the prairie and sagebrush site are not sensitive to assumptions about the resident community.

Many additional assumptions are contained in the model structure itself. The model I used assumes that the observed dynamics can be described by a survival process and a colonisation process operating on a grid of cells. To test whether the results are sensitive to such assumptions about demographic processes, Adler *et al.* (2006) repeated the analysis of the prairie community using a phenomenological multinomial regression model. While the values of the low-density growth rates changed, the conclusion that variability had a strong stabilising effect on coexistence did not. Similarly, an alternative analysis of the sagebrush steppe data, based on a model of individual genet demographics, confirmed the weak effect of temporal variability on the three grasses, but a slightly stronger effect on *A. tripartita* (Adler *et al.* 2010). These alternative analyses show that my qualitative conclusions are robust to specific model parameters and assumptions.

While such exercises provide some reassurance that the qualitative conclusions are robust, I will continue to work on building a better model. An improved model would produce more reliable predictions about the structure of resident communities under different environmental conditions or containing different combinations of species. The accurate description of these resident communities is critical for the quantification of invasion growth rates. Another desirable feature would be model parameters with direct demographic interpretations. In fact, with a cleverly designed model and recently developed theory (Chesson 2008), it should be possible to calculate the strength of the storage effect directly from the model parameters, avoiding most of the simulations I relied on for this analysis. Integral projection models (Ellner 2006, Ellner and Rees 2007), adapted for a multispecies context (Adler *et al.* 2010), could provide a promising solution to these challenges.

Towards a comparative coexistence programme

A general understanding of cross-system variation in the strength of the storage effect will require not only improved models but also advances in theory and data. Existing storage effect theory focuses on quantifying the strength of the effect in a particular community. While it clearly identifies the life-history traits and demographic rates essential to the mechanism, it does not predict where we should expect to find those traits. Making such predictions will require a synthesis of storage effect theory with the evolutionary ecology literature, which explains life-history evolution across environments. In particular, we need more information about the environmental cues that organisms rely on to adjust their demographic rates. To test new theories about cross-system patterns, we will need appropriate long-term data sets from many more communities. Now that we have the tools to quantify the strength of particular coexistence mechanisms in specific communities, we can conduct 'comparative coexistence' studies analogous to the comparative demography studies that have so successfully tested life-history theory (Silvertown *et al.* 1993, Franco and Silvertown 2004). As my use of historical censuses demonstrates, the necessary data may already be collected and waiting for analysis.

ACKNOWLEDGEMENTS

Jonathan Levine, Janneke HilleRisLambers, Phaedon Kyriakidis and Qingfeng Guan made important contributions to the original research summarised in this chapter. I thank Harmony Dalgleish for help with an earlier version of the manuscript, Colleen Kelly and Michael Bowler for suggestions that improved the final product, and Corey Moffet of the US Experiment Station (USDA-ARS) for access to the raw chart quadrat data. I was supported by National Science Foundation grant DEB-0614068 and the Utah Agricultural Experiment Station, Utah State University.

REFERENCES

Adler, P. B. and Drake, J. M. (2008). Environmental variation, stochastic extinction, and competitive coexistence. *American Naturalist* **172**, 186–95.

Adler, P. B., Ellner, S. P. and Levine, J. M. (2010). Coexistence of perennial plants: an embarrassment of niches. *Ecology Letters* **13**, 1019–1029.

Adler, P. B., HilleRisLambers, J., Kyriakidis, P., Guan, Q. and Levine, J. M. (2006). Climate variability has a stabilizing effect on coexistence of prairie grasses. *Proceedings of the National Academy of Sciences, USA* **103**, 12793–12798.

Adler, P. B., HilleRisLambers, J. and Levine, J. M. (2009). Weak effect of climate variability on coexistence in a sagebrush steppe community. *Ecology* **90**, 3303–3312.

Adondakis, S. and Venable, D. L. (2004). Dormancy and germination in a guild of Sonoran Desert annuals. *Ecology* **85**, 2582–2590.

Albertson, F. W. and Tomanek, G. W. (1965). Vegetation changes during a 30-year period in grassland communities near Hays, Kansas. *Ecology* **46**, 714–720.

Allan, R. P. and Soden, B. J. (2008). Atmospheric warming and the amplification of precipitation extremes. *Science* **321**, 1481–1484.

Angert, A. L., Huxman, T. E., Chesson, P. and Venable, D. L. (2009). Functional tradeoffs determine species coexistence via the storage effect. *Proceedings of the National Academy of Sciences, USA* **106**, 11641–11645.

Cáceres, C. E. (1997). Temporal variation, dormancy, and coexistence: a field test of the storage effect. *Proceedings of the National Academy of Sciences, USA* **94**, 9171–9175.

Chesson, P. (1990). Geometry, heterogeneity and competition in variable environments. *Philosophical Transactions of the Royal Society of London B* **330**, 165–173.

Chesson, P. (2000). Mechanisms of maintenance of species diversity. *Annual Review of Ecology and Systematics* **31**, 343–366.

Chesson, P. (2008). Quantifying and testing species coexistence mechanisms. In F. Valladares, A. Camacho, A. Elosegui *et al.* (eds), *Unity in Diversity: Reflections on Ecology after the Legacy of Ramon Margalef*. Bilbao, Spain: Fundación BBVA, pp. 119–164.

Chesson, P. and Huntly, N. (1989). Short-term instabilities and long-term community dynamics. *Trends in Ecology and Evolution* **4**, 293–298.

Chesson, P. L. and Warner, R. R. (1981). Environmental variability promotes coexistence in lottery competitive systems. *American Naturalist* **117**, 923–943.

Chesson, P., Gebauer, R. L. E., Schwinning, S. *et al.* (2004). Resource pulses, species interactions, and diversity maintenance in arid and semi-arid environments. *Oecologia* **141**, 236–253.

Clark, J. S. (2005). Why environmental scientists are becoming Bayesians. *Ecology Letters* **8**, 2–14.

Descamps-Julien, B. and Gonzalez, A. (2005). Stable coexistence in a fluctuating environment: an experimental demonstration. *Ecology* **86**, 2815–2824.

Ellner, S. and Rees, M. (2007). Stochastic stable population growth in integral projection models: theory and application. *Journal of Mathematical Biology* **54**, 227–256.

Ellner, S. R. (2006). Integral projection models for species with complex demography. *American Naturalist* **167**, 410–428.

Facelli, J., Chesson, P. and Barnes, N. (2005). Differences in seed biology of annual plants in arid lands: a key ingredient of the storage effect. *Ecology* **86**, 2998–3006.

Franco, M. and Silvertown, J. (2004). Comparative demography of plants based upon elasticities of vital rates. *Ecology* **85**, 531–538.

Hill, R. R. (1920). Charting quadrats with a pantograph. *Ecology* **1**, 270–273.

Hutchinson, G. E. (1941). Ecological aspects of succession in natural populations. *American Naturalist* **75**, 406.

Jain, S. L., Hoerling, M. and Eischeid, J. (2005). Decreasing reliability and increasing synchroneity of western North American streamflow. *Journal of Climate* **18**, 613–618.

Karl, T. R. and Trenberth, K. E. (2003). Modern global climate change. *Science* **302**, 1719–1723.

Kelly, C. K. and Bowler, M. G. (2002). Coexistence and relative abundance in forest trees. *Nature* **417**, 437–440.

Kelly, C. K., Bowler, M. G., Pybus, O. and Harvey, P. H. (2008). Phylogeny, niches, and relative abundance in natural communities. *Ecology* **89**, 962–970.

Lauenroth, W. K. and Adler, P. B. (2008). Demography of perennial grassland plants: survival, life expectancy and life span. *Journal of Ecology* **96**, 1023–1032.

Levine, J. and HilleRisLambers, J. (2009). The importance of niches for the maintenance of species diversity. *Nature* **461**, 254–257.

Pake, C. E. and Venable, D. L. (1995). Is coexistence of Sonoran Desert annuals mediated by temporal variability in reproductive success? *Ecology* **76**, 246–261.

Pake, C. E. and Venable, D. L. (1996). Seed banks in desert annuals: implications for persistence and coexistence in variable environments. *Ecology* **77**, 1427–1435.

Rosenzweig, M. L. and MacArthur, R. H. (1963). Graphical representation and stability conditions of predator-prey interactions. *American Naturalist* **97**, 209–223.

Salinger, M. (2005). Climate variability and change: past, present and future – an overview. *Climatic Change* **70**, 9–29.

Silvertown, J., Franco, M., Pisanty, I. and Mendoza, A. (1993). Comparative plant demography: relative importance of life-cycle components to the finite rate of increase in woody and herbaceous perennials. *Journal of Ecology* **81**, 465–476.

Warner, R. R. and Chesson, P. L. (1985). Coexistence mediated by recruitment fluctuations: a field guide to the storage effect. *American Naturalist* **125**, 769–787.

West, N. E., Rea, K. H. and Harniss, R. O. (1979). Plant demographic studies in sagebrush-grass communities of southeastern Idaho. *Ecology* **60**, 376–388.

MICK E. HANLEY AND REBECCA J. SYKES

5

Seedling herbivory and the temporal niche

5.1 SEEDLINGS AND THE TEMPORAL NICHE

The facilitation of plant coexistence via temporal variation in plant recruitment is increasingly studied (see this volume plus Pake and Venable 1996, Chesson and Huntly 1997, Kelly and Bowler 2002, Verhulst *et al.* 2008). For the most part however, corroborating studies have examined fluctuations in abiotic factors and the role of biotic agents has been largely overlooked. This omission is symptomatic of the plant coexistence literature in general; the role of predators, herbivores, pathogens and parasites in maintaining species coexistence is more often assumed than demonstrated (but see Kelly and Bowler 2009a). Nonetheless, while a number of agents, biotic and abiotic, result in the death of entire seedling cohorts, foremost among the factors limiting seedling recruitment is herbivory (Moles and Westoby 2004, Fenner and Thompson 2005). Herbivore attack has obvious effects on seedling demography (Lindquist and Carroll 2004, Maron and Crone 2006, Maron and Kauffman 2006), but even beyond population-level considerations, selective seedling removal also exerts long-lasting effects on plant community composition. We propose that temporal fluctuation in herbivore populations, and consequently variation in the intensity of herbivory experienced by plants during their regeneration phase, exerts a powerful influence over plant species contribution to the established community.

There are four necessary conditions of any temporal dynamic involving herbivory. First, seedling herbivores must be capable of moderating plant community composition in established vegetation.

Temporal Dynamics and Ecological Process, ed. C. K. Kelly, M. G. Bowler, G. A. Fox.
Published by Cambridge University Press.

Second, herbivores should select preferred seedlings on the basis of readily apparent ecophysiological characteristics. Third, and related to the previous assumption, any variation in seedling susceptibility to herbivore attack (i.e. defensive traits) will most probably correlate with competitive ability. Finally, herbivore populations must show fluctuations in numbers and therefore variation in their influence on regenerating plants. Consequently, before it is possible to develop any conceptual framework to explain how temporal variation in seedling herbivory influences species coexistence, we must first evaluate the evidence for these conditions.

5.2 SEEDLING HERBIVORY AND PLANT COMMUNITY COMPOSITION

At the outset it is worth noting that we use the term 'seedling' to mean any plant still dependent on stored nutrient reserves, usually the cotyledon, for their nutrition (see Hanley *et al.* 2004). However, we also use the term 'juvenile', which includes seedlings and young plants that have not yet reached maturity and are thus actively recruiting to the established community.

Hanley *et al.* (1995a) was one of the first studies investigating the effects of selective seedling herbivory on subsequent plant community development. In September 1993, seeds of six species were sown into 20 1-m^2 temperate grassland plots cleared of all above- and below-ground vegetation. The timing was coincidental with the main germination period for most plants in this habitat and also with the peak abundance of the principal seedling herbivore: terrestrial molluscs. Immediately after seeds were sown, molluscicide was applied to the periphery of 10 plots in order to exclude slugs and snails for the first month following germination. Subsequently, seedling number, followed by percentage cover estimates when the plant community had established, tracked the course of plant community development in plots with and without seedling predators. Significant treatment effects were found for both individual plant number and relative cover for all six species during the first 9 months of the experiment. Four species (*Agrostis capillaris*, *Stellaria graminea*, *Taraxacum officinale* and *Trifolium repens*) performed better in plots from which molluscs were excluded during seedling establishment; presumably because they were selectively removed by molluscs during the seedling stage. The two remaining species, *Ranunculus repens* and *Senecio jacobaea*, performed better in

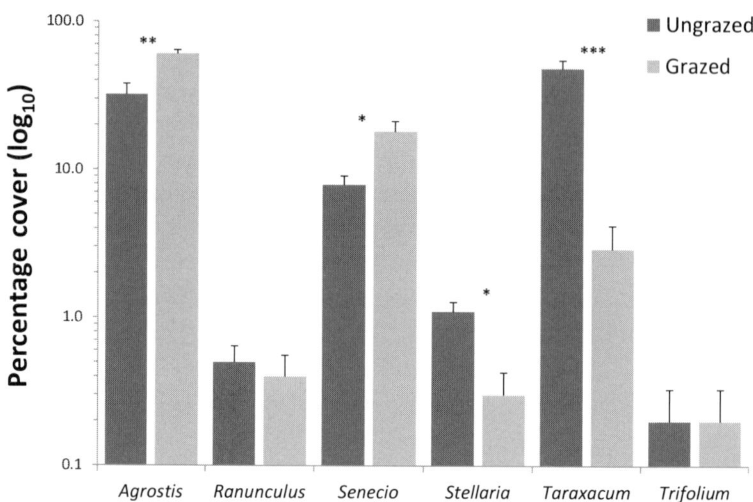

Figure 5.1 Relative cover of six temperate grassland plant species 34 months after seedlings in half of 20 experimental plots were subject to mollusc herbivory (see Hanley *et al.* 1995a for details). Results of one-way ANOVA on relative species cover shown as $^{*}\ p < 0.05$, $^{**}\ p < 0.01$, $^{***}\ p < 0.001$.

grazed plots. Relatively low rates of seedling attack on *Senecio* in particular, and the absence of competition from faster growing species preferentially removed by molluscs during the first month after germination, allowed *Senecio* to dominate the established vegetation of grazed plots.

The plots were monitored for a further 2 years, and percentage cover data obtained 34 months after the experiment began reinforced the long-term influence that seedling selection by molluscs had on the plant community (Figure 5.1). Two species, *Stellaria graminea* and *Taraxacum officinale*, retained significantly greater cover in plots wherein seedlings were exposed to mollusc herbivory during the regeneration stage. By contrast, along with *Agrostis capillaris*, which had successfully colonised by virtue of its capacity for vegetative reproduction, *Senecio jacobaea* continued to dominate the established vegetation of grazed plots.

This kind of impact of seedling herbivory on plant recruitment is apparent across a variety of ecosystems and through the actions of many different herbivores (Izahki and Ne'eman 1996, Green *et al.* 1997, Vasconcelos and Cherrett 1997, Lindquist and Carroll 2004, Asquith and Mejia-Chang 2005, Buschmann *et al.* 2005). The important point emerging from these varied examples is that the composition of the mature plant community shifts according to herbivory experienced during the regeneration stage. Moreover these effects vary from species to species, such that the recruitment and abundance of some plants are favoured by the presence of seedling herbivores while others apparently benefit from their absence. Precisely this effect was shown by Burt-Smith *et al.* (2003), who describe how the ranking of seedling resistance to herbivory in laboratory feeding trials (with the generalist grasshopper *Acheta domestica*) was negatively correlated with relative abundance of mature plants at Cedar Creek, Minnesota, USA. The six least acceptable species in the feeding trials comprised over 60% of the relative cover of mature plants in the synthesised prairie communities, while the most acceptable seedling species never contributed more than 5% to the established community. More recently, Strauss *et al.* (2009) reported a similar negative correlation between seedling palatability and relative abundance for five herbaceous plant species attacked by slugs in California.

These studies strongly suggest that seedling selection, and therefore relative abundance of different plant species in the established community, can be influenced greatly by seedling susceptibility to attack by generalist herbivores; plant species that produced unpalatable seedlings were consistently more abundant. Yet less common, but more palatable, species were also present in the community, suggesting the possibility that they are able to exploit a relaxation of seedling herbivory in order to recruit. Variation in seedling recruitment is fundamental to the assumptions of temporal niche models and evidently herbivory is a major cause of fluctuation in seedling establishment success. Thus seedling selection by herbivores can influence plant species composition in mature vegetation; but why are certain seedlings selected by herbivores while their neighbours are ignored?

5.3 PATTERNS OF SEEDLING SELECTION

A number of experiments have confirmed that plant species vary consistently in their relative acceptability to generalist seedling herbivores. Of the species used in Hanley *et al.*'s (1995a) study, for example, *Stellaria graminea* and *Taraxacum officinale* seedlings are amongst the most acceptable of all temperate grassland species, while *Senecio jacobaea* is consistently avoided (Hanley *et al.* 1995b, Hanley 2004). In the most wide-ranging study of its kind, Fenner *et al.* (1999) examined the relative seedling and adult palatability of 29 different northwest

European plant species. This work not only showed considerable variation in patterns of mollusc selection of herb and grass species, it also demonstrated that seedling palatability is very different from that displayed by adult conspecifics.

Because Fenner *et al.* (1999) quantified seedling palatability by means of plant extracts incorporated into agar gel they concluded that seedling selection by molluscs was driven largely by chemical cues. The assertion has subsequently been corroborated by Hanley *et al.* (2011). Thus, while factors such as age, phenology and neighbour environment are important (Hanley *et al.* 1995b, 1996a, 1996b, Hanley 2004), seedling selection is based primarily on the development and expression of anti-herbivore defences. In the case of *Senecio jacobaea*, for example, even from a very early age seedlings contain large concentrations of pyrrolizidine alkaloids (Schaffner *et al.* 2003, Elger *et al.* 2009). Other species, by contrast, have poorly developed chemical or structural defences during early seedling ontogeny, only developing anti-herbivore deterrents as they approach maturity (Boege and Marquis 2005, Barton 2007). Although there is a surprising lack of information on relative seedling defence, especially for plant species drawn from the same habitat, Elger *et al.* (2009) show that seedlings of English temperate grassland plant species vary greatly in the expression of different chemical defences.

Species-specific variation in the expression of anti-herbivore defence is pivotal in dictating the likelihood of seedling herbivory. Hanley and Lamont (2001) for example, exposed Proteaceae seedlings to herbivory in the Mediterranean-climate shrublands of Western Australia. Although there was variation between study sites, Hanley and Lamont (2001) reported a consistent negative relationship between rates of herbivory and seedling phenolic concentrations across the 12 species examined. A study with captive kangaroos also demonstrated a significant negative relationship between shoot phenolic content and kangaroo selection of 4-month old juvenile *Hakea* (Proteaceae) species (Rafferty *et al.* 2005). Beyond Hanley and Lamont (2001) and Rafferty *et al.* (2005), however, few studies have explicitly examined the link between seedling defence and herbivore selection. Nonetheless, it is apparent that significant species-specific variation in defence exists and it is also likely that this variation dictates patterns of selection by herbivores. Put simply, well-defended seedlings are less likely to be eaten.

If anti-herbivore defence is so beneficial to seedling establishment, why then do only some seedlings possess well-developed chemical or structural defences? Although there has been considerable debate on the topic, most theories assume that anti-herbivore defence will entail some kind of cost to the plant (Herms and Matson 1992, Stamp 2003, Agrawal and Fishbein 2006); an idea most succinctly encapsulated in Herms and Matson's (1992) dilemma for plants: 'to grow or defend'. In essence Herms and Matson's (1992) growth differentiation balance hypothesis (GDBH) suggests that where the resources required for plant growth are limiting and plant competition is intense, the allocation costs of anti-herbivore defence are such that investment in defence inevitably means a reduction in resource allocation elsewhere.

Although these so-called fitness costs are by no means established for all species (Koricheva 2002, Haring *et al.* 2008), where apparent, they are most manifest in reduced plant growth or fecundity (Fine *et al.* 2006, Glynn *et al.* 2007). However, as with many aspects of the plant defence and herbivory literature, there is a remarkable paucity of information regarding growth–defence tradeoffs for seedlings (Hanley *et al.* 2007b). Nonetheless, even at the recruitment stage, plants can be assumed to face the 'growth versus defence' dilemma (Herms and Mattson 1992, Boege and Marquis 2005, Kelly and Hanley 2005) and this tradeoff may go a long way towards explaining how selective seedling removal has so marked an effect on plant community composition.

Species that allocate resources to seedling defence can be at a significant advantage over faster growing, but poorly defended species when germination coincides with high rates of seedling herbivory (Hanley *et al.* 1995a). However, when seedling herbivory is relaxed, the faster maturing seedlings may dominate the plant community by virtue of superior competitive ability (Hanley *et al.* 1995a). Moreover, in the only systematic multispecies comparison, Kelly and Hanley (2005) showed a consistent negative association between seedling growth and palatability to the snail *Helix aspersa*, supporting an allocation tradeoff between anti-herbivore defence and seedling competitive ability. Species-specific variation in, and tradeoffs between, ecophysiological traits expressed during the recruitment stage are fundamental assumptions of our theory.

5.4 TEMPORAL VARIATION IN HERBIVORE POPULATIONS

Plant species coexistence due to temporal fluctuations in recruitment success is only possible when the factors limiting seedling recruitment vary through time. There is a wealth of evidence of year-to-year fluctuations in the abundance and activity of seedling herbivores. These include the classic studies on rodent population cycles conducted by

Charles Elton in the early part of the twentieth century through to more recent studies demonstrating yearly fluctuations in the intensity of vole herbivory on 1–2-year-old tree saplings in Finnish boreal forest (Elton 1924, Vehviläinen and Koricheva 2006). In fact, the many studies documenting population cycling of voles, lemmings and hares in boreal and temporal ecosystems (Elton 1927, Meserve 1971, Kendall *et al.* 1998, Lambin *et al.* 2006, Ims *et al.* 2007) offer by far the strongest support for the long-term temporal shifts in herbivore populations required to elicit temporal variation in the intensity of seedling herbivory.

Evidence of temporal population fluctuations is not restricted to rodents. Barnes and Weil (1944) and Symondson *et al.* (2002) for example demonstrate year-to-year changes in mollusc populations. Symondson *et al.*'s (2002) study of slug populations in an English arable system over a 5-year period showed important variation between years; e.g. between 1992 and 1996 the highest and lowest density of slugs recorded from standardised soil samples varied by up to four orders of magnitude, with at least an order of magnitude difference between any one monthly sample. Moreover, over a 3-year period, Barnes and Weil (1944) reported a five-fold difference in slug numbers recorded during the month of October. Slugs are one of the major seedling herbivores in temperate ecosystems (Crawley 1997), and October is the month during which the majority of autumn-germinating plant species in temperate northern hemisphere ecosystems pass through the seedling phase. Any year-to-year variation in the abundance of the organism responsible for the majority of seedling losses can be expected to have significant repercussions for plant species demography and plant community composition.

Much more work needs to be done on the interaction between herbivore population cycles and seedling recruitment before we can say definitively that temporal variation in the intensity of seedling herbivory influences plant species coexistence. Nevertheless, it is increasingly apparent that relaxation of herbivory via population cycling is the only way in which some plants are able to recruit to the mature community (Bryant *et al.* 2009, Lindquist *et al.* 2009).

5.5 TEMPORAL FLUCTUATIONS IN HERBIVORY AND SEEDLING ESTABLISHMENT

Given that temporal fluctuations in herbivore populations can theoretically combine with variation in seedling defence differentially to affect plant recruitment, we propose a conceptual framework to explain

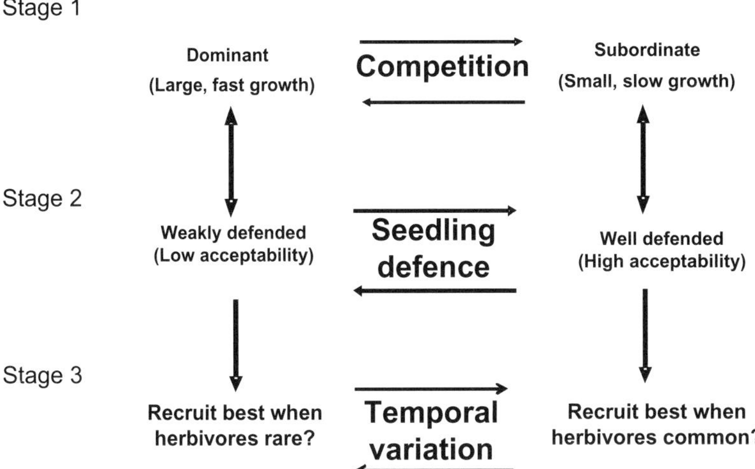

Figure 5.2 A conceptual framework linking seedling competitive ability, anti-herbivore defence (acceptability) and herbivore abundance. Temporal shifts in the intensity of herbivory interact with seedling acceptability and competitive ability to dictate which species pass through the establishment phase and dominate regeneration microsites.

how temporal variation in seedling herbivory helps maintain plant species coexistence (Figure 5.2).

All else being equal, established ecological theory predicts that a more competitive species will exclude a weaker competitor (Gause 1932). Thus, inherently faster growing or larger seedlings should capture resources more quickly and efficiently than smaller, slower growing neighbours and win the battle for dominance over establishment microsites (Figure 5.2, Stage 1). However, things are seldom equal and environmental perturbations frequently upset the balance of competition. One such perturbation is herbivory, and a large body of research also shows that well-defended (low acceptability) plants are less prone to herbivore attack than their weakly defended (high acceptability) neighbours (Figure 5.2, Stage 2). As we have seen, however, the growth–defence tradeoff increases the likelihood that competitively dominant plant species will be only weakly defended against herbivores. Consequently there is scope for the interaction between competition and herbivory to dictate the course of seedling recruitment: the removal of a dominant (poorly defended) competitor by herbivores will facilitate the survival of subordinate (well-defended) neighbours through the process of exploiter-mediated coexistence.

This concept is nothing new. Darwin (1859) briefly noted in *The Origin of Species* that the relaxation of mowing and browsing influenced plant community composition via this mechanism. However, we add a temporal component (Stage 3): through temporal fluctuations in the number of herbivores present during seedling regeneration, different species survive and establish to maturity depending on their allocation to anti-herbivore defence, their competitive ability, and the intensity of herbivory operating during the establishment phase. By virtue of their competitive ability, poorly defended seedlings have a significant establishment advantage when herbivory is absent or at low intensity. Their slow-growing but well-defended neighbours, by contrast, have the advantage when seedling herbivory is intense enough to remove the competition.

Although established theory predicts growth–defence tradeoffs at the seedling stage (Herms and Matson 1992, Boege and Marquis 2005), beyond Kelly and Hanley (2005) and Fine *et al.* (2006), there have been few systematic attempts to explore this relationship across a range of sympatric plant species. Second, there is only circumstantial evidence from herbivore exclusion experiments (e.g. Hanley *et al.* 1995a) that fluctuations in the intensity of herbivory at the seedling stage can facilitate plant species coexistence through the differential recruitment of species with different defence/growth trait combinations.

5.6 EVIDENCE FOR THE GROWTH–DEFENCE TRADEOFF IN SEEDLINGS

Although Herms and Matson's (1992) theory of a plant growth–defence tradeoff is widely accepted, the expected fitness costs of adopting anti-herbivore defence have proved difficult to elucidate (Bergelson and Purrington 1996, Koricheva 2002). This failure stems from a number of reasons, including shifting ontogenetic trends in plant defence (Boege and Marquis 2005, Elger *et al.* 2009) and the only recently recognised role of plant tolerance as an alternative means of coping with herbivory (Agrawal and Fishbein 2006, Hanley and Fegan 2007). For seedlings there is a further complication in that the GDBH actually predicts that initial inability to acquire resources from the external environment will effectively limit the production of anti-herbivore defences. Nonetheless, while recent evidence suggests that anti-herbivore defence may increase significantly with seedling age (Elger *et al.* 2009, Barton and Koricheva 2010), even very young seedlings exhibit anti-herbivore defences (Elger *et al.* 2009). Moreover these defences vary considerably between coexisting species (Hanley and Lamont 2001, 2002, Elger *et al.* 2009).

One of the main problems associated with elucidation of a growth–defence tradeoff has been identification of the relevant costs and benefits. This is perhaps a particular problem associated with studies of mature plants since the costs may be varied (shoot or root growth, flowering, seed size or number, etc.) and not apparent for many years after herbivory has occurred. For seedlings, matters are generally more straightforward. Herbivory usually kills the plant and there is no more obvious fitness cost than death. Consequently, not only is a better understanding of seedling growth–defence tradeoffs relevant to our plant species coexistence framework, it may provide a much more tractable way of examining the relationship between growth and defence than studies involving mature plants.

We examined growth–defence tradeoffs at the seedling stage using ten sympatric, herbaceous, chalk grassland plant species. Seeds of each species were collected from Weather Hill (51°15′N, 1°42′W), Salisbury Plain, England, during September 2004. We selected congeneric pairs on the basis that their shared evolutionary history confers a fundamental physiological similarity, increasing the likelihood and intensity of competition (Kelly and Bowler 2005, Kelly and Hanley 2005). Congeneric pairs are also likely to offer a chance to compare quantitative differences in seedling defence that represent simple additions or subtractions to that exhibited by its close relative (Harborne 1993). Seeds of each of the test species were set to germinate in 90-mm diameter plastic Petri dishes containing two layers of 90-mm diameter Whatman No. 1 filter paper and 5 ml of distilled water. The dishes were maintained in a dark incubator set at 15°C. Immediately following appearance of the radical, seedlings were transferred into 50-mm diameter plastic plant pots. The potting medium used was rendzina soil collected from Weather Hill, sieved through a 15-mm^2 mesh prior to use.

Since it is difficult to quantify all those aspects of a seedling's ecophysiology and morphology that would confer upon it any level of anti-herbivore defence, we took acceptability to a generalist herbivore as our surrogate for seedling allocation to defence. This approach is broadly similar to that used by Kelly and Hanley (2005), but here we examined herbivore selection of whole seedlings; this allowed us to integrate the effects of structural defences in addition to the chemical defences that were the focus in Kelly and Hanley's (2005) study with seedling extracts incorporated into agar discs.

We planted two 7-day-old seedlings of the same 'test' species 45 mm apart in 50-mm diameter plant pots filled with rendzina. These seedlings were grown on in greenhouse conditions (mean daily min temp = 17.3°C ± 0.2°C; max = 22.6°C ± 0.2°C; 12-h day:night) for 7 days, at which time two newly emerged lettuce seedlings (*cv* Tom Thumb) were planted 45 mm apart in the same pot, perpendicular to the 'test' seedlings. Lettuce seedlings, cultivated simultaneously in large plastic trays containing commercial potting compost, were used to ascertain the relative acceptability of the 'test' species and allow comparison between congeneric pairs with reference to the same 'index' species (Fenner *et al.* 1999). Rapid development of lettuce seedlings compared with the test species meant that 7-day-old seedlings were at approximately the same ontogenetic stage as 14-day-old test seedlings. When the test seedlings were 14 days old they were exposed to herbivory by snails (*Helix aspersa* Müller). Five replicate pots for each test species were sunk into large plastic propagator trays (350 mm × 215 mm × 70 mm deep) filled with commercial potting compost, such that the top of each pot was flush with the level of the compost. One pot was placed into the centre of each tray, with the remaining four pots located in the tray corners. This arrangement was replicated ten times for each test species. Four snails (*Helix aspersa*) of uniform size (circa 30 mm diameter) were then added to each tray and retained overnight (≈ 16 h) using a clear plastic propagator lid (350 mm × 215 mm × 115 mm deep). The total number of test species and lettuce (index) seedlings attacked by snails was determined for each replicate tray (all attacked seedlings suffered 100% aboveground tissue loss). These values were used to calculate an acceptability index (AI) for seedlings within individual trays; based on the formula given by Fenner *et al.* (1999), AI per tray is equal to

$$\frac{\text{Mean number of test seedlings attacked}}{\text{Mean number of test + index seedlings attacked}}.$$

Average AI for each test species was then calculated across all 10 replicate trays.

Alongside the seedling acceptability trial, seedling growth was quantified by planting one newly germinated seedling into the centre of a 50-mm diameter pot containing rendzina. Twelve seedlings of each of the ten 'test' species were grown on in the same greenhouse conditions until 14 days old before being removed from the pots, cleaned of any adhering soil and oven dried for 24 h at 60°C. We then quantified dry weight biomass for each seedling. Following the reasoning of Kelly and Hanley (2005), we used absolute size at 14 days old rather than other commonly employed measures of plant growth such as relative growth rate (RGR) to compare seedling competitive ability. Although absolute size naturally incorporates differences in RGR between species, it also allows for the effects of initial seed mass on seedling competitive ability: larger seeded species often produce larger, more competitive seedlings (Westoby *et al.* 1996, Hanley *et al.* 2007a).

A comparison of mean growth versus seedling acceptability (AI) between congeneric pairs shows that in four out of five comparisons the most acceptable species was also the larger (Table 5.1). When coupled with the results from Kelly and Hanley's (2005) study of growth–defence tradeoffs in seedlings of four other congeneric temperate herbaceous species pairs (five of five comparisons showed the same trend, although one of Kelly and Hanley's (2005) pairs also involved *T. pratense* and is excluded here), we find support for a growth–defence tradeoff at the seedling stage ($p = 0.02$; Siegal and Castellan, 1988).

5.7 EVIDENCE: VARIATION IN INTENSITY OF HERBIVORY AT SEEDLING STAGE

Assuming that the seedling growth–defence tradeoff holds for at least some plant species comparisons (see also Fine *et al.* 2006), our framework predicts that variation in herbivore pressure will differentially affect the recruitment success of two plant species with contrasting competitive and anti-herbivore defence capabilities. Thus at low intensities of seedling herbivory, weakly defended but more competitive species will dominate any new recruitment opportunities, while more intense herbivore pressure will favour well defended, but generally less competitive species. Of the four species pairs that supported the growth–defence tradeoff in our experiment with chalk grassland congeners, the *Trifolium* pair exhibited the greatest contrast between relative seedling acceptability and growth. *T. pratense* seedlings were much larger than *T. repens* seedlings (one-way ANOVA $F_{1,22} = 76.26$ $p < 0.0001$ following $\ln(x + 1)$ transformation), and also more susceptible to snail herbivory (one-way ANOVA $F_{1,18} = 5.22$, $p = 0.035$ following arcsine transformation). Consequently, we selected this congeneric pair to examine how fluctuation in herbivore pressure influenced seedling recruitment success.

Full details of this experiment are reported by Hanley and Sykes (2009), but a summary of the methods and results are as follows. Following germination in Petri dishes, *Trifolium pratense* and *T. repens*

Table 5.1 *Comparison of relative seedling growth and acceptability for five congeneric plant species native to the chalk grasslands of southern England. Acceptability to snails (*Helix aspersa*) at 14 days old was calculated as seedling consumption relative to a standard index species lettuce (*Lactuca sativa*). Seedling growth was quantified as absolute dry weight biomass. Evidence for a growth defence tradeoff (*sensu *Herms and Matson 1992) for each congeneric pair is also noted.*

Species pair	Seedling acceptability index	Seedling mass (mg) at 14 days old	Evidence for growth–defence tradeoff?
Leontodon autumnalis	0.48	1.2	No
Leontodon hispidus	0.28	1.9	
Ononis repens	0.64	9.1	Yes
Ononis spinosa	0.57	5.4	
Plantago lanceolata	0.63	5.7	Yes
Plantago media	0.57	2.3	
Rumex acetosa	0.48	6.2	Yes
Rumex crispus	0.59	6.6	
Trifolium pratense	0.64	5.3	Yes
Trifolium repens	0.33	2.2	

seedlings were transferred to 110-mm diameter pots containing rendzina. Eleven seedlings of each species were planted into the same regular stratified hexagonal array in each of the 24 pots. When the seedlings were 14 days old, each pot was sunk into a plastic box and exposed to 0, 2, 5 or 10 snails (*Helix aspersa*) overnight to simulate spatiotemporal variation in grazing intensity. Following snail removal, plants were cultivated for a further 106 days at which time all plants were harvested. The dry weight biomass of individual plants, and for the species as a whole, was taken as a measure of how interspecific competition between the two species was affected by snail herbivory at the seedling stage (Figure 5.3).

Whether measured by total species biomass or mean individual biomass, variation in the intensity of herbivory applied to the *Trifolium* seedling mixtures at 14 days old had a marked effect on the balance of competition. In the ungrazed treatment (zero snails), the pots were dominated by the more competitive *T. pratense*. In the two-snail treatment there appears to be codominance between the two *Trifolium* species. However, when the intensity of seedling herbivory was increased in the five-snail treatment, we observed community

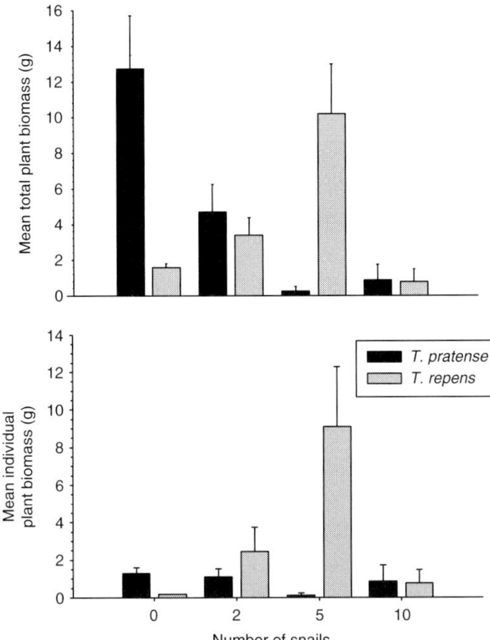

Figure 5.3 Mean total dry weight biomass (**a**) and mean individual plant biomass (**b**) of *Trifolium repens* and *T. pratense* plants 106 d following seedling exposure to snails (*Helix aspersa*). Eleven seedlings of each species were planted in hexagonal arrays in 110-mm diameter pots and at 14 days old subjected to herbivory by 0, 2, 5 or 10 snails overnight to simulate spatiotemporal variation in the intensity of seedling herbivory. Data shown are mean biomass values per pot averaged across all replicate pots (± SE). From Hanley and Sykes (2009).

dominance by the weaker competitor (*T. repens*), almost certainly by virtue of its relative unacceptability to snail attack. At the most intense level of herbivory (ten snails) nearly all plants were removed at the seedling stage. These results provide support for our hypothesis that fluctuations in herbivore pressure at the seedling stage differentially affect recruitment to the mature phase.

5.8 SYNTHESIS AND FUTURE DIRECTIONS

The results of our greenhouse experiment with *Trifolium* offer an insight into how fluctuation in seedling herbivory might affect interspecific competition and so influence community composition in established vegetation. We recognise that this particular aspect of the temporal niche is likely only one of many factors that shape the course of seedling recruitment. Nonetheless, seedling herbivory is a powerful selective filter and should be more widely examined as a mechanism promoting species diversity.

The potential for a link between our model and contemporary temporal niche theory is no better illustrated than by reference to one of the most recent, differential sensitivity storage theory (Kelly and Bowler 2002, 2005, 2009b). This theory predicts that for a pair of ecologically similar, common and rare species, the more abundant will be less sensitive to the environment and as a result show less fluctuation in recruitment from juvenile to adult stages. To enable persistence, the rarer species must be a stronger competitor than the more common species, possessing a competitive advantage in the form of faster growth. As the primary agent of seedling mortality (Moles and Westoby 2004) and because of ecophysiological differences in defence and growth traits exhibited at the seedling stage (Grime *et al.* 1997, Elger *et al.* 2009), it is not unreasonable to assume that shifts in herbivore pressure are a prime cause of fluctuation in plant species regeneration success. Thus with regards to differential sensitivity storage theory, one might reasonably suppose that sensitivity to seedling herbivory (weak defence) is balanced against an ability to grow quickly and outcompete neighbours for regeneration microsites. Seedling herbivory will normally reduce or eliminate recruitment of the stronger competitor because it is inherently more sensitive to herbivore attack. This leaves the less sensitive, but weaker competitor to monopolise regeneration opportunities and so dominate the established community. Occasionally, however, environmental fluctuations reduce the impact of seedling herbivory, so allowing the more sensitive but stronger competitor to establish and maintain a presence in the community. Interestingly a comparison of relative abundance for our two *Trifolium* species at Weather Hill reveals that *T. pratense* (the more herbivore sensitive but stronger competitor) was less common than *T. repens* (Hanley and Sykes unpublished).

In seeking to establish a role for temporal fluctuation in the intensity of seedling herbivory within plant species coexistence models, ecologists must consider several important issues. First, the growth–defence tradeoff may be confounded by differences in a plant's ability to cope with herbivory once it has occurred. The so-called third option of plant tolerance is now widely seen as an important way in which plants respond to herbivory (Tiffin 2000, Agrawal and Fishbein 2006, Leimu and Koricheva 2006). While tolerance is poorly explored for seedlings, studies conducted so far show that some species are inherently better able to cope with tissue removal at the seedling stage than their likely competitors (Hanley and May 2006, Hanley and Fegan 2007). Consequently, simply because a seedling survives herbivory does not mean it will escape significant fitness costs later on. While the characteristics that determine seedling tolerance are poorly understood at present, there is a strong possibility that these traits are linked to those governing growth, defence and other ecophysiological processes (Agrawal and Fishbein 2006). Thus species-specific variation in herbivore tolerance may affect seedling regeneration just as crucially as competitive ability and anti-herbivore defence in environments where herbivory shows spatiotemporal fluctuations.

A further consideration is timescale. Within many ecosystems an intense barrage of herbivory coincides with seedling emergence and during this time the vast majority of seedlings may be killed and only a few ever establish and survive to reproductive maturity (Hanley *et al.* 1996a, 1996b, Green *et al.* 2008). Nonetheless by virtue of their seedbanks, many plant species possess a long-term insurance against reproductive failure. It may take only 1 year out of many when environmental conditions or predator numbers conspire to reduce dramatically herbivore populations, but during these comparatively short windows of opportunity poorly defended seedlings can establish, ensuring persistence in the community. This possibility has been invoked for plant communities containing long-lived species such as trees (Green *et al.* 2008), but persistent seedbanks are a common feature of many temperate ecosystems such as grasslands and wetlands (Kalamees and Zobel 2002, Fenner and Thompson 2005). For woody plants where it is possible to age individuals, a relatively accurate picture of fluctuations in seedling regeneration success can be gained (Hanley and Lamont 2001, Kelly and Bowler 2002, 2005). Even then, however, matching recruitment to periods when the intensity of seedling herbivory has been relaxed is difficult. For perennial herbaceous species that cannot be aged, this possibility is even more remote. Potentially, therefore, the only way of demonstrating whether temporal fluctuations in seedling herbivory differentially affect plant species regeneration (and so community persistence) is to observe annual shifts in herbivore abundance, and seedling recruitment to the established community, over many years.

Anthropogenic disruption of normal temporal fluctuations in seedling herbivory may provide a means of examining these effects however. Reductions in tree species diversity as a result of logging are known to influence the abundance of rodents in boreal forests (Vehviläinen *et al.* 2007). Where the abundance of seedling herbivores has been reduced in this, or similar ways, ecologists may find evidence supporting a link between fluctuation in seedling herbivory, species-specific recruitment and plant species diversity.

This possibility raises a particularly pressing issue. Many ecologists are now reporting disruption of normal population cycles in seedling herbivores such as voles and hares. Such perturbations have been strongly linked to anthropogenic climate change (Aars and Ims 2002, Bierman *et al.* 2006, Ims *et al.* 2007). Any long-term disruption of the interaction between herbivore abundance and seedling herbivory will inevitably shift the balance between the recruitment of species whose seedlings are sensitive to herbivore attack and those that are able to avoid or tolerate it. Ironically, perhaps, we are beginning to see the pivotal role played by spatiotemporal shifts in seedling herbivory in the maintenance of plant species diversity just as human activities are likely to disrupt this process.

ACKNOWLEDGEMENTS

We thank Clare Phillips and Emma Fegan for technical assistance. Part of this work was funded via a British Ecological Society Grant (SEPG No. 2240) to RJS.

REFERENCES

Aars, J. and Ims, R.A. (2002). Intrinsic and climatic determinants of population demography: the winter dynamics of tundra voles. *Ecology* **83**, 3449–3456.

Agrawal, A. A. and Fishbein, M. (2006). Plant defense syndromes. *Ecology* **87**, S132–S149.

Asquith, N. M. and Mejia-Chang, M. (2005). Mammals, edge effects, and the loss of tropical forest diversity. *Ecology* **86**, 379–390.

Barnes, H. F. and Weil, J. W. (1944). Slugs in gardens: their numbers, activities and distribution: Part I. *Journal of Animal Ecology* **13**, 140–175.

Barton, K. E. (2007). Early ontogenetic patterns in chemical defense in *Plantago* (Plantaginaceae): genetic variation and trade-offs. *American Journal of Botany* **94**, 56–66.

Barton, K. E. and Koricheva, J. (2010). The ontogeny of plant defense and herbivory: characterizing general patterns using meta-analysis. *American Naturalist* **175**, 481–493.

Bergelson, J. and Purrington, C. B. (1996). Surveying patterns in the cost of resistance in plants. *American Naturalist* **148**, 536–558.

Bierman S. M., Fairbairn, J. P., Petty, S. J. *et al.* (2006). Changes over time in the spatiotemporal dynamics of cyclic populations of field voles (*Microtus agrestis* L.). *American Naturalist* **167**, 583–590.

Boege, K. and Marquis, R. J. (2005). Facing herbivory as you grow up: the ontogeny of resistance in plants. *Trends in Ecology and Evolution* **20**, 441–448.

Bryant J. P., Clausen, T. P., Swihart, R. K. *et al.* (2009). Fire drives transcontinental variation in tree birch defense against browsing by snowshoe hares. *American Naturalist* **174**, 13–23.

Burt-Smith, G. S., Grime, J. P. and Tilman, D. (2003). Seedling resistance to herbivory as a predictor of relative abundance in a synthesised prairie community. *Oikos* **101**, 345–353.

Buschmann, H., Keller, M., Porret, N., Dietz, H. and Edwards, P.J. (2005). The effect of slug grazing on vegetation development and plant species diversity in an experimental grassland. *Functional Ecology* **19**, 291–298.

Chesson, P. and Huntly, N. (1997). The roles of harsh and fluctuating conditions in the dynamics of ecological communities. *American Naturalist* **150**, 519–553.

Crawley, M. J. (1997). Plant-herbivore dynamics. In M. J. Crawley (ed.), *Plant Ecology*, 2nd edn. Oxford: Blackwell, pp. 401–474.

Darwin, C. R. (1859). *The Origin of Species*. London: John Murray.

Elger, A., Lemoine, D. G., Fenner, M. and Hanley, M. E. (2009). Plant ontogeny and chemical defence: older seedlings are better defended. *Oikos* **118**, 767–773.

Elton, C. R. (1924). Periodic fluctuations in the numbers of animals: their causes and effects. *British Journal of Experimental Biology* **2**, 119–163.

Elton, C. R. (1927). *Animal Ecology*. London: Sidgwick and Jackson.

Fenner, M., Hanley, M. E. and Lawrence, R. (1999). Comparison of seedling and adult palatability in annual and perennial plants. *Functional Ecology* **13**, 546–551.

Fenner, M. and Thompson, K. (2005). *The Ecology of Seeds*, 2nd edn. Cambridge: Cambridge University Press.

Fine, P. V. A., Miller, Z. J., Mesones, I. *et al.* (2006). The growth defense trade-off and habitat specialization by plants in Amazonian forests. *Ecology* **87**, S150–S162.

Gause, G. F. (1932). Experimental studies on the struggle for existence. I. Mixed population of two species of yeast. *Journal of Experimental Biology* **9**, 389–402.

Glynn, C., Herms, D. A., Orians, C. M., Hansen, R. C. and Larsson, S. (2007). Testing the growth differentiation balance hypothesis: dynamic responses of willows to nutrient availability. *New Phytologist* **176**, 623–634.

Green, P.T., Lake, P. S. and O'Dowd, D. J. (1997). Control of seedling recruitment by land crabs in rain forest on a remote island. *Ecology* **78**, 2474–2486.

Green, P. T., O'Dowd, D. J. and Lake, P. S. (2008). Recruitment dynamics in a rainforest seedling community: context-independent impact of a keystone consumer. *Oecologia* **156**, 737–385.

Grime, J. P., Thompson, K., Hunt, R. *et al.* (1997). Integrated screening validates primary axes of specialisation in plants. *Oikos* **79**, 259–281.

Hanley, M. E. (2004). Seedling herbivory and the influence of plant species richness in seedling neighbourhoods. *Plant Ecology* **170**, 35–42.

Hanley, M. E., Collins, S. A. and Swann, C. (2011). Advertising acceptability: is mollusk olfaction important in seedling selection? *Plant Ecology* **212**, 727–731.

Hanley, M. E., Cordier, P. K., May, O. C. and Kelly, C. K. (2007a). Seed size and seedling growth: differential response of Australian and British Fabaceae to nutrient limitation. *New Phytologist* **174**, 381–388.

Hanley, M. E. and Fegan, E. L. (2007). Timing of cotyledon damage affects growth and flowering in mature plants. *Plant, Cell and Environment* **30**, 812–819.

Hanley, M. E., Fenner, M. and Edwards, P. J. (1995a). An experimental field study of the effects of mollusc grazing on seedling recruitment and survival in grassland. *Journal of Ecology* **83**, 621–627.

Hanley, M. E., Fenner, M. and Edwards, P. J. (1995b). The effect of seedling age on the likelihood of herbivory by the slug *Deroceras reticulatum*. *Functional Ecology* **9**, 754–759.

Hanley, M. E., Fenner, M. and Edwards, P. J. (1996a). The effect of mollusc grazing on seedling recruitment in artificially created grassland gaps. *Oecologia* **106**, 240–246.

Hanley, M. E., Fenner, M. and Edwards, P. J. (1996b). Mollusc grazing and seedling survivorship of four common grassland species: the role of gap size, species and season. *Acta Oecologica* **17**, 331–341.

Hanley, M. E., Fenner, M., Whibley, H. and Darvil, B. (2004). Early plant growth: identifying the end point of the seedling phase. *New Phytologist* **163**, 61–66.

Hanley, M. E. and Lamont, B. B. (2001). Herbivory, serotiny and seedling defence in Western Australian Proteaceae species. *Oecologia* **126**, 409–417.

Hanley, M. E. and Lamont, B. B. (2002). Relationships between physical and chemical attributes of seedlings: how important is seedling defence? *Functional Ecology* **16**, 216–222.

Hanley, M. E., Lamont, B. B., Fairbanks, M. M. and Rafferty, C. M. (2007b). Plant structural traits and their role in anti-herbivore defence. *Perspectives in Plant Ecology, Evolution and Systematics* **8**, 157–178.

Hanley, M. E. and May, O. C. (2006). Cotyledon damage at the seedling stage affects growth and flowering potential in mature plants. *New Phytologist* **169**, 243–250.

Hanley, M. E. and Sykes, R. J. (2009). Impacts of seedling herbivory on plant competition and implications for species coexistence. *Annals of Botany* **103**, 1347–1353.

Harborne, J. B. (1993). *Introduction to Ecological Biochemistry*, 4th edn. London: Academic Press.

Haring, D. A., Huber, M. J., Suter, D., Edwards, P. J. and Luescher, A. (2008). Plant enemy derived elicitors increase the foliar tannin concentration of *Onobrychis viciifolia* without a trade-off to growth. *Annals of Botany* **102**, 979–987.

Herms, D. A. and Mattson, W. J. (1992). The dilemma of plants: to grow or to defend. *Quarterly Review of Biology* **67**, 283–335.

Ims, R. A., Henden, J.-A. and Killengreen, S. T. (2007). Collapsing population cycles. *Trends in Ecology and Evolution* **23**, 79–86.

Izhaki, I. and Ne'eman, G. (1996). The importance of porcupine and bast scale on Aleppo pine recruitment after fire. *Acta Oecologica* **17**, 97– 107.

Kalamees, R. and Zobel, M. (2002). The role of the seed bank in gap regeneration in a calcareous grassland community. *Ecology* **83**, 1017–1025.

Kelly, C. K. and Bowler, M.G. (2002). Coexistence and relative abundance in forest tree species. *Nature* **417**, 437–440.

Kelly, C. K. and Bowler, M.G. (2005). A new application of storage dynamics: differential sensitivity, diffuse competition and temporal niches. *Ecology* **86**, 1012–1022.

Kelly, C. K. and Bowler, M.G. (2009a). Investigating the role of enemies in temporal niche dynamics: differential sensitivity, competition, and stable coexistence. *Theoretical Population Biology* **76**, 278–284.

Kelly, C. K. and Bowler, M.G. (2009b). Temporal niche dynamics, relative abundance and phylogenetic signal in coexisting species. *Theoretical Ecology* **2**, 161–169.

Kelly, C. K. and Hanley, M. E. (2005). Juvenile growth and palatability in congeneric British herbs. *American Journal of Botany* **92**, 1586–1589.
Kendall, B. E., Prendergast, J. and Bjørnstad, O.N. (1998). The macroecology of population dynamics: taxonomic and biogeographic patterns in population cycles. *Ecology Letters* **1**, 160–164.
Koricheva, J. (2002). Meta-analysis of sources of variation in fitness costs of plant anti-herbivore defences. *Ecology* **83**, 176–190.
Lambin, X., Bretagnolle, V. and Yoccoz, N. G. (2006). Vole population cycles in northern and southern Europe: is there a need for different explanations for single pattern? *Journal of Animal Ecology* **75**, 340–349.
Leimu, R. and Koricheva, J. (2006). A meta-analysis of tradeoffs between plant tolerance and resistance to herbivores: combining the evidence from ecological and agricultural studies. *Oikos* **112**, 1–9.
Lindquist, E. S. and Carroll, C.R. (2004). Differential seed and seedling predation by crabs: impacts on tropical coastal forest composition. *Oecologia* **141**, 661–671.
Lindquist, E. S., Krauss, K. W., Green, P. T. *et al.* (2009). Land crabs as key drivers in tropical coastal forest recruitment. *Biological Reviews* **84**, 203–223.
Maron, J. L. and Crone, E. (2006). Herbivory: effects on plant abundance, distribution and population growth. *Proceedings of the Royal Society of London B* **273**, 2575–2584.
Maron, J. L. and Kauffman, M. J. (2006). Habitat-specific impacts of multiple consumers on plant population dynamics. *Ecology* **87**, 113–124.
Meserve, P. L. (1971). Population ecology of prairie vole, *Microtus ochrogaster*, in western mixed prairie of Nebraska. *American Midland Naturalist* **86**, 417–433.
Moles, A. T. and Westoby, M. (2004). What do seedlings die from and what are the implications for evolution of seed size? *Oikos* **106**, 193–199.
Pake, C. E. and Venable, D. L. (1996). Seed banks in desert annuals: implications for persistence and coexistence in variable environments. *Ecology* **77**, 1427–1435.
Rafferty, C., Lamont, B.B. and Hanley, M. E. (2005). Selective feeding by western grey kangaroos on seedlings of *Hakea* species varying in morphology and chemistry. *Plant Ecology* **177**, 201–208.
Schaffner, U., Vrieling, K. and van der Meijden, E. (2003). Pyrrolizidine alkaloid content in *Senecio*: ontogeny and developmental constraints. *Chemoecology* **13**, 39–46.
Siegal, S. and Castellan, N. J. (1988). *Nonparametric Statistics for the Behavioral Sciences*, 2nd edn. New York: McGraw-Hill.
Stamp, N. (2003). Out of the quagmire of plant defense hypotheses. *Quarterly Review of Biology* **78**, 23–55.
Strauss, S. Y., Stanton, M. L., Emery, N. C *et al.* (2009). Cryptic seedling herbivory by nocturnal introduced generalists impacts survival, performance of native and exotic plants. *Ecology* **90**, 419–429.
Symondson, W. O. C., Glen, D. M., Ives, A. R., Langdon, C. J. and Wiltshire, C. W. (2002). Dynamics of the relationship between a generalist predator and slugs over five years. *Ecology* **83**, 137–147.
Tiffin, P. (2000). Mechanisms of tolerance to herbivore damage: what do we know? *Evolutionary Ecology* **14**, 523–536.
Vasconcelos, H. L. and Cherrett, J. M. (1997). Leaf-cutting ants and early forest regeneration in central Amazonia: Effects of herbivory on tree seedling establishment. *Journal of Tropical Ecology* **13**, 357–370.

Vehviläinen, H. and Koricheva, J. (2006). Moose and vole browsing patterns in experimentally manipulated pure and mixed forest stands. *Ecography* **29**, 497–506.

Vehviläinen, H., Koricheva, J. and Ruohomaki, K. (2007). Tree species diversity influences herbivore abundance and damage: meta-analysis of long-term forest experiments. *Oecologia* **152**, 287–298.

Verhulst, J., Montaña, C., Mandujano, M. C. and Franco, M. (2008). Demographic mechanisms in the coexistence of two closely related perennials in a fluctuating environment. *Oecologia* **156**, 95–105.

Westoby, M., Leishman, M. R. and Lord, J. (1996). Comparative ecology of seed size and dispersal. In J. Silvertown, M. Franco and J. L. Harper (eds), *Plant Life Histories: Ecology, Phylogeny and Evolution*. Cambridge: Cambridge University Press, pp. 143–162.

NORMA L. FOWLER AND CRAIG M. PEASE

6

Temporal variation in density dependence in an herbaceous community

6.1 INTRODUCTION

Plant communities are structured in time as well as space. Their component species may differ in the timing of establishment, growth and reproduction both within and between years, creating temporal differences in relative abundances. Under appropriate conditions, temporal differences among co-occurring species can allow their coexistence (Grubb 1977, Chesson and Huntly 1997, Higgins *et al.* 2000, Kelly and Bowler 2002, 2005, Chesson *et al.* 2004, Schwinning *et al.* 2004, Adler *et al.* 2006). Whether or not temporal differences are responsible for the coexistence of any given pair or set of species, temporal niches are a fundamental part of community structure. Understanding this aspect of community structure is becoming even more important as we are called upon to interpret and to predict responses of communities to climate change (Dukes and Mooney 1999, Walther *et al.* 2002, Fischlin *et al.* 2007).

In this study we examine temporal variation in population dynamics in a set of eight co-occurring herbaceous perennials. In addition to examining temporal variation in population sizes (densities), we use simple population dynamic models to provide estimates of temporal variation in equilibrium population density. Because actual densities lagged equilibrium population densities, the latter were more useful for comparing the temporal niches of different species and for relating temporal variation in precipitation to population dynamics. The degree of lag itself can be interpreted as a measure of the intensity

Temporal Dynamics and Ecological Process, ed. C. K. Kelly, M. G. Bowler, G. A. Fox.
Published by Cambridge University Press.

of density-dependent population regulation, and variation in the degree of lag as another aspect of a species' temporal niche.

The community we studied occurs in central Texas, where precipitation is highly variable from month to month and year to year. Because this community is so conspicuously affected by temporal variation in precipitation, one of our initial hypotheses was that all the species would simply track precipitation together. Alternatively, the temporal variability and lack of autocorrelation of precipitation in this region may allow each species to have a unique temporal niche.

6.2 METHODS

The community and climate

The analyses described here are based upon 16 annual censuses of permanent 0.5-m^2 quadrats established as part of a long-running demographic study in Pedernales Falls State Park, Texas (30.33°N, 98.26°W; Fowler 1984, 1995, Fowler *et al.* 2006). The site was a savanna, with clusters of woody plants (*Quercus fusiformis*, *Juniperus ashei*, *Berberis trifoliolata* and other woody species) scattered in a grassy matrix. We are concerned here only with the matrix vegetation, that is, the community of plants that grew in open areas outside of the clusters of woody plants. This community was dominated by perennial grass species. Eight species were abundant enough to use in the present study (Table 6.1).

This community experienced frequent, severe, and unpredictable water shortages. In Johnson City, 16 km away, average July and August temperatures were 28°C; average January temperature was 9°C; and precipitation averaged 861 mm per year (calculated from NOAA National Climatic Data Center data 1980–2008). Almost all precipitation is rain. Annual PET (potential evapotranspiration) is approximately twice as high as precipitation. Average May precipitation is higher than that of other months, but there is no month that is not rainless in some years (Figure 6.1). Species must therefore be able to tolerate drought at any time of the year. There is little correlation in precipitation between subsequent months in same year ($r = 0.14$). The perennial grasses in this community are all capable of dying back in response to low soil moisture and quickly re-growing during periods of higher soil moisture. The study site has a red clay soil (rhodustalf, Hensley series; Dittemore and Allison 1979) with a low infiltration rate.

Table 6.1 *The eight most abundant vascular plant species in the permanent plots. Bunchgrasses have very little rhizomatous or stoloniferous spread (i.e. little or no vegetative reproduction).* Bothriochloa ischaemum *is the only non-native species. Plant sizes are for the study site only.* $N_{t+1}/N_t \rightarrow (R+1)$ *as* $N_t \rightarrow 0$ *and the population approaches purely geometric growth.*

Species	Life form	Estimated $(R + 1)$
Bouteloua rigidiseta	C_4 bunchgrass, 5–20 cm tall	1.41
Aristida longiseta	C_4 bunchgrass, 5–20 cm tall	1.32
Nassella leucotricha	C_3 bunchgrass, 5–20 cm tall	1.45
Eragrostis intermedia	C_4 bunchgrass, 5–20 cm tall	1.26
Bothriochloa ischaemum	C_4 bunchgrass, 20–30 cm tall	1.55
Hilaria belangeri	C_4 stoloniferous grass, 5–20 cm tall	1.32
Evolvulus sericeus (Convolvulaceae)	prostrate or scrambling forb, 5–15 cm long stems	2.40
Sida abutifolia (Malvaceae)	prostrate forb, 5–15 cm long stems	2.11

Data sets

Censuses were conducted every year in May and June for 16 years (1982–97). Censuses of *Bouteloua rigidiseta* were made every year. Censuses of each of the other species were made in every year except 1987. The number of quadrats from which data were collected varied among years from 6 to 36. In each quadrat included in a census, the location of every individual of every perennial plant species was recorded on a sheet of clear plastic placed on a mapping table directly over the quadrat (Fowler 1995). A non-destructive measure of the size of each mapped individual was also recorded: number of tillers of each graminoid, number of stems of each forb. For each species–quadrat–census combination, density (N) was calculated as the number of tillers or stems m^{-2}. Because species differed in morphology, differences among species in density do not necessarily reflect differences among species in aboveground biomass. In particular, *Bothriochloa ischaemum* has much larger tillers than the other grasses. Six quadrats were unmanipulated; the rest received manipulations of *Bouteloua rigidiseta* or *Aristida longiseta* density at various times. For a more complete description of the design, see Fowler (1995) and Fowler and Pease (2010).

From the census data, we constructed, for each of the eight species in Table 6.1, a data set of (N_t, N_{t+1}) pairs for each quadrat, where t and $t + 1$ are sequential years and N is the density of that species in a given quadrat. Data from eight perennial species are analysed here.

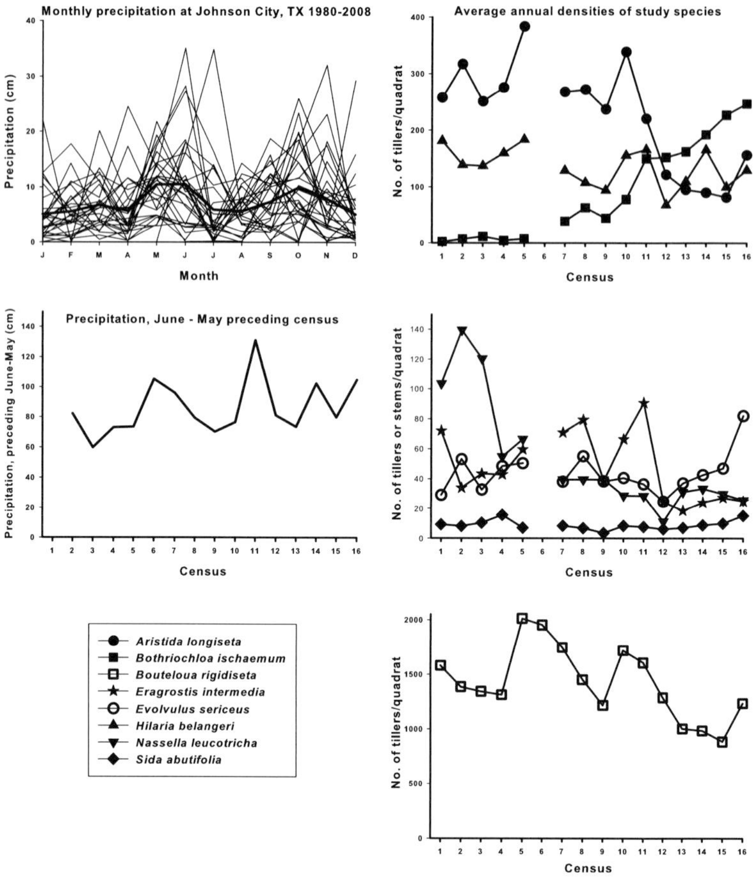

Figure 6.1 Left-hand column. Upper: monthly precipitation at Johnson City, Texas, 1980–2008; thick line is the average monthly precipitation during this period. Centre: annual precipitation during the 12 months preceding each census. Lower: key to symbols. Right-hand column. Average density (tillers or stems per quadrat) of each species at each census in the unmanipulated quadrats. Only *Bouteloua rigidiseta* was included in the census of year 6. Note different *y*-axis scales.

The other perennial species were too sparse, or grew in too few quadrats, to analyse. There were a total of 174 (*Bouteloua rigidiseta*) or 168 (the other seven species) (N_t, N_{t+1}) pairs, each pair representing two sequential censuses of one quadrat. For each (t, $t + 1$) pair of sequential years, data from at least six quadrats (hence, six (N_t, N_{t+1}) pairs) were available. Because the same value was sometimes used as N_{t+1} in one pair and as N_t in the next pair, the (N_t, N_{t+1}) pairs are not fully independent of each other.

Models

To analyse these data sets, we used a hyperbolic model of density-dependent population growth,

$$N_{t+1} = (1 + R)N_t/(1 + N_t R/K) + \varepsilon, \tag{6.1}$$

and estimated R and K by fitting each model to a data set. N_t and N_{t+1} are observed densities as described above. R and K are fitted coefficients. This version of the model has a single value of R and a single value of K, and will be called the 'single-K model'. Both R and K have biological interpretations. If no additional terms are added to the model, K represents the equilibrium population size. Note that, although K is often called the carrying capacity, its value is determined by competition from other species, density-dependent disease effects, and density-dependent predation impacts, as well as by resource levels.

Mathematically, as K approaches infinity, $1 + R$ approaches the potential finite rate of increase of the population, the population grows geometrically, and the model above simplifies to the 'geometric growth model',

$$N_{t+1} = (1 + R)N_t + \varepsilon. \tag{6.2}$$

Note that these models are difference equations, not differential equations. Population growth is being modelled as a discontinuous process. $1 + R$ is the analogue of e^r, not of r, of the common logistic model of population growth (Case 2000).

We used a hyperbolic model, rather than the more common discrete logistic or Ricker models (Case 2000), because previous analyses of the *Bouteloua rigidiseta* data set identified this model as providing excellent fits to the data (Fowler and Pease 2010). It has also been shown to work well for other plants (Watkinson 1980). It has the desirable property that when $N >> K$ it predicts a smooth decline in population size, rather than a crash (Case 2000), and is therefore more realistic for most perennial plant populations. It does not produce oscillations. If one integrates the continuous logistic function, $\delta N/\delta t = rN(K - N)/K$ over one time step, the result is the hyperbolic function above (Case 2000, p. 108, after re-arrangement).

To identify temporal patterns of population regulation, we expanded the single-K model to allow K to take different values in different annual intervals. We will call this the 'multi-K model'. Previous analysis of the *B. rigidiseta* data set had revealed that this produced better fits to the data than allowing R to take different values in different years (Fowler and Pease 2010). We also followed our earlier analyses of *B. rigidiseta* in substituting the geometric growth model for a hyperbolic model in those annual intervals in which it was required. Annual intervals in which a geometric growth model is required are intervals in which K_t is so much larger than N that it is best treated as infinite, as discussed below.

We used the SAS (version 9.1, SAS Institute, Cary, NC, USA) procedure NLMIXED to find a maximum likelihood solution for all the parameters of a model simultaneously. The single-K models had three fitted parameters: R, K and σ^2_ε. The variance of ε, written as σ^2_ε, represents variation in N_{t+1} not otherwise accounted for by the model (ε was assumed to be normally distributed with mean zero). Multi-K models had up to 15 fitted parameters: in addition to R and σ^2_ε, a value of K_t for each non-geometric growth annual interval was fitted. Although σ^2_ε is a single parameter for the purposes of AIC calculation, a separate estimate of ε was provided by PROC NLMIXED for each quadrat–interval combination. The estimates of R were similar, but not identical, between the single-K and multi-K models. Annual intervals best modelled by geometric growth were identified when NLMIXED failed to converge properly (usually because the Hessian matrix was not positive definite) and reported an extremely large estimate of K_t for one or more annual intervals. Geometric growth in those intervals was confirmed when substituting geometric growth for hyperbolic growth in those intervals solved the convergence problems. After adjusting the multi-K model so that it was a geometric growth model for those intervals, if any, for which geometric growth was more appropriate, we compared it to the corresponding single-K model. The multi-K model was considered to be a better fit than the corresponding single-K model if the AIC_c (corrected Akaike Information Criterion) value of the multi-K model was at least 2.0 less than the AIC_c value of the corresponding single-K model.

Models were fitted to each species separately. For *Bouteloua rigidiseta*, models were fitted to a data set that included all 15 annual intervals. Only 13 annual intervals were available for the other species: the lack of census data for those species from year 6 eliminated interval 5 (year 5 to year 6) and interval 6 (year 6 to year 7). *Bothriochloa ischaemum* was so sparse in year 1 that interval 1 (year 1 to year 2) had to be dropped, leaving 12 annual intervals in the data set of this species.

While the improvement in AIC_c values of the multi-K models over the corresponding single-K model for seven of the eight species indicates that K differed among annual intervals for these seven species, our data did not allow precise estimates of individual values of K_t to be obtained even for non-geometric growth intervals. We therefore emphasise general patterns, rather than individual estimates of K_t, in the Results (section 6.3) and Discussion (section 6.4). Standard errors give some indication of this uncertainty of individual estimates of K_t, although the lack of independence of sequential (N_t, N_{t+1}) pairs from the same quadrat prevents the calculation of confidence intervals. Standard errors of estimates of K_t averaged 20% of the estimated values of K_t for *Bouteloua rigidiseta*, 30–40% for *Evolvulus sericeus*, *Aristida longiseta* and *Bothriochloa ischaemum* and more than 50% for *Hilaria belangeri*, *Eragrostis intermedia* and *Nasella leucotricha*.

Correlations

To quantify the similarity in temporal patterns between pairs of species, Spearman non-parametric correlation coefficients were calculated for N, K and $(K - N)$. For each species, the observed densities at the end of each annual interval (N_{t+1}) in the unmanipulated quadrats were averaged. These averages were used in the calculation of $(K - N)$ for each annual interval, and in the calculations of correlation coefficients. K was set equal to an arbitrarily large constant value for all intervals of geometric growth. Spearman non-parametric correlation coefficients were also calculated for correlations with precipitation.

Because sequential values of N from the same quadrat are not statistically independent, ordinary significance tests of correlation coefficients are not valid. For the set of all possible species pairs, the null hypothesis of no relationships was tested by assuming that positive and negative relationships were equally likely and calculating the probability of getting by chance a result at least as extreme as the observed number of positive and negative correlations.

6.3 RESULTS

Temporal variation in density

The densities of each of the eight study species varied widely during the 16 years of this study (Figure 6.1). Precipitation during the preceding 12 months was not well correlated with the densities of any of the eight study species (Table 6.2). For example, following the peak in rainfall between censuses 10 and 11, *Eragrostis intermedia* had a relatively high density at census 11, but no other species did. *Aristida longiseta* and *Bouteloua rigidiseta* had peaks at census 10, which followed a relatively dry 12-month period. The average correlation between density and precipitation was 0.04.

Table 6.2 *Spearman correlation coefficients. First column: correlation between average density and precipitation during the 12 months (June–May) preceding the census. Second column: correlation between precipitation and estimated values of* K_t. *Upper right: correlations between densities of pairs of species. Lower left after precipitation columns: correlations between estimated values of* K_t. Sida abutifolia *was best fit by a model with a single value of* K, *so no correlations between estimated* Ks *for pairs of species are reported for this species.*

Species	ppt v *N*	ppt v *K*	*Al*	*Bi*	*Br*	*Ei*	*Ev*	*Hb*	*Nl*	*Sa*
Aristida longiseta	−0.29	−0.03	–	−0.81	0.82	0.63	0.32	0.43	0.57	−0.07
Bothriochloa ischaemum	0.47	0.32	−0.15	–	−0.59	−0.48	−0.06	−0.31	−0.83	0.03
Bouteloua rigidiseta	0.18	0.48	0.41	0.27	–	0.81	0.04	0.45	0.28	−0.15
Eragrostis intermedia	0.01	0.01	0.22	0.50	0.56	–	0.03	0.31	0.23	−0.07
Evolvulus sericeus	0.16	0.35	0.51	−0.32	0.12	−0.17	–	0.20	0.25	0.24
Hilaria belangeri	0.18	0.27	0.41	0.26	0.71	0.82	0.14	–	0.28	0.37
Nassella leucotricha	−0.50	0.27	0.40	0.07	0.25	0.46	0.29	0.49	–	0.10
Sida abutifolia	0.11	–	–	–	–	–	–	–	–	–

Towards the end of the study *Bothriochloa ischaemum* increased in density, as did *Evolvulus sericeus*, while *Aristida longiseta*, *Eragrostis intermedia* and *Bouteloua rigidiseta* all declined (Figure 6.1). *Nassella leucotricha*'s decline had begun by census 7. As a result, *B. ischaemum* density was negatively correlated with the densities of the other grasses (Table 6.2, upper right). The only other negative correlations between species involved *Sida abutifolia*. The average correlation in density between pairs of species was 0.11.

Temporal variation in equilibrium population size

Every species except *Sida abutifolia* was better fit by a model that allowed K to vary among annual intervals (multi-K model) than by a single-K model. The relatively poor performance of the multi-K *Sida abutifolia* model may be due simply to its low densities (Figure 6.1) and to the fact that the number of stems is a low-precision estimator of aboveground biomass for this species. Among the seven species best fit by multi-K models, only *Evolvulus sericeus* did not have any annual intervals that were best fit by geometric growth. For the other six species, two to four annual intervals were best fit by geometric growth (Figure 6.2). Overall, out of 107 species–interval combinations for which a model could be fitted, 16 (15%) were best modelled by geometric growth. *Bothriochloa ischaemum*, the invading species, was best modelled by geometric growth in only three intervals, and these intervals occurred when it was at low densities early in the study (Figures 6.1 and 6.2).

There was some synchrony between species in equilibrium population sizes (Figure 6.2, lower left after precipitation columns). Four of 13 annual intervals (intervals 3, 8, 11, and 12) had no species growing geometrically, and intervals 9 and 10 had half of the species growing in a fashion not distinguishable from geometric growth. The average correlation between the estimated equilibrium population densities of a pair of species was 0.30 (Table 6.2, $N = 21$ pairs of species).

However, there were also substantial differences among species in the way their estimated equilibrium population sizes (K_t) varied over time (Figure 6.2). In addition to differing in the timing of geometric growth years, species differed temporally in the magnitudes of estimable values of K_t. For example, *Eragrostis intermedia* apparently had a particularly good year (very large estimated K_t) in the second annual interval, *Aristida longiseta* in the third, and *Bouteloua rigidiseta*, *Hilaria belangeri* and *Nassella leucotricha* in the fourth.

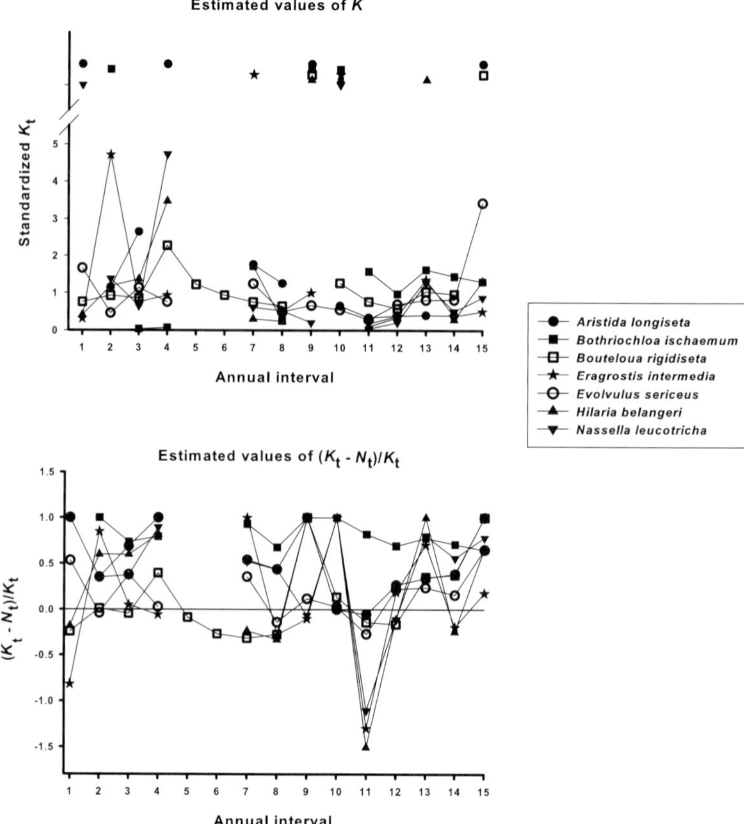

Figure 6.2 (a) Temporal variation in estimated equilibrium population densities. To facilitate comparisons of species, estimated values of K_t of each species (lower portion of the graph, symbols connected by lines) have been standardised by dividing each estimated value of K_t by the average value of all estimated K_t of that species. Values of K_t that were too large to be estimated (i.e. effectively infinite; these are the geometric growth intervals) are shown across the top portion of this graph (symbols not connected by lines). **(b)** Temporal variation in the tracking of equilibrium population size. Values shown are standardised differences from the equilibrium population density of the particular species, $(K_t - N_t)/(K_t)$. If a species was best modelled as growing geometrically during an annual interval, the value 1.0 was assigned to that species for that interval. Points above the horizontal line represent intervals in which density (N_t) lagged estimated equilibrium density (K_t); points below the horizontal line represent intervals in which density was greater than estimated equilibrium density. Note that the *x*-axes of both graphs in this figure represent annual intervals, not censuses. For example, interval 1 began at census 1 and ended at census 2.

The correlation between estimated K_t and precipitation during the preceding 12 months ranged from −0.03 to +0.48, with an average value of 0.24; only one of the seven values was negative ($r_s = -0.03$, *Aristida longiseta*).

The correlations between species pairs seen in observed densities (N_t) were different than those seen in estimated values of equilibrium population sizes (K_t). The negative correlations with *Bothriochloa ischaemum* seen in N were weaker or absent from K_t. *Hilaria belangeri* was more strongly correlated with *Bouteloua rigidiseta* ($r_s = 0.71$ for estimated K_t values, up from 0.45 between observed densities) and *Eragrostis intermedia* ($r_s = 0.82$, up from 0.31), while the correlation between the latter two species weakened ($r_s = 0.56$, down from 0.81). There were only three negative correlations in K_t (between *Bothriochloa ischaemum* and *Aristida leucotricha*, between *B. ischaemum* and *Evolvulus sericeus* and between *E. sericeus* and *Eragrostis intermedia*). The probability of having three or fewer negative values out of 21 is 0.0007.

Temporal variation in the tracking of equilibrium population size

In most annual intervals most of the species were below their equilibrium densities (K_t) (Figure 6.2). There were substantial overshoots ($N_t >> K_t$) by three species (*Nassella leucotricha*, *Eragrostis intermedia* and *Hilaria belangeri*) in the 11th interval. A total of 27 values of ($K_t - N_t$) were negative (out of 93 points in Figure 6.2b; 29%), indicating some degree of overshoot ($N_t > K_t$). Only *Bothriochloa ischaemum* and *Sida abutifolia* never overshot K_t. In the case of *B. ischaemum*, this may reflect its ongoing invasion.

The correlation between ($N_t - K_t$) and precipitation during the preceding 12 months ranged from −0.13 to +0.48, with an average value of 0.21; only one of the eight values was negative ($r_s = -0.13$, *Aristida longiseta*). The probability of getting positive values of r_s six or more times out of seven events is 0.05.

Only three of 21 pairs of species had negatively correlated values of ($K_t - N_t$), although the average correlation was only 0.21. The probability of getting three or more negative values out of 21 is 0.001. Negative correlations in values of ($K_t - N_t$) occurred between *Aristida longiseta* and *Bothriochloa ischaemum*, *A. longiseta* and *Eragrostis intermedia*, and *Bothriochloa ischaemum* and *Evolvulus sericeus*. The largest positive correlation in values of ($K_t - N_t$) was between *Bouteloua rigidiseta* and *Hilaria belangeri* ($r_s = 0.79$).

6.4 DISCUSSION

The shortgrass community of this study seems well suited to temporal niche differentiation among species. Plant growth is frequently water-limited; rainless periods of a month or more can occur at any time of year (Figure 6.1); there is very little correlation in rainfall between sequential months; and total precipitation per year also varies widely (Figure 6.1). If different temporal patterns of precipitation favour different plant species, there are certainly enough different precipitation patterns for each species to have a different temporal niche. If species are specialised in this way, we would expect to see each species behaving in an individualistic fashion over time. An alternative hypothesis is that water shortages affect all the species in unison. This latter hypothesis was supported by observations of the site 'greening up' after rain and 'browning off' in dry periods, sometimes several times in one year. These transitions were very conspicuous in the study site and in similar sites across the region. The hypothesis of synchronous species behaviour is also supported by the fact that, across taxa and climates, positive temporal correlations are far more common than negative ones (Houlahan *et al.* 2007).

The quantitative results presented here support the hypothesis of individualistic species behaviour and temporal niche differentiation more strongly than they support the hypothesis of synchronous species behaviour. While most correlations between pairs of species were positive, they were no more than moderate in magnitude if estimated equilibrium population sizes (K_t values) are used to calculate them ($r_s = 0.30$) and even less ($r_s = 0.11$) if observed densities (N_t values) are used. The relatively weak correlations between precipitation and either K_t or N_t ($r_s = 0.24$ and 0.04, respectively) are also consistent with individualistic species behaviour. The individualist behaviour we observed is similar to the differences among species reported for prairie grasses (Adler *et al.* 2006), but not to the synchrony of a California Mediterranean-climate grassland community (Elmendorf and Harrison 2009, but see Fox *et al.* 2006, Levine *et al.* 2008). Temporal niche separation of a type expected to lead to coexistence has been described in desert annuals (Pake and Venable 1995, 1996, Adondakis and Venable 2004) and desert shrubs (Verhulst *et al.* 2008), herbaceous species of Mediterranean-climate shrubland (Facelli *et al.* 2005), tropical forest trees (Kelly and Bowler 2002, 2005, Newbery and Lingenfelder 2009) and herbaceous species of a temperate mountain grassland (Herben *et al.* 2003), among others.

Differences among co-occurring species in their responses to temporal differences in the environment can only be a basis for their coexistence if a number of restrictive conditions are met (Chesson and Huntly 1997, Higgins *et al.* 2000, Kelly and Bowler 2002, Chesson *et al.* 2004, Levine and Rees 2004). We do not yet know which pairs of the eight species discussed here, if any, meet all of the necessary conditions for coexistence by temporal niche separation, but it seems likely that the temporal niche differences we found (Figure 6.2) at least contribute to the coexistence of the five native grasses in the study site and elsewhere in the region. *Nassella leucotricha*, the only C_3 species among the six grasses of this study, is known to grow more quickly than some C_4 grasses in cooler weather in this region (T. Basham, pers. comm.) but has not been compared to the other grass species of this study. *N. leucotricha* may be favoured in years with more rainfall in cooler periods. *Bouteloua rigidiseta* and *Aristida longiseta* seedlings germinate whenever the soil first remains moist, sometime in October, November, December or January (Fowler 1988). For successful establishment these two species need a period of substantial rainfall during those months, followed by a combination of cool-enough temperatures and sufficient precipitation for soil moisture to remain high enough for seedlings to survive. *Hilaria belangeri*, the only clonal grass, sets little seed. It might be favoured after a series of years that have not met the requirements for *B. rigidiseta* and *A. leucotricha* germination and establishment, reducing competition from these species. But note that this discussion of temporal variation in *K* among species implicitly assumes that temporal variation in equilibrium population sizes (Figure 6.2) reflects consistent temporal differences among species. Purely stochastic temporal variation in *K* may have little effect on coexistence (Turelli 1978, 1981) although under some conditions it can reduce average population size (Chesson 1991).

Bouteloua ischaemum, the tallest and largest of the six grasses, and the only non-native species, was apparently not coexisting with any of the native grasses, despite some temporal differentiation from them. It increased in density during the study (Figure 6.1) and after the period of this study continued to increase in abundance in the study site, with concomitant decreases in the other grass species (N. Fowler, pers. obs.). *B. ischaemum* frequently forms near-monospecific stands in the region (Gabbard and Fowler 2007). *Evolvulus sericeus*, one of the two dicot species, increased in density some years after the first large increases in *B. ischaemum* density. *E. sericeus* can use grass tillers for support as a scrambling vine would, and therefore is the only study species not over-topped by *B. ischaemum*.

When evidence suggests that species are responding differently to temporal differences in precipitation, as it does here, we would like to identify the rainfall patterns that favour each species. Unfortunately, doing so presents some formidable analytical challenges. If the climate has enough autocorrelation between months to justify an a priori division of the year into two seasons, an analysis of the effect of each season's rainfall is justified (e.g. Fox *et al.* 2006, Adler and HilleRisLambers 2008, Lima *et al.* 2008). However, in central Texas, plant growth is possible almost year-round and we feel that there is not enough correlation among months to justify the a priori pooling of months into seasons. Nor does the climate have clear-cut pulse (after rainfall) and inter-pulse (drought) periods (Goldberg and Novoplansky 1997, Novoplansky and Goldberg 2001). As a result, the number of potential independent variables is large (12 if only a simple linear effect of each month in the preceding year is considered, but pooling weeks into months may conceal important variation, effects need not be linear, and previous years may be relevant). Even a simple categorical classification of the preceding 12 months as wet or dry yields $2^{12} = 4096$ possible permutations. In our experience, it is easy to find statistical relationships between monthly precipitation and any measure of plant density or performance, but we cannot be confident that these relationships are not spurious. For example, we know that chance negative correlations in precipitation between months during the study period produce probable artefacts such as a negative correlation between plant density and precipitation in one month followed by a positive correlation the following month. We conclude that 16 censuses are simply too few, given a climate with so little month-to-month and year-to-year predictability in precipitation. For this reason we have not included a more detailed analysis of the effects of precipitation on population regulation here.

The individualistic behaviour of each species observed in this study may in part reflect the effects of competing species and of density- or weather-dependent pathogens and herbivores (Adler and HilleRisLambers 2008). A weather pattern that would otherwise be favourable may become unfavourable if it fosters the spread of a fungal pathogen, or if it favours a competitor more. In the study community, any weather than favours *Bothriochloa ischaemum* would presumably disfavour the native grasses, which over the years have been replaced by this larger and taller non-native grass. A second example might be *Hilaria belangeri*, discussed above. More generally, these factors may account for the relatively small correlations between the equilibrium population sizes of different species.

The analyses used in this study separated observed changes in density (N) into changes in equilibrium population size (K) and changes in the failure of species to track their shifting equilibria ($K - N$). Both components contributed substantially to changes in N. By disaggregating these two components of temporal variation in N, we were able to detect differences among species not otherwise apparent. These differences among species had been concealed by the ongoing invasion of the site by the non-native invasive species *Bothriochloa ischaemum* during the course of the study and the corresponding declines in the other perennial grass species (Figure 6.1). Had the invasion of *B. ischaemum* occurred uniformly over the entire site, our analyses might not have been able to detect other patterns. However, *B. ischaemum* spread across this site as advancing fronts from relatively few centres. Because we calculated (N_t, N_{t+1}) for each quadrat separately, much of the data set was not affected by the invasion in any given year.

Densities often did not track changes in K very well. Lags ($K > N$) were common, and there were also substantial overshoots ($K < N$; Figure 6.2). Overshoots were due to rapid drops in equilibrium population size, and lags were often due to increases in equilibrium population size. As a result, the relationship between N and K, and hence the strength of density dependence, was largely driven by temporal variation in K. Species differed in the timing and magnitude of overshoots and lags just as they did in the timing of changes in their equilibrium population sizes.

ACKNOWLEDGEMENTS

We thank the many people over the years who have helped collect, transcribe, assemble and manage the census data, especially Rick Miller, Doug Brown, Don Campanella and Deb Overath. We thank the staff and volunteers at Pedernales Falls State Park, who have welcomed us at the Park and facilitated this project. We thank Peter Chesson for a useful discussion. Parts of the data were collected and assembled with funding from the National Science Foundation.

REFERENCES

Adler, P. B. and HilleRisLambers, J. (2008). The influence of climate and species composition on the population dynamics of ten prairie forbs. *Ecology* **89**, 3049–3060.

Adler, P. B., HilleRisLambers, J., Kyriakidis, P. C., Guan, Q. and Levine, J. M. (2006). Climate variability has a stabilizing effect on the coexistence of prairie grasses. *Proceedings of the National Academy of Sciences, USA* **103**, 12793–12798.

Adondakis, S. and Venable, D. L. (2004). Dormancy and germination in a guild of Sonoran Desert annuals. *Ecology* **85**, 2582–2590.

Case, T.J. (2000). *An Illustrated Guide to Theoretical Ecology*. Oxford: Oxford University Press.

Chesson, P. (1991). Stochastic population models. In J. Kolasa and S.T.A. Pickett (eds), *Ecological Heterogeneity* (Ecological Studies: Analysis and Synthesis 86). New York: Springer-Verlag, pp. 123–143.

Chesson, P., Gebauer, R. L. E., Schwinning, S. *et al.* (2004). Resource pulses, species interactions, and diversity maintenance in arid and semi-arid environments. *Oecologia* **141**, 236–253.

Chesson, P. and Huntly, N. (1997). The roles of harsh and fluctuating conditions in the dynamics of ecological communities. *American Naturalist* **150**, 519–553.

Dittemore, W. H. and Allison, J. E. (1979). *Soil survey of Blanco and Burnet Counties, Texas*. Washington DC: Soil Conservation Service, US Department of Agriculture.

Dukes, J. S. and Mooney, H. A. (1999). Does global change increase the success of biological invaders? *Trends in Ecology and Evolution* **14**, 135–139.

Elmendorf, S. C. and Harrison, S. P. (2009). Temporal variability and nestedness in California grassland species composition. *Ecology* **90**, 1492–1497.

Facelli, J. M., Chesson, P. and Barnes, N. (2005). Differences in seed biology of annual plants in arid lands: a key ingredient of the storage effect. *Ecology* **86**, 2998–3006.

Fischlin, A., Midgley, G. F., Price, J. T. *et al.* (2007). Ecosystems, their properties, goods, and services. In M.L. Parry, O.F. Canziani, J.P. Palutikof, P.J. van der Linden and C.E. Hanson (eds), *Climate Change 2007: Impacts, Adaptation and Vulnerability. Contribution of Working Group II to the Fourth Assessment Report of the Intergovernmental Panel on Climate Change*. Cambridge: Cambridge University Press, pp. 211–272.

Fowler, N. L. (1984). Patchiness in patterns of growth and survival of two grasses. *Oecologia* **62**, 424–428.

Fowler, N. L. (1988). What is a safe site? Neighbor, litter, germination date, and patch effects. *Ecology* **69**, 947–961.

Fowler, N. L. (1995). Density-dependent demography in two grasses: a five-year study. *Ecology* **76**, 2145–2164.

Fowler, N. L., Overath, R. D. and Pease, C. M. (2006). Detection of density dependence requires density manipulations and calculation of λ. *Ecology* **87**, 655–664.

Fowler, N. L. and Pease, C. M. (2010). Temporal variation in the carrying capacity of a perennial grass population. *American Naturalist* **175**, 504–512.

Fox, L. R., Steele, H. N., Holl, K. D. and Fusari, M. H. (2006). Contrasting demographies and persistence of rare annual plants in highly variable environments. *Plant Ecology* **183**, 157–170.

Gabbard, B.L. and Fowler, N. L. (2007). Wide ecological amplitude of a diversity-reducing invasive grass. *Biological Invasions* **9**, 149–160.

Goldberg, D. and Novoplansky, A. (1997). On the relative importance of competition in unproductive environments. *Journal of Ecology* **85**, 409–418.

Grubb, P. J. (1977). The maintenance of species-richness in plant communities: the importance of the regeneration niche. *Biological Reviews* **52**, 107–145.

Herben, T., Krahulec, F., Hadincova, V., Pechackova, S. and Wildova, R. (2003). Year-to-year variation in plant competition in a mountain grassland. *Journal of Ecology* **91**, 103–113.

Higgins, S. I., Bond, W. J. and Trollope, W. S. W. (2000). Fire, resprouting and variability: a recipe for grass-tree coexistence in savanna. *Journal of Ecology* **88**, 213–229

Houlahan, J. E., Currie, D. J., Cottenie, K. *et al.* (2007). Compensatory dynamics are rare in natural ecological communities. *Proceedings of the National Academy of Sciences, USA* **104**, 3273–3277.

Kelly, C. K. and Bowler, M. G. (2002). Coexistence and relative abundance in forest tree species. *Nature* **417**, 437–440.

Kelly, C. K. and Bowler, M. G. (2005). A new application of storage dynamics: differential sensitivity, diffuse competition, and temporal niches. *Ecology* **86**, 1012–1022.

Levine, J. M., McEachern, A. K. and Cowan, C. (2008). Rainfall effects on rare annual plants. *Journal of Ecology* **96**, 795–806.

Levine, J. M. and Rees, M. (2004). Effects of temporal variability on rare plant persistence in annual systems. *American Naturalist* **164**, 350–363.

Lima, M., Morgan Ernest, S. K., Brown, J. H., Belgrano, A. and Stenseth, N. C. (2008). Chihuahuan desert kangaroo rats: nonlinear effects of population dynamics, competition, and rainfall. *Ecology* **89**, 2594–2603.

Newbery, D. M. and Lingenfelder, M. (2009). Plurality of tree species responses to drought perturbation in Bornean tropical rain forest. *Plant Ecology* **201**, 147–167.

Novoplansky, A. and Goldberg, D. E. (2001). Effects of water pulsing on individual performance and competitive hierarchies in plants. *Journal of Vegetation Science* **12**, 199–208.

Pake, C. E. and Venable, D. L. (1995). Is coexistence of Sonoran Desert annuals mediated by temporal variability in reproductive success? *Ecology* **76**, 246–261.

Pake, C. E. and Venable, D. L. (1996). Seed banks in desert annuals: implications for persistence and coexistence in variable environments. *Ecology* **77**, 1427–1435.

Schwinning, S., Sala, O. E., Loik, M. E. and Ehleringer, J. R. (2004). Thresholds, memory, and seasonality: understanding pulse dynamics in arid/semi-arid ecosystems. *Oecologia* **141**, 191–193.

Turelli, M. (1978). Does environmental variability limit niche overlap? *Proceedings of the National Academy of Sciences, USA* **75**, 5085–5089.

Turelli, M. (1981). Niche overlap and invasion of competitors in random environments. I. Models without demographic stochasticity. *Theoretical Population Biology* **20**, 1–56.

Verhulst, J., Montaña, C., Mandujano, M.C. and Franco, M. (2008). Demographic mechanisms in the coexistence of two closely related perennials in a fluctuating environment. *Oecologia* **156**, 95–105.

Walther, G.-R., Post, E., Convey, P. *et al.* (2002). Ecological responses to recent climate change. *Nature* **416**, 389–395.

Watkinson, A. R. (1980). Density-dependence in single species populations of plants. *Journal of Theoretical Biology* **83**, 345–357.

D. LAWRENCE VENABLE AND SARAH KIMBALL

7

Population and community dynamics in variable environments: the desert annual system

7.1 INTRODUCTION

Desert annual plants are frequently used to illustrate the principles of adaptation to variable environments, the population dynamic functions of dispersal and dormancy, and how temporal variation may promote species coexistence (Cohen 1966, Venable and Lawlor 1980, Shmida and Ellner 1984, Chesson and Huntly 1988, 1989, Philippi and Seger 1989, Venable *et al.* 1993, Chesson 2000). All of these topics involve ecological and evolutionary responses to environmental variability. High levels of environmental variation driven by rainfall are a signature characteristic of hot deserts (Frank and Inouye 1994, Davidowitz 2002). Desert annuals have provided useful conceptual models because they have very simple life cycles and respond on a rapid time scale to such environmental variation (Patten 1975, Venable and Pake 1999, Venable 2007). 'Good wildflower years', when showy-flowered annuals blanket the desert, often occur in association with abrupt desert annual population increases. Such years are correlated with greater than average germination season rainfall and global climatic cycles, such as El Niño and Pacific Decadal Oscillations, in the case of US southwestern deserts (Cayan *et al.* 1999, Venable and Pake 1999, Bowers 2005). Desert annuals spend most of their lives as seeds and may even go unnoticed during their normal flowering season in years of little germination or high mortality. Persistent seed-banks play an important role in population and community dynamics, and it is not uncommon for species to reappear following years of absence.

Temporal Dynamics and Ecological Process, ed. C. K. Kelly, M. G. Bowler, G. A. Fox. Published by Cambridge University Press.

While desert annuals are small and short lived, they occur as members of mature, persistent communities. This means that it is relatively easy to monitor multiple generations during the course of a single long-term project. Thus, in addition to being good conceptual models, desert annuals make good empirical models for exploring ecological and evolutionary dynamics in variable environments. Here, we present the results of our work combining the collection of long-term population dynamic data with several short-term focused approaches to understanding the ecology of Sonoran Desert winter annuals. We begin by introducing our study system and long-term data set, demonstrating how our data provide evidence for bet hedging and coexistence via the storage effect. Next, we describe a fundamental functional tradeoff that structures the dominant members of our community and determines the degree of interannual variation in fecundity. Finally, we explain long-term trends in response to climate change.

7.2 STUDY SYSTEM

Water is usually the critical resource in deserts and its availability is pulsed and irregular (Schwinning *et al.* 2004). In the northeastern Sonoran Desert, approximately half of all precipitation occurs during the summer monsoon season, from July to September, with the rest falling sporadically from October to April. The summer and winter growing seasons have associated annual floras and a smaller set of plants can grow during either season. About three quarters of Sonoran Desert annuals in Arizona are winter annuals. Their abundance and growth are heavily determined by variation in the timing and amount of precipitation during the winter growing season.

To understand biotic responses to environmental variability, we need long-term population dynamic data. Since 1982, we have followed the population dynamics of the winter annual plant species at the University of Arizona's Desert Laboratory at Tumamoc Hill (Venable and Pake 1999, Venable 2007). Our study site is a gently sloping alluvial plain dominated by *Larrea divaricata* (creosote bush) that has been protected from grazing since 1906, shortly after the property became the Carnegie Institute's Desert Botanical Laboratory in 1903. We have 72 permanent plots in the shrub–open-space matrix along a 250-m transect. We visit all plots following each rainfall event to map individual plants at germination, noting the species, and subsequently to record death and reproduction at bi-weekly to monthly intervals until all plants die. Since 1989, we have taken 180 23-cm^2 soil cores (approximately 4 cm deep; the vast majority of seeds are in the top 2 cm) each year following germination but prior to reproduction. From these soil cores, we float out any seeds and determine the number of viable seeds of each species that was left dormant following germination. We have calculated germination fraction for each species as the number of seedlings germinated per m^2 each year divided by the number of seedlings plus the number of viable seeds per m^2 remaining dormant in the soil (Pake and Venable 1996). As a result of this monitoring, we have a long-term data set with the number of seedlings, the germination fraction, per capita survivorship (the proportion of germinated seedlings that survived to reproduce), and per capita fecundity (the average number of seeds produced by each plant that reproduced, Figures 7.1–7.3). We calculated per germinant fecundity as per capita survivorship multiplied by per capita fecundity (Venable and Pake 1999, Venable 2007).

We have plots located both under creosote shrubs and in the open since this is the dominant habitat mosaic at our site. Weather stations record air temperature and precipitation. We monitor environmental conditions both under shrubs and in the open, including measurements of soil moisture and soil temperature at two different depths, the amount of photosynthetically active radiation and air temperature. These additional measurements indicate that under-shrub habitats are less extreme in terms of air temperature and are more mesic than open habitats.

7.3 BET HEDGING

Organisms in variable, unpredictable environments are expected to evolve adaptations to reduce risk (Philippi and Seger 1989). Evolutionary bet hedging is a reduction in short-term fitness in favour of long-term reduction of risk (Venable 2007). It is an adaptive strategy that buffers organisms from temporal variation and is thought to be widespread in nature. Some traits that have been considered bet hedging adaptations are invertebrate silk-ballooning dispersal (Bell *et al.* 2005), the perennial life cycle (Tuljapurkar 1990), diapause in many types of animals, from crustaceans (Hairston and Munns 1984) and mosquitoes (Andreadis 1990) to sponges (Fell 1995) and fish (Martin 1999), persistent seedbanks of weeds (Roberts and Feast 1972, Swanton and Weise 1991), variable offspring size in frogs (Crump 1981) and disease dynamics involving persistent sporulation (Henderson *et al.* 1999, Stumpf *et al.* 2002).

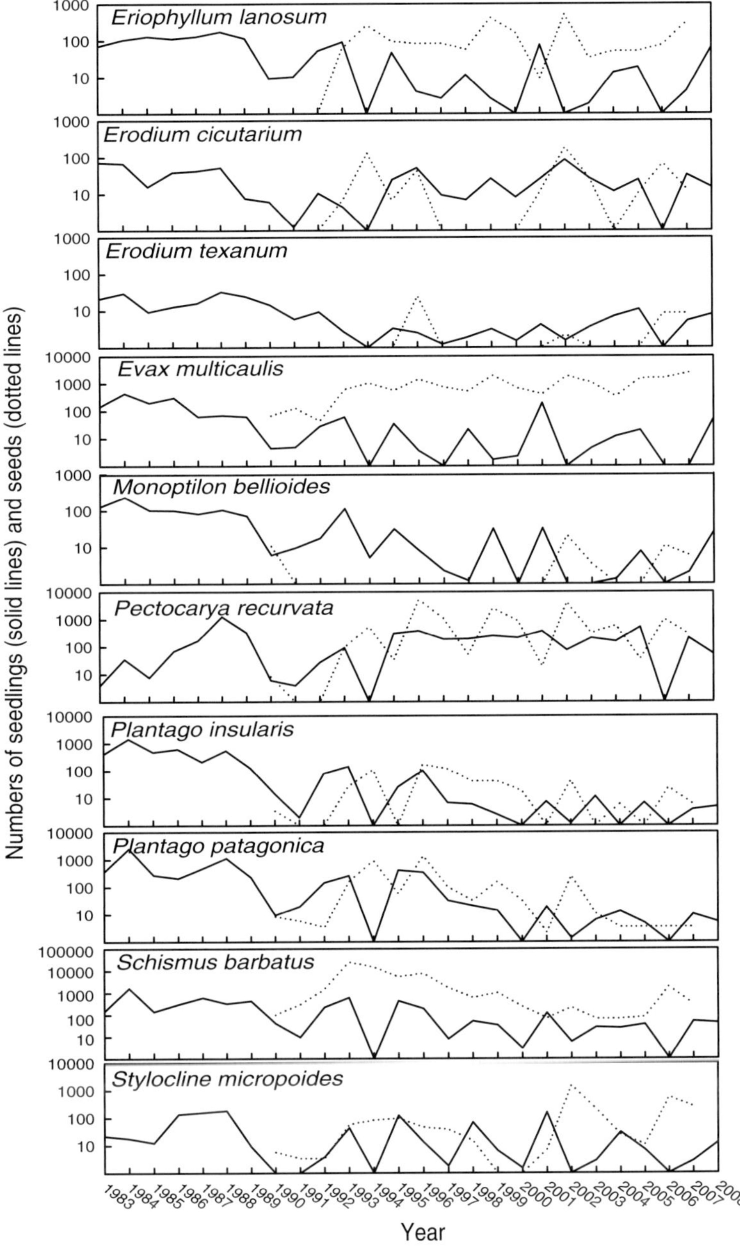

Figure 7.1 Abundance of several dominant species through time based on density per m^2 at germination and the density of viable seeds remaining in the seedbank after the end of the germination season (in February). Viable seed data are not available before 1990.

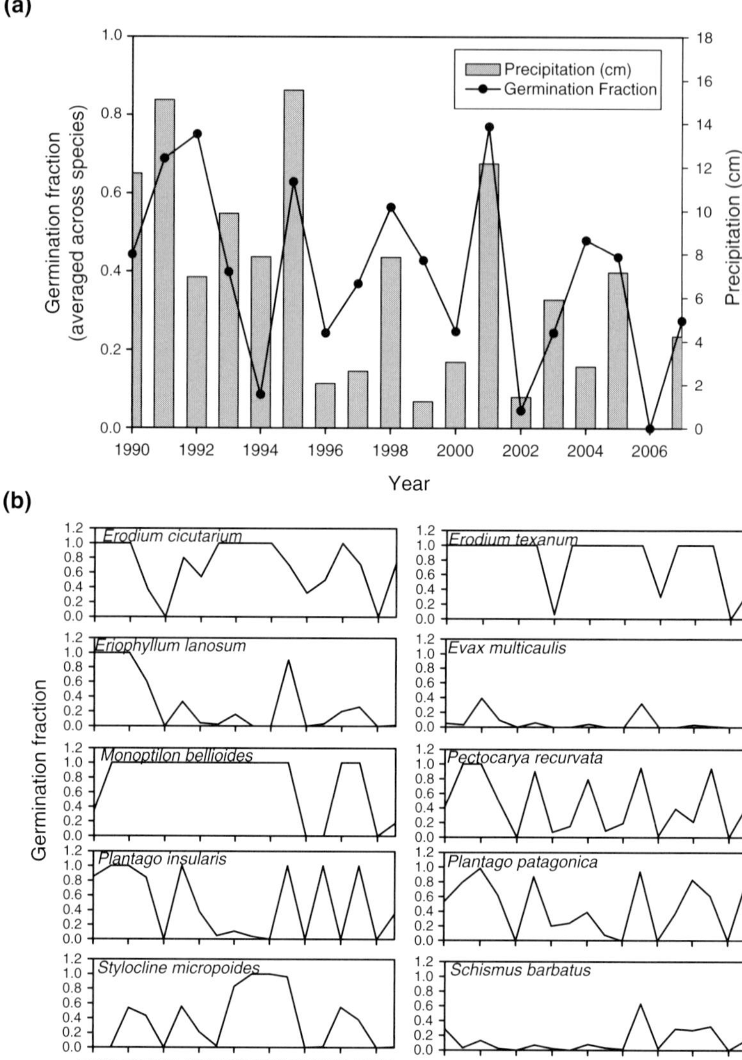

Figure 7.2 **(a)** Germination in the community of desert annuals is driven by the amount of rain during the winter germination season (usually October–January, but can be as late as March). For each year, germination season rain was calculated as the total precipitation starting with the first germination inducing rain and ending with the last germination inducing rain (so actual calendar dates bracketing germination season varied between years). This graph indicates the total amount of precipitation that triggered germination during those months. **(b)** Individual species' germination responses. Germination fraction is calculated as the density of seedlings emerging in natural field sites divided by the total number of viable seeds, calculated as germinated plus ungerminated viable seeds in the seedbank at the end of the germination season (February).

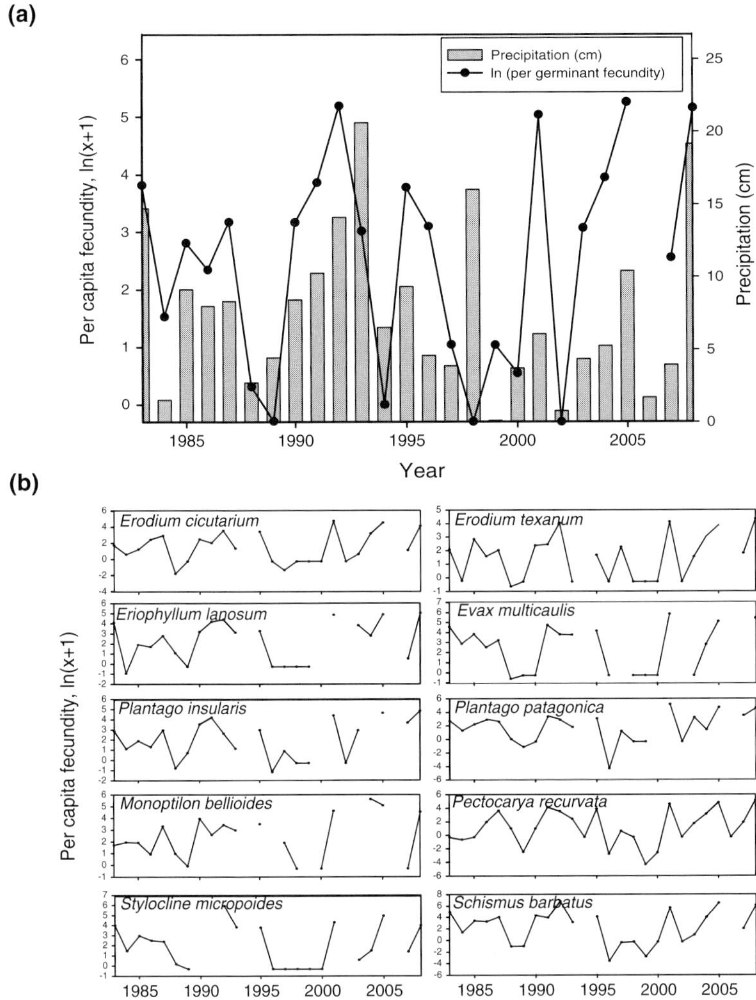

Figure 7.3 **(a)** Per capita fecundity (per germinant seed yield) varies among years and is largely driven by growing season precipitation (here indexed as January–March precipitation). **(b)** However, species have individualistic survival and fecundity responses (per germinant seed yields). Unconnected points are due to zero germination which results in years with undefined per germinant seed yield.

The concept is usually defined in terms of a tradeoff between mean and variance of population growth, financial profit, or Darwinian fitness (Seger and Brockmann 1987). Organisms with a bet-hedging trait should have higher long-term average fitness.

In our desert system precipitation is highly variable, so that a germinating seed may or may not receive adequate conditions for growth and reproduction. If a seed germinates, it may fail completely or it may successfully grow and produce many more new seeds for the next year. If it remains dormant, the best-case scenario is to survive to be just one seed in the next year. However, in the dormant state, a seed presumably has a good chance of surviving years when it would have experienced reproductive failure had it germinated. By producing some seeds that persist in the soil seedbank, a plant hedges against this possibility of dying with little or no seed production. Theory predicts that the amount of bet hedging, measured as the proportion of viable seeds that forgo germination, should be related to the year-to-year variation in fitness (Ellner 1985).

Desert annuals produce potentially long-lived seedbanks through delayed germination. We demonstrated this directly for one of our species with carbon dating of the seedbank via tandem accelerator mass spectrometry (TAMS). Aboveground nuclear bomb tests caused atmospheric ^{14}C levels to peak in 1963. Their subsequent gradual decline provided a signal for aging seedbanks with TAMS. The soil seedbank for *Pectocarya recurvata* was shown to have an age structure that declined exponentially with age, with the oldest of the 53 seeds sampled estimated to be 5 years old (Moriuchi *et al.* 2000). This was a species with a relatively high germination fraction (see Figure 7.4) and most of those 5 years had adequate rainfall for germination of non-dormant seeds. To test bet-hedging theory we used our long-term demographic data on natural germination fractions and the 28-year sequence of data on temporal variation in per germinant fecundity. As predicted, species that experienced greater variation in per germinant fecundity had lower average historic field germination fractions (Figure 7.4). In other words, species with reproductive success that varies greatly from year to year have a higher proportion of dormant seeds.

Previous studies have described adaptations that may function for bet hedging (Glennon *et al.* 2006, Matsuo 2006) and some of these have described the putative risk factors (Evans and Dennehy 2005). A few studies have related variable strength of a risk factor with variable expression of a bet-hedging trait (Philippi 1993, Clauss and Venable 2000). Other interesting studies have used demographic data to parameterise simulations which suggest that specific bet-hedging traits are adaptive (Philippi *et al.* 2001, Evans *et al.* 2007). In an intriguing recent study of bacteria under variable artificial selection in the laboratory, bet hedging was seen to evolve and the molecular mechanisms were

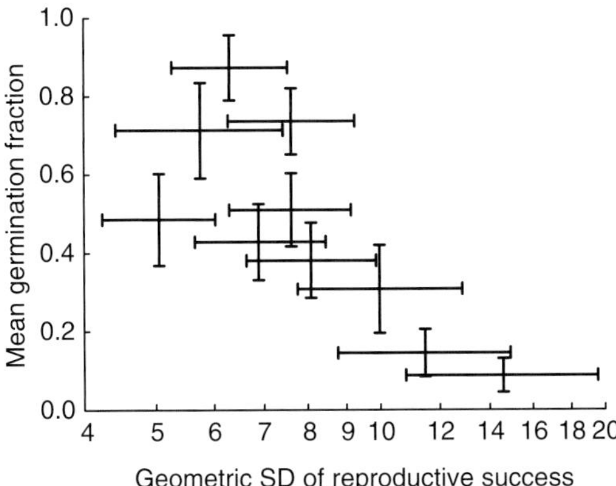

Figure 7.4 Mean germination fraction in relation to variation in the per capita reproductive success. Germination is averaged over 14 years. Demographic variation (22 years) is given as the geometric standard deviation, exp(SD(ln(average number of seeds per germinating seed))). $r = -0.78$; $p < 0.005$. Species and year bootstraps highly robust; error bars are +/− one bootstrapped SD. From Venable (2007).

described (Beaumont *et al.* 2009). The study of desert annuals described here has quantified actual, natural long-term fitness variance for a guild of species and related it to their observed long-term average dormancy in that same undisturbed field setting (Venable 2007).

7.4 FUNCTIONAL UNDERPINNINGS OF POPULATION DYNAMICS

The data presented above demonstrate the life-history pattern predicted for bet hedging. However, they also demonstrated a spectrum of life histories for plants coexisting in a community. This raises the question as to why species growing together in the same habitat have different demographic variances. The simplest answer, demonstrated with the long-term data, is that species with greater variance in per germinant fecundity were the ones whose reproductive success was more sensitive to the amount of winter precipitation (Venable 2007). This suggests that functional traits related to water use and growth rate may be important in determining population dynamics in this system.

To explore the connection between functional traits and population dynamics, we conducted detailed ecophysiological measurements on the common species of this winter annual community (Angert *et al.* 2007, Huxman *et al.* 2008). During the 2004–05 growing season, we harvested plants every 2 weeks from germination until death to determine growth and changes in biomass allocation. Leaves were scanned to determine leaf area and plant tissues were separated into roots, leaves, stems and reproductive material to determine dry mass partitions. From this, we calculated RGR for each species and investigated biomass allocation patterns in response to rain pulses. We also analysed carbon isotope ratios of leaves. Isotope ratios were used to calculate Δ, a measure of integrated intrinsic water-use efficiency (WUE, Ehleringer 1993, Dawson *et al.* 2002). These measurements revealed that the species in this community exhibit a tradeoff between growth rate and WUE. Furthermore, position along this RGR–WUE tradeoff correlates with variance in per germinant fecundity. Species with high demographic variance are fast-growing and have low WUE while the demographically buffered species are slow-growing and have high WUE (Figure 7.5, Angert *et al.* 2007, Huxman *et al.* 2008).

The amount of new leaf growth following a large rain pulse during the 2004–05 growing season varied depending on the position of a species along the tradeoff. The demographically variable, fast-growing species produced large amounts of leaf material following a large rain pulse late in the season, while the demographically buffered, slow-growing species did not exhibit this plastic leaf response (Angert *et al.* 2007, 2009). These results demonstrated that growth characteristics of individual species were important in determining among-species variation in demographic response to environmental variation.

We dissected how WUE is attained in these winter annual plants. Instantaneous WUE is the ratio of the rate of carbon assimilation through photosynthesis to the rate of water lost by stomatal conductance, and it is usually correlated with integrated WUE as measured with carbon isotope analyses (Ehleringer 1993). Somewhat surprisingly, our measure of integrated WUE did not correlate with instantaneous measures of stomatal conductance. Instead, water-use efficient species tended to have high carbon assimilation rates (Huxman *et al.* 2008). At the height of the growing season we used a Li-Cor 6400 gas exchange system to measure carbon and water flux under varying levels of CO_2 (A–C_i curves). To see what aspects of the carbon assimilation rate are responsible for the greater assimilation rate for water-use efficient plants, we calculated J_{max} (light-saturated rate of electron transport) and V_{Cmax} (maximum rate of carboxylation) from the A–C_i curves. We discovered that demographically buffered plants with high WUE had high investment in J_{max} rather than V_{Cmax}. The ratio of J_{max} to V_{Cmax} was consistently higher for these plants than for values typical

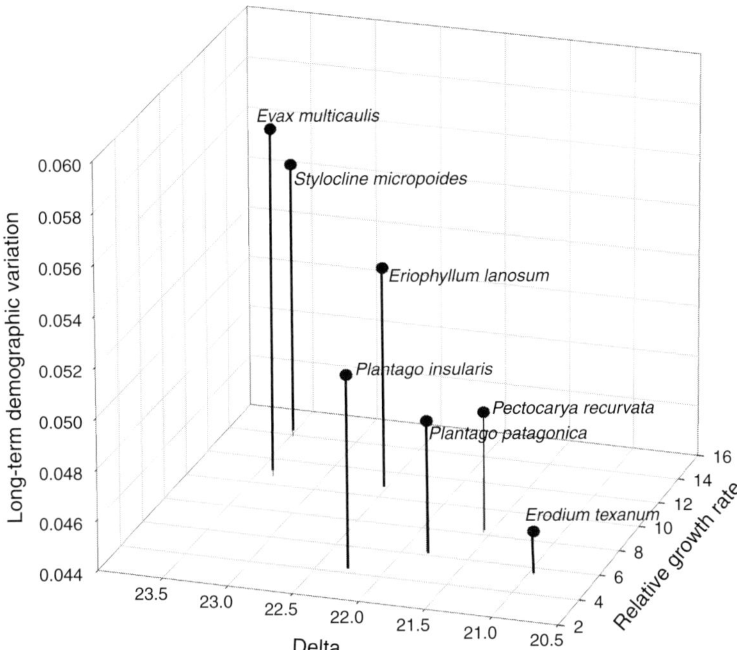

Figure 7.5 Carbon isotope discrimination (delta) is an indicator that decreases with higher integrated water-use efficiency (WUE). It is strongly related to relative growth rate (RGR) consistent with a tradeoff between growth rate and water-use efficiency. Faster growing, lower water-use efficiency species had highly variable per germinant fecundity (plotted as geometric SD of per germinant fecundity), while slower growing, higher water-use efficiency species had more buffered per germinant fecundity.

of a range of previously studied angiosperms (Wullschleger 1993). It seems paradoxical that plants in such high light environments would have unusually high allocation to photosynthetic light-capturing machinery. However, the enzymes involved in light capture are quite temperature sensitive, suggesting that this unusually high allocation may be an adaptation to increase photosynthesis at low temperatures (Harley *et al.* 1992). This may make sense considering that winter annuals escape the most water-stressed conditions of the desert by completing their life cycle during the coldest months of the year. Winter annuals are characteristic of warm but not cold deserts (Comstock and Ehleringer 1992), suggesting that low temperature provides a severe challenge to this life history. Our more water-use efficient, low variance species appear to be especially specialised to low-temperature photosynthesis. This led us to hypothesise that they may also differ in the phenology of their annual growth cycle (Huxman *et al.* 2008), and that such differences may help to maintain species diversity.

7.5 SPECIES DIVERSITY

We have identified over 70 species of winter annuals in our study plots over the last 28 years. Species diversity of germinated plants has changed from year to year as population sizes vary and all species do not emerge in great numbers in every year (Figure 7.6a). Species richness and Shannon diversity index for seedlings were both positively related to precipitation (Figure 7.6a). Nonetheless, there has been a long-term trend of declining precipitation that was not accompanied by a long-term decline in species diversity. Rather, there has not been any sustained tendency for diversity to increase or decrease (Pake and Venable 1995, Venable 2007). This long-term stability in species diversity despite highly variable population dynamics suggests that strong coexistence-promoting mechanisms are at work in this community.

Since desert annual species may be present in a community but not germinate in a given year, a more meaningful measure of species diversity would include persistent dormant seeds. Our long-term data set on the seedbank enabled us to include these dormant individuals in the calculation of species diversity. The total representation of a species is the sum of the seedlings that germinated and the viable seeds that remained dormant. Interestingly, diversity measured this way had a lower Shannon index from 1990–95 than subsequently (Figure 7.6b). During a late 1980s drought there were several years of high germination followed by reproductive failure. This depleted the numbers of seeds and seedlings by 1990. Small-seeded *Schismus barbatus* demonstrated a very strong demographic response to the increasingly wet growing seasons from 1990–93 (Figure 7.3) so that from 1990–95 it dominated the seedbank and accounted for more than 50% of seeds and seedlings (cf. Figure 7.1). After 1996, this species was a less prominent member of the seedbank, so species evenness was higher, increasing diversity (Figure 7.6b). Throughout this response to the late 1980s population declines, diversity of seedlings remained high (Figure 7.6a). As one would expect in a system with bet hedging in the form of prolonged seed dormancy, diversity of both seedlings and seeds was less variable and less related to precipitation than diversity of germinated seedlings alone.

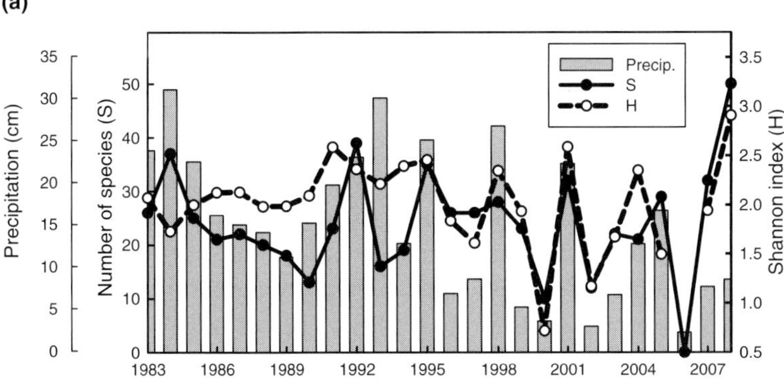

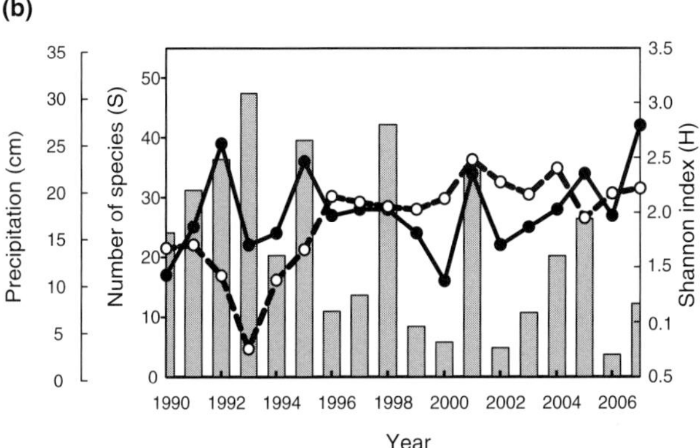

Figure 7.6 **(a)** Diversity of germinated seedlings and total growing season precipitation (October through May) for each year of our long-term study. S is the number of species and H is the Shannon diversity index. **(b)** Diversity of germinated seedlings plus ungerminated seeds in the seed bank for 1990–2007.

7.6 TEMPORAL VARIATION AND COEXISTENCE

Ecologists have explored and catalogued the potential mechanisms by which competing species stably coexist (Tilman and Pacala 1993, Chesson 2000). Niche-based mechanisms for stable coexistence rely on species differences that allow species to recover from low density. Some of these coexistence mechanisms operate independently of environmental variation. These include differential utilisation of multiple limiting resources (Schoener 1974, Tilman 1988) and frequency-dependent predation (Gendron 1987). Other scenarios require environmental variation for stable coexistence (Chesson 2000). Under this category are competition–colonisation tradeoffs in a metapopulation disturbance regime (Levins and Culver 1971, Hastings 1980), relative nonlinearity of competition (Armstrong and McGehee 1980) and the storage effect (Chesson and Warner 1981, Chesson 1994). The storage effect combines species-specific responses to the environment (temporal niche) and population dynamic buffering due to persistent life-history stages (such as seedbanks) to generate positive average low-density growth rate for each species.

Three coexistence-promoting mechanisms are likely to be important in structuring of this winter annual community. First, resource partitioning is certainly involved and is reflected in habitat and range differences for coexisting plants. Species are known to occur at different positions along salinity and moisture gradients and to differ in occurrence between bajada and flat environments as well as to have preferential occurrence on north versus south facing slopes and in the open or under-shrub habitats. Much information on ranges and habitats exists at the natural history level and is recorded in local floras (Felger 1980, Bowers and Turner 1985, Rondeau 1996). This type of range and habitat information documents resource partitioning and often hints at the mechanisms involved. Our long-term data set provides information on differences between open and under-shrub habitats (Pake and Venable 1995). For example, the timing of life-cycle events varied depending on whether plants were in the open or under shrubs (Kimball *et al.* 2011). Early-germinating species germinated first in the open, while later germinating species germinated first under shrubs. Senescence also differed depending on microhabitat, although it was not related to timing of germination (Kimball *et al.* 2011). Such microhabitat differences suggest the importance of resource partitioning in promoting coexistence in this system.

Another mechanism of potential, but unproven, importance is frequency-dependent seed predation. In this scenario, high density makes the seeds of a particular species more attractive to seed predators, which reduces the growth rate of common species. Little work has been published on this potential mechanism. However, granivory is an important ecosystem process in deserts, involving a great variety of rodents, birds and ants (Brown *et al.* 1979). Preliminary evidence from seed-tray feeding trials which systematically varied seed densities of different species suggests that rodents do prefer seeds of species that are more common (J. L. Horst and D. L. Venable, unpublished data).

Third, temporal environmental variation creates storage effect coexistence when (1) demographic independence of species arises from partially uncorrelated responses to temporal environmental variation, (2) some aspect of the life history provides 'storage' or resistant buffering to the combined negative effects of competition and unfavourable phases of the environmental variation (Chesson 1994, 2000) and (3) the degree of competition experienced by each species covaries positively with its population dynamic response to the environment to a greater extent when it is common than when it is rare. When these three elements are combined, each species will have positive long-term average growth rates when at low density, resulting in stable coexistence. The storage effect is widely thought to be the predominant temporal variation-dependent coexistence mechanism and it has been investigated in communities such as coral reef fishes (Chesson and Warner 1981), freshwater zooplankton (Cáceres 1997), tropical trees (Runkle 1989, Kelly and Bowler 2002), prairie grasses (Adler *et al.* 2006) and desert annuals (Pake and Venable 1995).

Our long-term data set demonstrates that species have differences in their germination responses to environmental variation (Figure 7.2) and in their survival and fecundity responses to environmental variation (Figure 7.3). Such species-by-year interactions for germination and per germinant fecundity provide the important first element for a storage effect: the partially uncorrelated responses to environmental variation that separate species niches temporally. We have also documented the second element required for a storage effect, in the form of a persistent seedbank which provides a buffer against unfavourable conditions (Moriuchi *et al.* 2000, Venable 2007). This seedbank allows species to persist when they are not able to germinate and reproduce successfully every year. When a species experiences favourable conditions, so that its germination and potential for growth and reproduction are high, it will contribute more to competition and tend to limit its own growth. If it is at low density, this effect will be weak. However, when a species is at high density, conditions favouring germination and growth will result in self-limiting competition. This means that species have greater covariance between competition and environment when common than rare (the third element required for a storage effect). When a species is at low density, competition with other species will be more important, but since species have partially uncorrelated responses to environmental variation, this will not always be a problem. Thus when at low density a species will have times when it is has a favourable environment and low competition, and so population size can increase strongly. This results in storage effect coexistence (Chesson 2000).

We used long-term demographic data to quantify the community average storage effect. This involved calculating the species-by-time interaction for germination and for per germinant fecundity (data like

Figures 7.2 and 7.3) as well as the average germination fraction and average survival of ungerminated seeds (Angert *et al.* 2009). For our system, the low-density advantage due to germination variation alone increases low-density long-term per capita population growth rate by $\approx$0.052. The low-density advantage due to per germinant fecundity variation alone adds 0.025 to the low-density growth rate. An additional increase of 0.027 comes from the covariance of germination fraction and reproductive variation, giving a total low-density growth rate advantage due to the storage effect $\approx$0.103. At our field site there are millions of individuals of most species. Thus any positive recovery rate will be adequate to result in indefinite coexistence (Chesson and Warner 1981, Chesson 1994). The observed storage effect is substantial, equivalent to a doubling time of 7 years for a species recovering from low density. This is the first detailed quantification of the storage effect using long-term demographic data on vital rates (Angert *et al.* 2009).

7.7 ECOPHYSIOLOGICAL TRAITS AND COEXISTENCE

Species-by-year interactions for germination and per germinant fecundity provide the quantitative measurement of the partially uncorrelated responses to environmental variation that separate species' niches temporally. Together with the average germination fraction and average survival of ungerminated seeds which determine the buffering effects of seedbanks, these interactions determine the magnitude of the storage effect reported above.

Species-by-year interactions for germination fractions result from species differences in germination physiology. Such physiological differences have been demonstrated in experiments in which conditions reflective of natural variation in the germination environment were manipulated in growth chambers (Adondakis and Venable 2004). We systematically varied the over-summering temperature and rainfall that seeds experienced. We also varied day length and day and night temperatures to reflect different months during the normal germination season (October–December). Finally, we conducted the germination experiments at different calendar dates bracketing the natural germination season (August–January). Species differed in their germination response to these growth chamber manipulations and also strongly differed in how the experimental variables interacted. Analysis of temperature, rainfall and calendar date when the different species have germinated in the field over the last 28 years also revealed strong species differences in germination niche (Kimball *et al.* 2010). These results provide information on how variation in germination biology gives rise to the species-by-year interaction for germination fraction, which determines the magnitude of storage effect coexistence.

Above, we showed that growth and water-use functional traits were related to the magnitude of temporal variance in per germinant fecundity. It is reasonable to assume that these same functional traits might also be involved in determining the magnitude of the species-by-year interactions in per germinant fecundity. To summarise the variation patterns of the numerous functional traits measured, we performed a principal component analysis of the physiological variables underlying the RGR–WUE tradeoff. Specific leaf area, leaf mass ratio, RGR plasticity, ratio of J_{max} to V_{Cmax} and leaf nitrogen content were all included in the analysis. The majority of variation of these five variables (54%) was captured by the first principal component axis, which differentiated slow-growing, high WUE species from fast-growing, low WUE species. We then developed a statistical test to determine if species that were more similar in their growth and water-use traits (as reflected in this principal component score) were more similar in their species-by-year interaction terms calculated from differences in reproductive success over 25 years. Functional trait similarity was highly correlated with similarity in species-by-year interactions for per germinant fecundity, thus completing the link from environmental variation to functional traits to population variation to community dynamics (Angert *et al.* 2009).

7.8 TIMING OF LIFE-CYCLE EVENTS

The growth and water-use traits that we have been able to link to population variation and storage effect coexistence relate to the ability to photosynthesise at low temperatures. Thus we hypothesised that population dynamic decoupling of species may relate to species differences in phenological specialisation made possible by these functional differences. Are high WUE, demographically buffered species more active during periods of cool temperatures? Temperatures are lower immediately following rain events. They are also lower later in the October to January germination season, and earlier in the spring growing season. We used our long-term demographic data set and finer scale short-term phenological data to investigate timing of germination, reproduction, and senescence of our core study species (Kimball *et al.* 2011). We found that, in any given year, buffered, high WUE species germinated and reproduced earlier in the season than variable species, while variable species germinated and reproduced later (Figure 7.7).

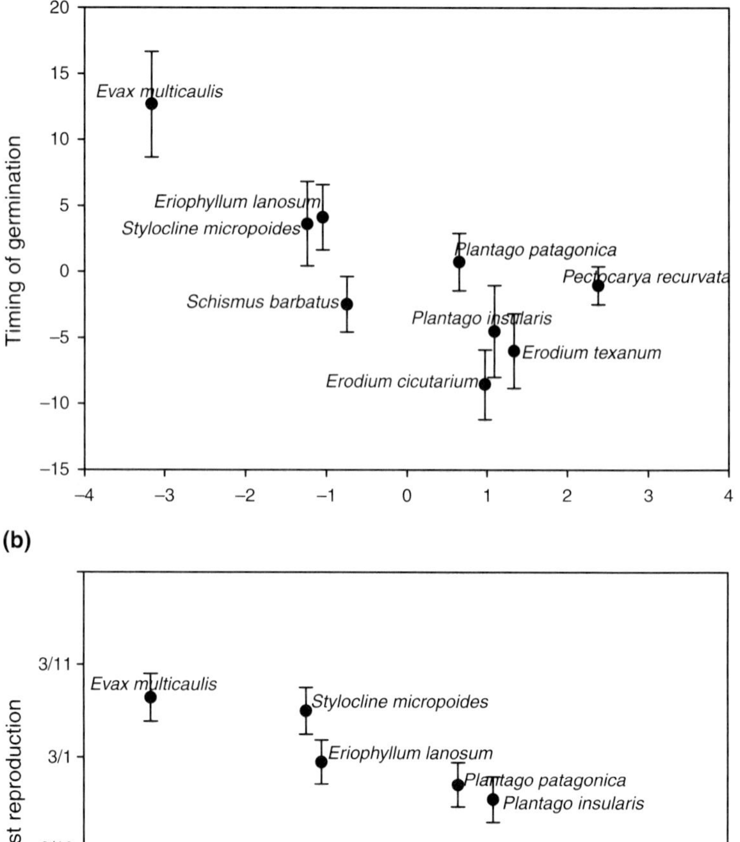

Figure 7.7 **(a)** Relationship between physiological PC score (position on the RGR–WUE tradeoff) and timing of germination from 1983–2008 ($r = -0.872$, $p = 0.002$). Species with low physiological PC scores have high RGR and low WUE, while those with high physiological PC scores have low RGR and high WUE. The number graphed on the *y*-axis is the average number of days by which each species deviated from the annual average germination date. For example, over the 25 years of the study, *E. multicaulis* germinated an average of 12.7 d after the average germination date (0) of all individuals of all species in any given year. **(b)** Relationship between physiological PC score and average date on which reproductive material was first produced during the 2004–2005 growing season ($r = -0.810$, $p = 0.008$).

While earlier reproduction exposed the cold-adapted species to cooler growing conditions in spring, earlier germination during autumn should, in principle, expose them to warmer growing conditions as seedlings. However, as will be explained below, over the last 28 years, the buffered, high WUE species have, on average, germinated under colder conditions, due to an increase in the abundance of these species and progressively later dates of germination over the years (Kimball *et al.* 2010). These differences in phenology interact with the variable pulsed timing of precipitation to favour different species in different years (species-by-year interaction).

7.9 LONG-TERM TRENDS

Long-term ecological data sets have recently been used to investigate the response of biotic communities to global climate change. Over the last 25 years, the Sonoran Desert has been getting warmer and drier (Weiss and Overpeck 2005). Climate models predict that this trend will continue and that the Southwestern United States will be affected more by climate change than other parts of the United States (McAfee and Russell 2008). Our weather data indicate that precipitation during the winter growing season has decreased an average of 6.6 mm/year and average temperatures have increased an average of 0.05°C/year since we began our long-term monitoring (Figure 7.8). Germination-triggering rain events have been arriving later in the year, during December rather than October. This means that germination has been occurring under colder temperatures, since temperatures in December are lower than in October (Figure 7.8, Kimball *et al.* 2010).

We used our long-term data set to determine how the winter annual community has responded to these changes in weather, and found that the abundance of all winter annuals has decreased over the last 25 years (Figure 7.9). Some species were affected more than others, leading to a change in community composition (Figures 7.1 and 7.9). We calculated the average physiology (average principal component explained above) of the plant community over time and found that species composition has shifted in favour of the demographically buffered, low RGR, high WUE cold-adapted species.

The shift in community composition could have been caused by changes in the germination, survival or fecundity of species in the community (Figure 7.10). In order to determine what stages in the life cycle of our plants were responsible for interannual changes in abundance, we performed a series of regressions of different stage

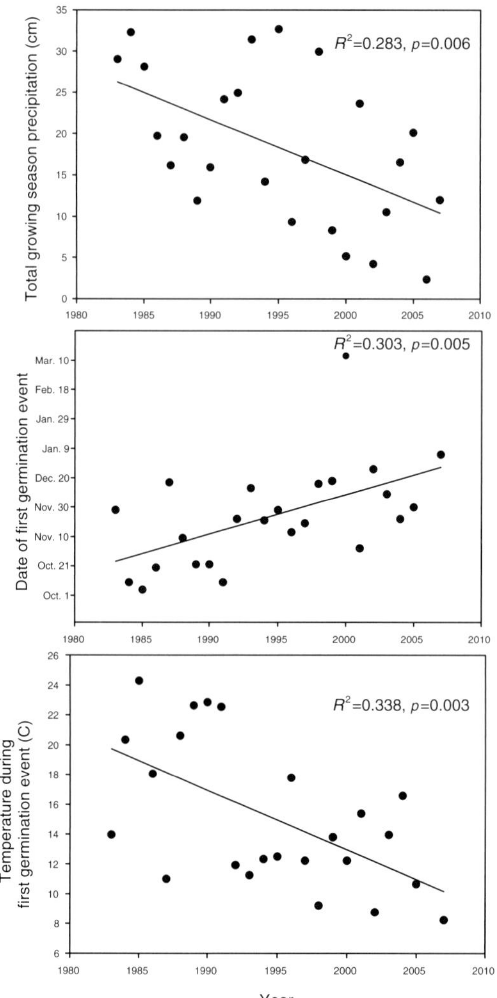

Figure 7.8 Growing season (October–April) precipitation, date of first germination and temperature during and for 5 days following first germination at the Desert Laboratory from 1983 to 2007.

transitions against total annual change in population size (Kimball *et al.* 2010). We used the long-term data set to calculate the size of annual transitions in the germination, survivorship and fecundity of each species, as well as an overall change in the number of seedlings from one year to the next. Linear regression and hierarchical partitioning analyses indicated that the transition from seeds produced in the previous year to seeds germinating in the current year (seed mortality and

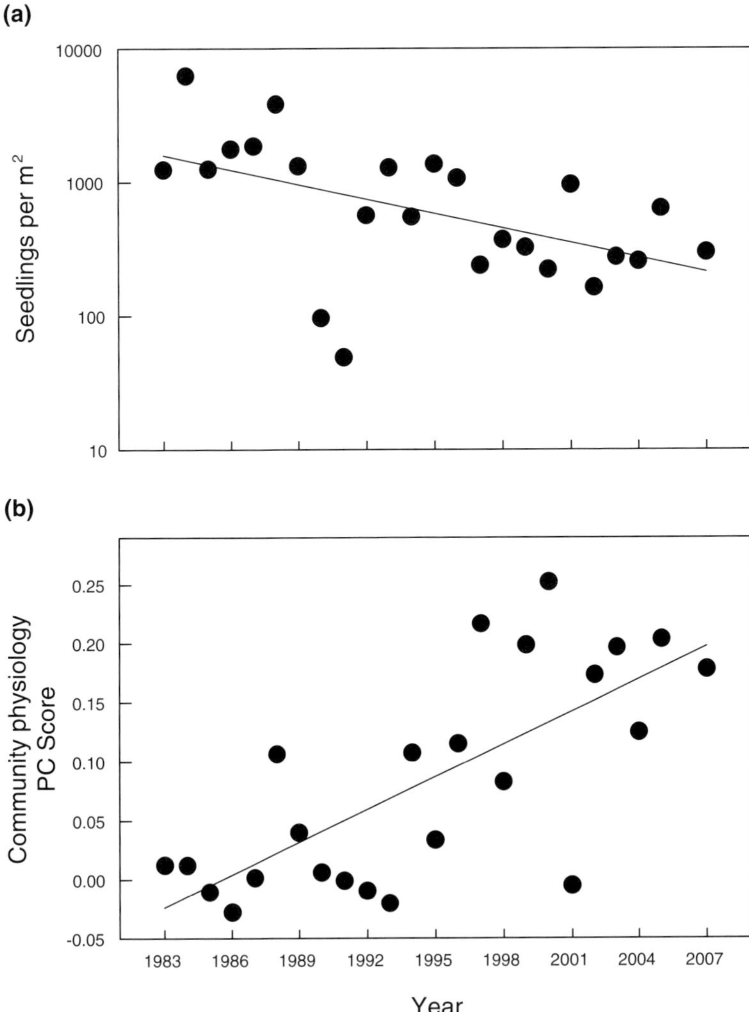

Figure 7.9 Patterns of abundance from 1983–2007. **(a)** Seedlings/m^2 over time. Each dot represents 1 year. The line is a linear regression. $R^2 = 0.272$, $p = 0.011$ (Figure 7.1 shows the trends in abundance for individual species). **(b)** The average physiological PC scores of seedlings over time with a linear regression line. $R^2 = 0.528$, $p < 0.0001$. Species with low physiological PC scores have high RGR and low WUE, while those with high physiological PC scores have low RGR and high WUE.

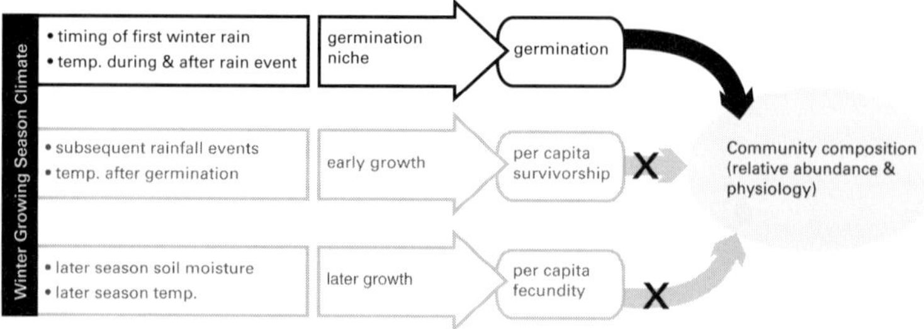

Figure 7.10 The role of three life-history stages (germination, survivorship and fecundity) in responding to climate and influencing community composition. These are three possible, not mutually exclusive, pathways by which winter growing season climate can effect community composition. Our analysis of vital rate shifts and their correlations with yearly population shifts showed that germination has been the most important factor for seven of nine species. Over the last 25 years, the winter rains have been arriving later in the year, resulting in decreased temperatures during germination. There have been fewer subsequent rainfall events, and later season temperatures have been getting warmer. Community composition has shifted to favour seedlings that germinate well under cold temperatures.

germination) was the life-history stage primarily responsible for changes in the abundance of seven of our nine focal species (Kimball *et al.* 2010). These results are a reminder that the response of biotic communities to climate change may be unexpected. Heating and drying have caused delays in germination-triggering rains that have led to lowered temperatures during germination, favouring an increase in cold-adapted species.

7.10 CONCLUSION

Combining the collection of long-term ecological data with a diversity of more short-term focused approaches is a powerful way to gain synthetic insights in ecology (Rees *et al.* 2001). We are using a 28-year data set on demographic vital rates along with more short-term investigations of life history, species interactions and ecophysiology to ask questions about bet hedging, fluctuation-dependent species coexistence mechanisms and responses to climate change (Huxman *et al.* 2013). With this system we can pinpoint the details of dynamics responsible for coexistence and long-term shifts in community composition and determine the functional biology responsible for these differential responses to the environmental variation. Desert annuals combine fast dynamics with mature, persistent communities in which probability distributions that are critical to fluctuation-dependent theories can be measured during the life of a single long-term project, analogous to several thousand years of data for shrubs or trees. Long-term data sets on the vital rates of individual species can provide an important window on coexistence and community response to climate change. The synthesis of ecophysiology with long-term demographic data provides advances in the fields of evolutionary ecology, population ecology and community ecology.

REFERENCES

Adler, P. B., HilleRisLambers, J., Kyriakidis, P. C., Guan, Q. F. and Levine, J. M. (2006). Climate variability has a stabilizing effect on the coexistence of prairie grasses. *Proceedings of the National Academy of Sciences, USA* **103**, 12793–12798.

Adondakis, S. and Venable, D. L. (2004). Dormancy and germination in a guild of Sonoran Desert annuals. *Ecology* **85**, 2582–2590.

Andreadis, T. G. (1990). Observations on installment egg hatching in the brown salt-marsh mosquito, *Aedes cantator*. *Journal of the American Mosquito Control Association* **6**, 727–729.

Angert, A. L., Horst, J. L., Huxman, T. E. and Venable, D. L. (2010). Phenotypic plasticity and precipitation response in Sonoran Desert winter annuals. *American Journal of Botany* **94**, 405–411.

Angert, A. L., Huxman, T. E., Barron-Gafford, G. A., Gerst, K. L. and Venable, D. L. (2007). Linking growth strategies to long-term population dynamics in a guild of desert annuals. *Journal of Ecology* **95**, 321–331.

Angert, A. L., Huxman, T. E., Chesson, P. and Venable, D. L. (2009). Functional tradeoffs determine species coexistence via the storage effect. *Proceedings of the National Academy of Sciences, USA* **106**, 11641–11645.

Armstrong, R. A. and McGehee, R. (1980). Competitive exclusion. *American Naturalist* **115**, 151–170.

Beaumont, H. J. E., Gallie, J., Kost, C., Ferguson, G. C. and Rainey, P. B. (2009). Experimental evolution of bet hedging. *Nature* **462**, 90–U97.

Bell, J. R., Bohan, D. A., Shaw, E. M. and Weyman, G. S. (2005). Ballooning dispersal using silk: world fauna, phylogenies, genetics and models. *Bulletin of Entomological Research* **95**, 69–114.

Bowers, J. E. (2005). El Niño and displays of spring-flowering annuals in the Mojave and Sonoran deserts. *Journal of the Torrey Botanical Society* **132**, 38–49.

Bowers, J. E. and Turner, R. M. (1985). A revised flora of Tumamoc Hill, Tucson, Arizona. *Madroño* **32**, 225–252.

Brown, J. H., Reichman, O. J. and Davidson, D. W. (1979). Granivory in desert ecosystems. *Annual Review of Ecology and Systematics* **10**, 201–227.

Cáceres, C. E. (1997). Temporal variation, dormancy, and coexistence: A field test of the storage effect. *Proceedings of the National Academy of Sciences, USA* **94**, 9171–9175.

Cayan, D. R., Redmond, K. T. and Riddle, L. G. (1999). ENSO and hydrologic extremes in the western United States. *Journal of Climate* **12**, 2881–2893.

Chesson, P. (1994). Multispecies competition in variable environments. *Theoretical Population Biology* **45**, 227–276.

Chesson, P. (2000). Mechanisms of maintenance of species diversity. *Annual Review of Ecology and Systematics* **31**, 343–366.

Chesson, P. L. and Huntly, N. (1988). Community consequences of life-history traits in a variable environment. *Annales Zoologici Fennici* **25**, 5–16.

Chesson, P. and Huntly, N. (1989). Short-term instabilities and long-term community dynamics. *Trends in Ecology and Evolution* **4**, 293–298.

Chesson, P. L. and Warner, R. R. (1981). Environmental variability promotes coexistence in lottery competitive systems. *American Naturalist* **117**, 923–943.

Clauss, M. J. and Venable, D. L. (2000). Seed germination in desert annuals: an empirical test of adaptive bet hedging. *American Naturalist* **155**, 168–186.

Cohen, D. (1966). Optimizing reproduction in a randomly varying environment. *Journal of Theoretical Biology* **12**, 119–129.

Comstock, J. P. and Ehleringer, J. R. (1992). Plant adaptation in the Great Basin and Colorado Plateau. *Great Basin Naturalist* **52**, 195–215.

Crump, M. L. (1981). Variation in propagule size as a function of environmental uncertainty for tree frogs. *American Naturalist* **117**, 724–737.

Davidowitz, G. (2002). Does precipitation variability increase from mesic to xeric biomes? *Global Ecology and Biogeography* **11**, 143–154.

Dawson, T. E., Mambelli, S., Plamboeck, A. H., Templer, P. H. and Tu, K. P. (2002). Stable isotopes in plant ecology. *Annual Review of Ecology and Systematics* **33**, 507–559.

Ehleringer, J. R. (1993). Gas-exchange implications of isotopic variation in aridland plants. In J. A. C. Smith and H. Griffiths (eds), *Water Deficits: Plant Responses from Cell to Community*. Oxford: Bios Scientific Publishers, pp. 265–284.

Ellner, S. (1985). ESS germination strategies in randomly varying environments. 1. Logistic-type models. *Theoretical Population Biology* **28**, 50–79.

Evans, M. E. K. and Dennehy, J. J. (2005). Germ banking: bet-hedging and variable release from egg and seed dormancy. *Quarterly Review of Biology* **80**, 431–451.

Evans, M. E. K., Ferrière, R., Kane, M. J. and Venable, D. L. (2007). Bet hedging via seed banking in desert evening primroses (*Oenothera*, Onagraceae): demographic evidence from natural populations. *American Naturalist* **169**, 184–194.

Felger, R. S. (1980). Vegetation and flora of the Gran Desierto, Sonora, Mexico. *Desert Plants* **2**, 87–114.

Fell, P. E. (1995). Deep diapause and the influence of low-temperature on the hatching of the gemmules of *Spongilla lacustris* (L) and *Eunapius fragilis* (Leidy). *Invertebrate Biology* **114**, 3–8.

Frank, D. A. and Inouye, R. S. (1994). Temporal variation in actual evapotranspiration of terrestrial ecosystems: patterns and ecological implications. *Journal of Biogeography* **21**, 401–411.

Gendron, R. P. (1987). Models and mechanisms of frequency-dependent predation. *American Naturalist* **130**, 603–623.

Glennon, V., Chisholm, L. A. and Whittington, I. D. (2006). Three unrelated species, 3 sites, same host monogenean parasites of the southern fiddler ray, *Trygonorrhina fasciata*, in South Australia: egg hatching strategies and larval behaviour. *Parasitology* **133**, 55–66.

Hairston, N. G. and Munns, W. R. (1984). The timing of copepod diapause as an evolutionarily stable strategy. *American Naturalist* **123**, 733–751.

Harley, P. C., Thomas, R. B., Reynolds, J. F. and Strain, B. R. (1992). Modeling photosynthesis of cotton grown in elevated CO_2. *Plant Cell and Environment* **15**, 271–282.

Hastings, A. (1980). Disturbance, coexistence, history, and competition for space. *Theoretical Population Biology* **18**, 363–373.

Henderson, I. R., Owen, P. and Nataro, J. P. (1999). Molecular switches: the ON and OFF of bacterial phase variation. *Molecular Microbiology* **33**, 919–932.

Huxman, T. E., Barron-Gafford, G., Gerst, K. L. *et al.* (2008). Photosynthetic resource-use efficiency and demographic variability in desert winter annual plants. *Ecology* **89**, 1554–1563.

Huxman, T. E., Kimball, S., Angert, A. L., Gremer, J. R., Barron-Gafford, G. A. and Venable, D. L. (2013). Understanding past, contemporary, future dynamics of plants, populations, and communities using Sonoran Desert winter annuals. *American Journal of Botany* **100**, 1369–1380.

Kelly, C. K. and Bowler, M. G. (2002). Coexistence and relative abundance in forest trees. *Nature* **417**, 437–440.

Kimball, S., Angert, A. L., Huxman, T. E. and Venable, D. L. (2010). Contemporary climate change in the Sonoran Desert favors cold-adapted species. *Global Change Biology* **16**, 1555–1565.

Kimball, S., Gremer, J. R., Angert, A. L., Huxman, T. E. and Venable, D. L. (2011). Differences in the timing of germination and reproduction relate to growth physiology and population dynamics of Sonoran Desert winter annuals. *American Journal of Botany* **98**, 1773–1781.

Levins, R. and Culver, D. (1971). Regional coexistence of species and competition between rare species. *Proceedings of the National Academy of Sciences, USA* **68**, 1246–1248.

Martin, K. L. M. (1999). Ready and waiting: delayed hatching and extended incubation of anamniotic vertebrate terrestrial eggs. *American Zoologist* **39**, 279–288.

Matsuo, Y. (2006). Cost of prolonged diapause and its relationship to body size in a seed predator. *Functional Ecology* **20**, 300–306.

McAfee, S. A. and Russell, J. L. (2008). Northern annular mode impact on spring climate in the western United States. *Geophysical Research Letters* **35**, L17701.

Moriuchi, K. S., Venable, D. L., Pake, C. E. and Lange, T. (2000). Direct measurement of the seed bank age structure of a Sonoran Desert annual plant. *Ecology* **81**, 1133–1138.

Pake, C. E. and Venable, D. L. (1995). Is coexistence of Sonoran Desert annuals mediated by temporal variability in reproductive success? *Ecology* **76**, 246–261.

Pake, C. E. and Venable, D. L. (1996). Seed banks in desert annuals: Implications for persistence and coexistence in variable environments. *Ecology* **77**, 1427–1435.

Patten, D. T. (1975). Phenology and function of Sonoran Desert annuals in relation to environmental changes. In *USABP Desert Biome Research Memorandum 75010*. Logan, UT: Utah State University, pp. 109–116.

Philippi, T. (1993). Bet-hedging germination of desert annuals: Variation among populations and maternal effects in *Lepidium lasiocarpum*. *American Naturalist* **142**, 488–507.

Philippi, T. and Seger, J. (1989). Hedging ones evolutionary bets, revisited. *Trends in Ecology and Evolution* **4**, 41–44.

Philippi, T. E., Simovich, M. A., Bauder, E. T. and Moorad, J. A. (2001). Habitat ephemerality and hatching fractions of a diapausing anostracan (Crustacea : Branchiopoda). *Israel Journal of Zoology* **47**, 387–395.

Rees, M., Condit, R., Crawley, M., Pacala, S. and Tilman, D. (2001). Long-term studies of vegetation dynamics. *Science* **293**, 650–655.

Roberts, H. A. and Feast, P. M. (1972). Fate of seeds of some annual weeds in different depths of cultivated and undisturbed soil. *Weed Research* **12**, 316–324.

Rondeau, R. J. (1996). Flora and vegetation of the Tucson Mountains. *Desert Plants* **12**, 1–47.
Runkle, J. R. (1989). Synchrony of regeneration, gaps, and latitudinal differences in tree species diversity. *Ecology* **70**, 546–547.
Schoener, T. W. (1974). Resource partitioning in ecological communities. *Science* **185**, 27–39.
Schwinning, S., Sala, O. E., Loik, M. E. and Ehleringer, J. R. (2004). Thresholds, memory, and seasonality: understanding pulse dynamics in arid/semi-arid ecosystems. *Oecologia* **141**, 191–193.
Seger, J. and Brockmann, H. J. (1987). What is bet-hedging? *Oxford Surveys in Evolutionary Biology* **4**, 182–211.
Shmida, A. and Ellner, S. P. (1984). Coexistence of plant species with similar niches. *Vegetatio* **58**, 29–55.
Stumpf, M. P. H., Laidlaw, Z. and Jansen, V. A. A. (2002). Herpes viruses hedge their bets. *Proceedings of the National Academy of Sciences, USA* **99**, 15234–15237.
Swanton, C. J. and Weise, S. F. (1991). Integrated weed management: the rationale and approach. *Weed Technology* **5**, 657–663.
Tilman, D. (1988). *Plant Strategies and the Dynamics and Structure of Plant Communities*. Princeton, NJ: Princeton University Press.
Tilman, D. and Pacala, S. (1993). The maintenance of species richness in plant communities. In R. E. Ricklefs and D. Schluter (eds), *Species Diversity in Ecological Communities*. Chicago, IL: University of Chicago Press, pp. 13–25.
Tuljapurkar, S. (1990). Delayed reproduction and fitness in variable environments. *Proceedings of the National Academy of Sciences, USA* **87**, 1139–1143.
Venable, D. L. (2007). Bet hedging in a guild of desert annuals. *Ecology* **88**, 1086–1090.
Venable, D. L. and Lawlor, L. (1980). Delayed germination and dispersal in desert annuals: escape in space and time. *Oecologia* **46**, 272–282.
Venable, D. L. and Pake, C. E. (1999). Population ecology of Sonoran Desert annual plants. In R. H. Robichaux (ed.), *Ecology of Sonoran Desert Plants and Plant Communities*. Tucson, AZ: University of Arizona Press, pp. 115–142.
Venable, D. L., Pake, C. E. and Caprio, A. C. (1993). Diversity and coexistence of Sonoran Desert winter annuals. *Plant Species Biology* **8**, 207–216.
Weiss, J. L. and Overpeck, J. T. (2005). Is the Sonoran Desert losing its cool? *Global Change Biology* **11**, 2065–2077.
Wullschleger, S. D. (1993). Biochemical limitations to carbon assimilation in C_3 plants: A retrospective analysis of the A/C_i curves from 109 species. *Journal of Experimental Botany* **44**, 907–920.

SUSANNE SCHWINNING, GORDON A. FOX
AND COLLEEN K. KELLY

8

Temporal niches, ecosystem function and climate change

8.1 INTRODUCTION

This chapter differs from the main current of this volume – the identification and quantification of coexistence mechanisms associated with temporal niche dynamics – in exploring the ramifications of these processes for ecosystem ecology. The intimate link between niches and ecosystem function has long been recognised, at least in the general sense that more species, representing a greater diversity of 'life-styles', make more complete use of available resources and thus achieve higher levels of productivity (e.g. Preston 1948, Odum 1953, MacArthur 1955, May 1975). This broadly stated principle has been unpacked in numerous models that are more specific, for example in resource-ratio niche theory (Tilman 1982) and various forms of spatial niche theories (Loreau 1998). However, the role of temporal niches in the ecosystem context is somewhat less well developed, but critical to understanding ecological responses to climate change.

Two main features characterise worldwide, anthropogenic climate change: a general warming trend that is strongest at low latitudes and weakest at high latitudes, and complex changes in precipitation patterns, currently predicted to include reductions in precipitation at the poleward fringe of the subtropical dry belt at midlatitudes (IPCC 2007, Scheff and Frierson 2012). Both temperature and precipitation shifts, as well as their interactions, have the potential to alter environmental heterogeneity. For example, the onset of spring/summer growing seasons could be advanced (Menzel *et al.* 2006) and the frequency and amplitude of extreme hydrological events such as drought

Temporal Dynamics and Ecological Process, ed. C. K. Kelly, M. G. Bowler, G. A. Fox.

and flooding increased (Huntington 2006). Both temperature trends and precipitation variability are important factors in structuring temporal niches, for example by functioning as triggers of life-history events (Kelly *et al.*, this volume, Venable and Kimball, this volume) or by controlling competitive interactions through their effects on primary production (Haxeltine and Prentice 1996). Simultaneous changes in seasonal temperature and precipitation patterns may have complex effects on populations and their interactions. Predicting such effects, and their feedbacks on climate, is one of the premiere challenges of earth system science and, in our opinion, cannot be adequately tackled without a more complete understanding of temporal niche dynamics and its role in ecosystem function.

At the most basic level, ecosystem function is measured in terms of primary productivity, the sum total of the productivity contributed by all plants present. For convenience, ecologists often group plant species into 'plant functional types' that are *alike* in their interactions with the ecosystem and as a group markedly *different* from other functional types (Figure 8.1). Classification schemes vary widely (Box 1996, Lavorel *et al.* 2007) but in general they involve structural and physiological distinctions that determine their participation in material cycles (e.g. capacity for nitrogen fixation, rooting depth, woodiness) and in energy flows (e.g. albedo, surface roughness) (Westoby and Wright 2006). Species belonging to different plant functional types are assumed to have more complementary ecosystem functions and less overlap in habitat and resource requirements. Species that are very similar in structure and physiology are called functionally redundant and assumed to be largely interchangeable in terms of their ecosystem functions (Figure 8.1; Walker 1992, Naeem 1998, Yachi and Loreau 1999).

Even though this conceptual framework does not specifically address how species richness is maintained, it is natural to assume that plant functional types coexist because they have little niche overlap and weak, if any, competitive interactions, or may even interact facilitatively (Dawson 1993, Schwinning and Parsons 1996, Chapin *et al.* 1998, Cardinale *et al.* 2002, Hooper *et al.* 2005). For example, deep-rooted shrubs may provide a good microhabitat for perennial grasses. By extension, competition between functionally redundant species can be expected to be stronger than, on average, with random community members, but differences in sensitivity to environmental factors variable in space and time (precipitation patterns, soil type, aspect, etc.) could provide opportunities for stable coexistence (Loreau 2000).

In this framework, ecosystem productivity and adaptability to climate change depends in part on direct effects of climate on plant

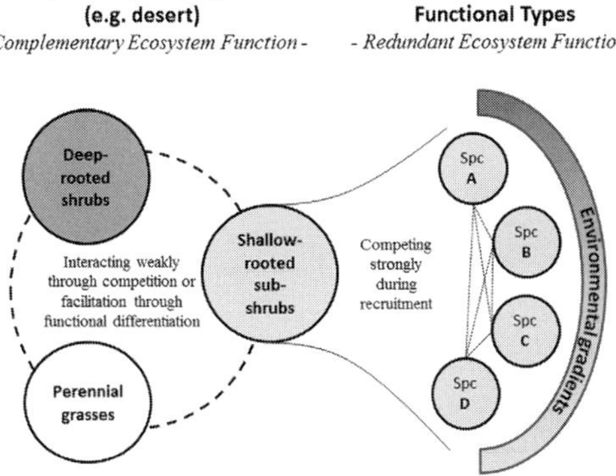

Figure 8.1 The presumed roles of complementarity and redundancy in ecosystem function. Left: Plant functional types, differentiated by numerous structural and physiological traits, giving them broad niche separation and complementarity in ecosystem function. Right: Functionally redundant species within a functional type, differentiated primarily by their sensitivities to environmental conditions, with high potential for competitive interactions. Functional richness is expected to elevate average ecosystem performance through synergistic interactions; redundancy is expected to reduce variance of ecosystem function in the face of environmental heterogeneity (= insurance function).

function – for example, aridification is relatively more adverse to the productivity of deeper-rooted species due to less deep infiltration – and in part on the environmental niche structure of redundant species, about which much less is known as a rule. Does a functional type with many species operate at a consistently higher level of primary production? Does it occupy a wider range of environmental niches? Can richness in redundant species buffer more efficiently against species loss or climate change? In sections 8.2–8.4, we attempt to address these questions from the viewpoint of community theory and the evolutionary origins of temporal niche structure.

8.2 PLANT FUNCTIONAL TYPES, REDUNDANCY AND PRODUCTIVITY

Agricultural experiments have long ago shown that crop mixtures can overyield, that is, produce more biomass, when intermixed than when grown separately on the same land (Vandermeer 1992). If carefully selected, mixtures can even produce more biomass per land area than any monoculture of the component species. This has been called

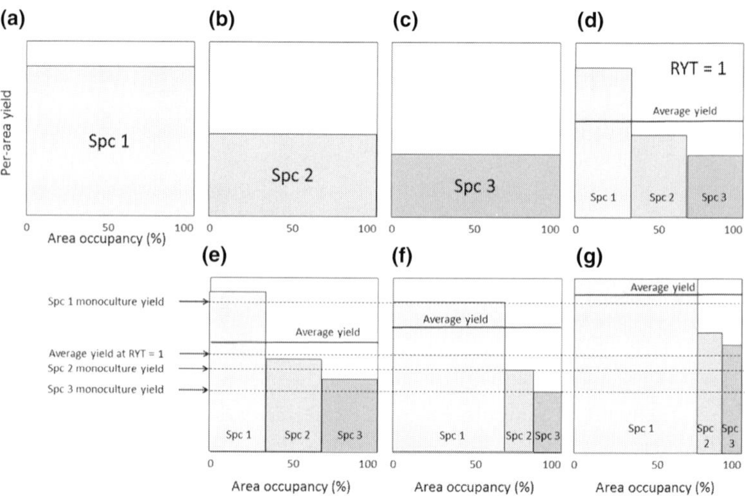

Figure 8.2 Overyielding. To quantify overyielding, the yield of polycultures is compared to the mean and maximal yields of the component species' monocultures. **(a–c)** Monoculture yields for three species. **(d)** Polyculture yield corresponding to Relative Yield Total = 1, in which area-based species yields are exactly as in the monocultures. **(e)** Non-transgressive overyielding through the complementarity effect. In this example, all three species yield more per area than in monoculture. **(f)** Non-transgressive overyielding through the selection effect. Here, the most productive species 1 increased area occupancy, while area-specific yields stayed the same. **(g)** Transgressive overyielding achieved through complementarity and selection effects.

transgressive overyielding, while *non-transgressive overyielding* is the term used when crop mixtures yield more than the average yield of monocultures, but no more than the highest yielding monoculture.

Theoretically, polycultures can overyield in two distinct ways (Loreau 2000; Figure 8.2). One or more species could increase yield-per-unit-cover (or per seed sown), so that species' yield reductions due to sharing space with other species is less than proportional to their reduction in cover or density (Tilman 1999). Overyielding through this mechanism has been called the 'complementarity effect' and is consistent with the idea that species have distinct resource needs, so when they are grown together they capture a greater fraction of the available resource per unit land. However, the complementarity effect does not necessarily result in transgressive overyielding if species have large differences in monoculture yields, because the negative effect of reducing the cover of the most productive species weighs against the positive effect of increasing individual yields-per-unit-cover. The second way in which polycultures can overyield is through the selection effect, which depends on the more productive species becoming dominant. This effect does not require synergistic relationships between species, only that species that are more productive competitively displace those that are less productive. While the selection effect was originally seen as evidence against the hypothesis of a productivity-enhancing effect of species richness (Huston 1997, Wardle 1999, Schwartz *et al.* 2000), it is now interpreted as a relevant mechanism through which ecosystems increase productivity in nature (Hector *et al.* 2002).

A second presumed benefit of species richness is the reduction of variance in ecosystem function in the presence of environmental variability. There are two ways to think about this 'insurance' function of biodiversity (Yachi and Loreau 1999). One way is in the sense of a 'portfolio effect': if species in a community have a component of independent yield variation, the variance around the community's average performance should decrease with the number of species simply as a consequence of statistical averaging (Doak *et al.* 1998). At the same time, the mean performance of the mixture could be above, below or the same as the average across monocultures, depending on species selection vis á vis the frequency of environmental conditions favouring them. The other way is in terms of 'compensatory dynamics' (Tilman 1996), in which better-adapted species grab resource opportunities vacated by less-well-adapted species. At the very least, this would provide even more stability of ecosystem function. But if some species fail where others succeed for ecologically significant reasons – e.g. because they have different environmental optima – a community of functionally redundant species could also regularly achieve higher average productivity, as long as species increase under conditions in which they are more productive and vice versa.

Which of these effects have actually been observed in experiments? The following is a brief summary for terrestrial ecosystems (mostly grassland experiments) derived from meta-analyses and informal syntheses (Hooper *et al.* 2005, Balvanera *et al.* 2006, Cardinale *et al.* 2006, 2007, 2011, Isbell *et al.* 2009, Zhang *et al.* 2012):

1. The great majority of mixtures overyield, but transgressive overyielding is uncommon;
2. Productivity responds more strongly to functional diversity than to species richness;
3. Mixtures overyield in part because they contain the most productive species, indicative of a 'selection effect';

4. In addition, there are positive effects of having multiple species in the community; indicative of a 'complementarity effect';
5. The magnitude of the complementarity effect and the degree of overyielding increases with the length of the biodiversity experiment or its spatial scale;
6. Mixtures that are more diverse are more stable over time and more resilient to some (e.g. nutrient stress, invasions) but not all types of perturbation (e.g. temperature stress, drought).

In general, these patterns suggest that plant species tend to have large differences in area–yield not easily overcome by complementary resource use and so transgressive overyielding is not the norm. In addition, the establishment of a maximally productive community takes time, presumably involving a slow process of sorting of species with respect to each other and environmental heterogeneity in space and time (Mouquet *et al.* 2002, Hillebrand *et al.* 2008, Zhang *et al.* 2012). Apart from a positive effect of including legumes, which, through symbiotic nitrogen fixation, enrich soil fertility and therefore the productivity of most other species (Fornara and Tilman 2009, Hooper *et al.* 2012), there is little further clarification of the ecological and biological mechanisms responsible for producing diversity effects on ecosystem function. The strong focus on grassland ecosystems in biodiversity experiments may limit the mechanisms available for complementary resource use, since herbaceous species are relatively restricted in morphological and physiological differentiation, e.g. compared to plants in forest communities (Zhang *et al.* 2012).

It is also noteworthy that experimental grassland communities appear to be poorly buffered against environmental factors most directly related to climate change – temperature and water. Since temperature and water are two of the most fundamental constraints on terrestrial plant photosynthesis, this is not a surprise, but raises questions about the generality of conclusions drawn from this system regarding the effects of species richness in climate change adaptation.

8.3 THEORETICAL EXPECTATIONS

Many authors have found it useful to explore heuristic models to determine the theoretical underpinnings of diversity–productivity relationships (Doak *et al.* 1998, Loreau 1998, 2004, Tilman 1999, Yachi and

Loreau 1999, 2007, Norberg *et al.* 2001, Mouquet *et al.* 2002, Loreau *et al.* 2003, Beckage and Gross 2006, Marquard *et al.* 2009). Here, we examine the properties of two very general types of models representing interactions between species with distinct functional niches, which do not require environmental heterogeneity to coexist, and between species that are functionally identical and coexist because of differential sensitivity to fluctuating environmental conditions.

Arguably, the quintessential model of coexistence between species with distinct resource niches is the Lotka–Volterra competition model. It requires neither temporal nor spatial variation for coexistence, and can be unpacked into more mechanistic consumer–resource models with two different kinds of resources (for two species) that are required in different proportions by each species (MacArthur and Levins 1967, Tilman 1982). For the purpose of illustration, we use the Lotka–Volterra model with explicit mortality terms:

$$\frac{dN_1}{N_1 dt} = r_1 N_1 \left(1 - \frac{\alpha_1 N_1 + \beta_1 N_2}{K_1}\right) - d_1 N_1 \quad (8.1a)$$

$$\frac{dN_2}{N_2 dt} = r_2 N_2 \left(1 - \frac{\alpha_2 N_2 + \beta_2 N_1}{K_2}\right) - d_2 N_2, \quad (8.1b)$$

where the r_i are the maximal growth rates of species, K_i the carrying capacities, d_i the per capita mortality rates, α_i the intraspecific competition coefficients and β_i the interspecific competition coefficients. For species to be in equilibrium, their rates of growth (the first terms in the equations) must equal their rates of mortality (the second terms), and the absolute magnitudes of both rates are adequate measures of productivity, or the ecosystem carbon and nutrient flux required to maintain the population at a given density. Thus, the total productivity or ecosystem flux F required for maintaining any combination of standing biomass for the two species is

$$F = d_1 N_1 + d_2 N_2. \quad (8.2)$$

In Figure 8.3, we superimposed contour lines for F-values onto the classic zero-growth isocline representation of two-species interaction models, so that ecosystem fluxes associated with monocultures can be readily compared with those for stable equilibria. Fixing values for d_1 and d_2, we varied the remaining parameters to generate a set of case examples.

Figure 8.3a shows that both non-transgressive and transgressive overyielding is possible under stable coexistence between competitors.

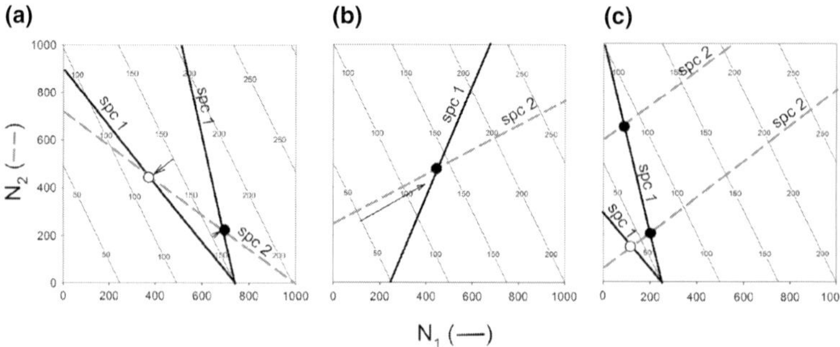

Figure 8.3 Coexistence and productivity in the Lotka–Volterra model. Shown are zero-growth isoclines (thick) and contour lines (fine) that indicate the total productivity required to maintain any combination of standing biomass (i.e. $F = d_1N_1 + d_2N_2$). Shown are only cases of stable coexistence. Transgressive overyielding at equilibrium is indicated by solid symbols, non-transgressive overyielding by open symbols. (a) Competition. By assumption, the productivity of species 1 in monoculture is higher than that of species 2. Depending on the slope of the species 1 isocline relative to the slope of the productivity contour lines, stable mixtures either transgressively or non-transgressively overyield. (b) Mutualism. A stable mixture of mutualists always has higher productivity than either monoculture. (c) Exploitation. If the exploited species (here, species 1) in monoculture, has higher productivity than species 2 in monoculture, stable mixtures either transgressively or non-transgressively overyield, depending on the slope of the species 1 isocline. If the exploiting species 2 has higher productivity in monoculture, stable mixtures necessarily transgressively overyield.

Transgressive overyielding happens if the slope of the higher-yielding species' isocline is steeper than the slope of the contour lines for ecosystem flux, mathematically:

$$\frac{\beta_1}{\alpha_1} < \frac{d_2}{d_1}. \tag{8.3}$$

Thus, transgressive overyielding does not occur when interspecific relative to the intraspecific competitive effect on species 1 is greater than the mortality rate of species 2 relative to species 1. Suppose that species 1 has the higher mortality rate, so that it must assimilate more resources to maintain a unit of standing biomass. It is then easy to see that a relatively steep reduction of its standing biomass by adding the competitor can only lower the combined resource uptake of the mixture. Biologically, this violates no known tradeoffs; it only requires that a faster growing species can be strongly competitively suppressed by a slower growing species.

By changing the sign of the competition coefficients, the Lotka–Volterra model can also express mutualistic (+,+) and exploitative (+,−) relationships between two species. Mutualistic interactions between primary consumers can arise if both species make the environment more benign for the other species, for example, one species may produce a more favourable microenvironment for the other species, while the other species may enhance soil fertility. Interactions between primary consumers can be exploitative if one species makes more resource available (e.g. a nitrogen-fixing legume), while the other species only competes for mutually limiting resources (Schwinning and Parsons 1996).

A stable mutualism always transgressively overyields (Figure 8.3b). This is a direct consequence of the premise that both species increase standing biomass in the presence of the other species. An exploitative interaction either transgressively or non-transgressively overyields (Figure 8.3c). If the exploited species (1) has the higher yield in monoculture, then, as in the case of competition, the isocline slope relative to the contour line slope determines the degree of overyielding. If, on the other hand, the exploiting species (2) has higher monoculture yield, an equilibrium exists only if the species 1 isocline is steeper than the contour lines, and the mixture necessarily transgressively overyields.

Thus we can conclude quite generally that linear interactions between coexisting species in a constant environment, if the productivity of the mixture is a linear combination of species' standing biomass, always overyield. All types of species interactions (e.g. (−,−), (+,+), (+,−)) can produce transgressive overyielding, but only mutualistic interactions (+,+) must transgressively overyield.

Different results obtain for nonlinear interactions. For example, Loreau (2004) observed that stable mixtures of competing species could underyield in a generalised Lotka–Volterra competition model with concave-up isoclines. Strong interference competition can produce this effect by lowering resource-use efficiencies in both competitors, as has been observed for two *Drosophila* species (Gilpin and Justice 1972). This may be an unlikely case for plants, however, since plants tend to become more efficient, not less efficient, under resource scarcity or competition.

We now turn to the lottery model (Chesson and Warner 1981, Chesson *et al.* 2001) to represent the interactions of functionally redundant species that do not vary in resource requirements, only in sensitivity to environmental fluctuations. A fundamental assumption of the lottery model with coexistence through the storage effect is that populations divide into at least two distinct life-history classes, of which one engages in high stakes lottery competition (e.g. seedlings) and the others are relatively persistent stages for surviving seedlings (e.g. perennial adults). Mathematically, the model is

$$\sum_{j=1}^{k} \beta_j(t) n_j(t) > 0 :$$

$$n_i(t+1) = n_i(t) * (1 - d_i) + \sum_{j=1}^{k} d_j n_j(t) * \frac{c_i(t)\beta_i(t)n_i(t)}{\sum_{j=1}^{k} c_j(t)\beta_j(t)n_j(t)} \quad (8.4a)$$

$$\sum_{j=1}^{k} \beta_j(t) n_j(t) = 0 : n_i(t+1) = n_i(t) * (1 - d_i), \quad (8.4b)$$

where k is the number of suitable sites (occupied or empty) for members of the community, n_i is the proportion of adults of species i relative to k, d_i is the per capita death rate of adults, β_i is the per capita seed production and c_i is a parameter that expresses temporal variation in competitiveness between seedlings of different species (Chesson and Warner 1981, Chesson *et al.* 2001). Equation (8.4a) expresses what happens in a year when at least one species recruits. The first term describes the number of plants that survive from one year to the next and the second term adds the number of new recruits, as determined by lottery competition for open sites. Equation (8.4b) expresses what happens in years when no species recruits. By expressing community dynamics in terms of proportions relative to the number of suitable sites, the model allows for potentially large variation in total site number (or establishment opportunities) from year to year, as may be the case in highly water-limited environments. However, establishment opportunities are equally available for all species. Thus, absolute establishment success of species may be highly correlated over time, and variation between species comparatively small, but for coexistence via temporal niches, only the independently variable fraction of establishment fluctuations matters.

An arbitrarily large number of species can stably coexist under this model, if each species has positive average growth rate at low density. Furthermore, species with average fitness differences can coexist if the stabilising effect generated by a lessening of intraspecific competition at low density is sufficiently strong (Chesson 1994).

In keeping with the definition used previously, we define the ecosystem function of this community as the matter flux required to grow and maintain adults, and in this case, to grow seeds as well. Since this is a model for functionally redundant species, we assume that species are identical in all metabolic costs for seed production and growth and maintenance of adults, and have the same mortality rates. We ignore the productivity of all seedlings unsuccessful in recruitment. Setting k constant, the long-term average annual ecosystem flux of the community is

$$F = C_{adult} d\overline{\sum_{J=1}^{k} n_J(t)} + C_{seed}\overline{\sum_{J=1}^{k} n_J(t)\beta_J(t)}\ , \tag{8.5}$$

where C_{adult} and C_{seed} are the average annual per capita metabolic costs associated with adult growth and maintenance, and with seed production, respectively. Here we set both cost parameters to unity and scale β so that the average cost of community-wide seed production is equivalent to the cost of maintaining full site occupancy by adults. The relative annual cost of seed production versus the growth and maintenance of adults of course varies between groups of species, although in most plants appreciable growth versus reproduction tradeoffs suggest that neither cost is negligibly small (Obeso 2002, Herben *et al.* 2012).

We consider three variations of the lottery model. For models 1 and 2, we assume that every year is a recruitment year (so new recruits immediately occupy vacated sites), as in the original model proposed by Chesson and Warner (1981). With adult numbers constant, variation in ecosystem function is caused only by varying seed production. However, seed production need not vary at all between species for stable coexistence to occur (i.e. all variation could be in c_i; Equation (8.4a)). Thus, one possible outcome for communities coexisting through temporal niche separation is that species richness has no effect on community productivity. In the following analysis, we assume the opposite extreme: that all recruitment variation stems from differences in seed production.

In model 1, year types arrive at random, so that species cannot take over in the community by being consistently favoured over several reproductive cycles. In model 2, environmental conditions remain constant for several reproductive cycles before they switch to a random, new regime. Model 3 is a version of the lottery model examined by Chesson *et al.* (2001), in which each species requires a different year type for recruitment. A consequence of this assumption is that communities with fewer species recruit less often than those with more species, and total adult densities decline with species numbers. Parameter values for the three models are shown in Table 8.1, and the relationships between species richness and yield under the different models are shown in Figure 8.4.

In model 1, more species-rich mixtures overyield on average, but no mixture transgressively overyields, and some mixtures, if they are composed of the lower-yielding species, underyield (Figure 8.4a). Since year types arrive in random order, years of maximal per capita

Table 8.1 *Parameter settings for the lottery models in Figure 8.3. Shown are seed production values (β_i) for five species across 5-year types. Mortality rate is* d = *0.1 unless otherwise specified and the competition parameter* $c_i = 1$ *for all species across all simulations. All year types have equal probability.*

	Models 1 and 2					Model 3				
Year type	Sp 1	Sp 2	Sp 3	Sp 4	Sp 5	Sp 1	Sp 2	Sp 3	Sp 4	Sp 5
1	1.13	0.68	0.45	0.23	2.25	0	0	0	0	4.28
2	2.25	1.13	0.68	0.45	1.13	5.63	0	0	0	0
3	1.13	2.25	1.13	0.68	0.45	0	5.41	0	0	0
4	0.68	1.13	2.25	1.13	0.23	0	0	4.95	0	0
5	0.45	0.23	0.45	2.25	0.23	0	0	0	4.73	0
Mean	1.13	1.08	0.99	0.95	0.86	1.13	1.08	0.99	0.95	0.86

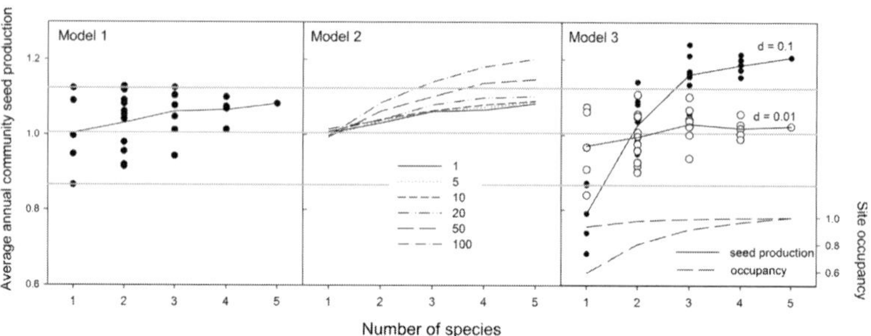

Figure 8.4 Biodiversity–productivity relationships in three lottery models. In models 1 and 2, all sites are always occupied and differences in productivity are solely due to differences in seed production. In model 1, year types arrive randomly; in model 2, year types repeat 1–100 times as indicated in the legend. In model 3, all species have a specific year type in which only they reproduce and results are shown for $d = 0.1$ (solid symbols) and $d = 0.01$ (open symbols). In model 3, site occupancy can be < 1 and is indicated by the broken lines. All panels show as grey lines the minimal, average and maximal seed production of monocultures at full patch occupancy. A mixture underyields if it yields below the average, overyields non-transgressively if it yields above the average but below the monoculture maximum, and overyields transgressively if it yields above the monoculture maximum.

reproduction are just as likely to arrive when a species is at high as when it is at low frequency. Thus, the contribution of any species to average seed yield just depends on its average frequency, and since species that are fitter on average achieve higher frequencies, species mixtures tend to yield more than the average of monocultures. Thus,

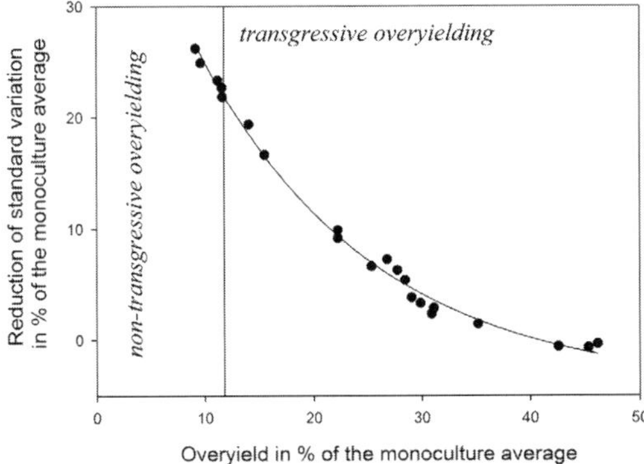

Figure 8.5 Transgressive overyielding at the cost of increased seed yield variability. Values are shown for model 2 with all five species and year-type repeat values between 1 (top left) and 350 (bottom right). Above the overyield percentage of 11.8%, the mixture yields more than the highest yielding monoculture.

overyielding is caused by the selection effect, and depends on the magnitude of fitness differences among species.

Model 2 does exhibit transgressive overyielding because species favoured by the environment have time to spread in the community, so that over time, high per capita seed yields are positively correlated with higher than average adult frequencies across species. However, immediately after an environmental regime shift, in the process of switching dominance from one species to another, the community is poorly adapted to the environment and underyields. The degree of overyielding and the magnitude of variance reduction are therefore negatively correlated in model 2 (Figure 8.5); a higher degree of overyielding occurs at the cost of higher temporal yield variance. The exact relationship depends on the time a community takes to switch dominance between species relative to the frequency of regime shift. For example, the existence of stable refugia could help recovery by limiting how rare a species can get, and large fitness differences between species under any stable regime (as in model 3) help the currently favoured species to rapidly displace all other species.

Transgressive overyielding also occurs in model 3, which exhibits an especially high degree of sensitivity to species numbers if adults have low survivorship. This is because diversity affects not only average per capita seed production, but also total adult densities. However, if adult mortality is low, in other words if adults are long-lived compared to average recruitment rates, the model behaves similarly to model 1, since most sites remain occupied at all times.

Thus, both transgressive and non-transgressive overyielding is possible, but not necessary in groups of functionally redundant species with distinct temporal niches. Contrary to what is sometimes expressed in the literature (Loreau 2000, Gonzalez and Loreau 2009, Hector *et al.* 2010), temporal niches may have no effect on ecosystem productivity or stability, if niche dynamics play out entirely through germination and seedling dynamics. This is assuming, of course, that seedling dynamics have negligible effects on ecosystem fluxes, as would be the case for long-lived species with adults that have a much larger combined biomass than seedlings, for example forest trees. Furthermore, groups of functionally redundant species could have highly variable production as a whole, but if these are caused by shared environmental responses, they cannot be buffered by temporal niches.

If temporal niche dynamics have a component that does correlate with adult performance, for example through seed production, non-transgressive overyielding can occur and only requires that species have average fitness differences and that species with higher fitness on average are more frequent on average. Transgressive overyielding, while possible, requires more stringent conditions. In species that are not recruitment-limited, transgressive overyielding can only occur if environmental variation has intermittent periods of relative stability that last through several reproductive cycles, so that the currently fittest species can spread. On the time scales of experiments, we might observe transgressive overyielding in annuals or biennials responding differentially to El Niño Southern Oscillation (ENSO) cycles. In tree populations, we would need centennial time scales to see this, making direct observation unlikely, but long-term cycles of population fluctuation in temporal niche partners found by Kelly and Bowler (2002) are consistent with this. Conversely, if adult densities are recruitment-limited, transgressive overyielding can occur if each species occupies a distinct specialised recruitment niche and therefore has partially additive effects on total adult abundance. Desert winter annuals may be a good example for this case, but at the same time, suggesting that the contribution of such ephemeral species groups to overall ecosystem productivity could be small.

Clearly, it matters whether adults or juveniles are driving niche dynamics. By definition, functional niches, which involve adult phenotypes and distinct resource use by adults, will necessarily leave imprints on ecosystem function and response to climate. For temporal storage effect niches driven by seed and seedling traits, the ecosystem effects are not so obvious. However, as a start, we can look at how different recruitment niches are associated with the evolution of distinct environmental sensitivities.

8.4 EVOLUTION OF FUNCTIONAL REDUNDANCY

As a fundamental consequence of the gradualist nature of Darwinian evolution, closely related species are, by and large, highly similar (Darwin 1859). With similar resource requirements, consumers and climate sensitivities, closely related species are integrated into the material and energy flows of ecosystems in similar ways, so that redundancy in ecosystem function is often generated by close relatives (Woodward and Kelly 1997). Furthermore, increasing evidence indicates that the line between functional redundancy and complementarity is drawn generally at the genus level (Kelly *et al.* 2008, Queenborough *et al.* 2009, Kelly *et al.* 2010, Burns and Strauss 2011). It is therefore of interest that congeneric species comprise on average 30% of woody communities worldwide (Figure 4 in Kelly and Bowler 2005). In the widespread distribution around this average, the hyper-diversity of tropical forests produces both higher percentages and much higher numbers of co-occurring congeners (e.g. México: 86 of 191 spp.; Panama: 121 of 220 spp.), signifying considerable potential redundancy in ecosystem function and services. We suggest that coexisting congeners are often the product of sympatric/parapatric speciation enabled by temporal niche dynamics, with implications for the action of redundancy in ecosystem function.

There is a developing consensus that sympatric and parapatric speciation predominate in biodiversity hotspots, although the prevalence of specific paths to speciation may differ with geography (Couvreur *et al.* 2011, Warren *et al.* 2011, Keller and Seehausen 2012, Hughes *et al.* 2013). Sympatric speciation requires environmental heterogeneity and genetic variability, with opportunities for genotypes of differing sensitivity to this environmental heterogeneity to become reproductively isolated while maintaining coexistence (Dieckmann and Doebeli 1999). This contrasts with allopatric speciation, produced when a physical barrier of distance or geography disrupts gene flow between two subpopulations, with subsequent adaptive divergence (Mayr 1942). Divergence in sensitivity is not an essential outcome of physical separation, but can evolve secondarily by coevolutionary character displacement when reproductively independent sister species are geographically reunited (Schluter 2001).

The same sort of differential sensitivity to the environment that can allow sympatric speciation also supports temporal niche dynamics, but temporal dynamics alone do not lead to sympatric speciation. An additional requirement is that differential sensitivity be linked to a mechanism of reproductive isolation (Doebeli and Dieckmann 2003). For example, anti-correlated seed production between years could stabilise the coexistence of two species, but would not promote sympatric speciation unless flower production is also anti-correlated to prevent gene flow between different sensitivity types. However, flowering in forest trees is notably correlated across years (Kelly 1994), for a number of good reasons (e.g. Iwasa *et al.* this volume, Satake *et al.* this volume), making this an unlikely route for reproductive isolation. On the other hand, within-year differences in the timing of reproduction, correlated with resource-use differences, have been shown to support sympatric speciation (Savolainen *et al.* 2006, Rymer *et al.* 2010). The necessary conditions for sympatric speciation have been more tightly connected to resource use in the tropical tree genus *Bursera*, where the association of ecological sensitivity and reproductive isolation is evident in the potentially pleiotropic control of environmental response traits (germination and vegetative bud-break) and flowering period, which in turn are reflected in local distribution patterns (Kelly *et al.* this volume).

Irrespective of the mechanism for reproductive isolation, the evolution of functional redundancy rests fundamentally on the divergence of seed/seedling niches, without concomitant change in adult ecology. The evolutionary record suggests a flexibility in seedling traits that exceeds that of adult traits. While the latter are very closely tied into phylogenetic groups and the majority of functionally relevant traits, seedling characters are not (Wright *et al.* 2000). In addition, we commonly see distinct recruitment traits, involving such factors as seed size, dispersal syndromes, and germination requirements, in groups of coexisting congeneric species (e.g. *Vaccinium* in Newfoundland tundra, Vander Kloet and Hill 2000; *Acacia* in Australian eucalypt forests, Brown *et al.* 2003; *Ephedra* in North America, Loera *et al.* 2012). Taken together, these lines of evidence signify that different seedling types can produce similar adults.

Combining the larger pattern of clade development documented in Williams and Kelly (unpublished manuscript) with the model of speciation in Kelly *et al.* (this volume) builds a picture of the physical directionality of evolutionary change. In the tropical dry forest of México, gradient analysis shows a proliferation of lineages of increasingly derived species from more to less benign conditions both locally and across the landscape (Williams and Kelly 2013, Williams and Kelly unpublished manuscript). In Kelly *et al.*'s speciation model, sympatric speciation into distinct temporal niches occurs most readily by way of a sensitive parent species giving rise to a more resistant daughter species. This occurs because recruits of the sensitive type are unable to survive harsh years, but the resistant type is not physiologically excluded from years that are more benign. Therefore, the resistant type has competition-free intervals in which it can grow from a low number but the competitive type does not. The directional selection on tolerance that leads to speciation also leads to greater resistance in the daughter species, supporting year-type specialisation and continued sympatry, but also allowing the daughter species to expand into new, harsher habitats.

Kelly *et al.*'s (this volume) model, together with the distributional data from Méxican dry forest, thus point to a specific chain of events in the process of speciation from temporal dynamics. That is, when a new, more tolerant species first arises, it owns a fundamental recruitment niche that overlaps with the parent species', but also includes conditions that are harsher than the parent can withstand. The *realised* recruitment niche of this new species is likely to be narrower than that of its parent, however, due to the greater competitive ability of the parent in most of its fundamental niche. Persistent exclusion of the daughter species from more benign conditions may lead eventually to further evolutionary separation and specialisation to the conditions of the realised niche, but the extent to which this leads to a narrower fundamental niche is an open question (Grime 1994).

In the larger context of climate change at the scale of glaciation cycles and tectonic uplift, this model meshes well with the frequently observed pattern of increased speciation rates coinciding with increased aridity (Axelrod 1972, Axelrod and Raven 1985, Becerra 2005, Hampe and Petit 2005, Valente *et al.* 2010, Metcalf and Nash 2012). A turn to more arid conditions would provide more exclusive habitat for incipient species and accelerate the establishment of secure population sizes. Patterns of current distributions in North American *Ephedra* are consistent with this scenario of climate-responsive radiation in the context of temporal niches (Loera *et al.* 2012), as are a number of genera from Méxican dry forest (*Bursera*, *Caesalpinia*, *Lonchocarpus* and *Croton*; Williams and Kelly unpublished manuscript) and African rainforest (*Isolona* and *Monodora*; Couvreur *et al.* 2011). Within these groups of congeners, extant sister species predominantly share the same climate bracket, with overlapping or nested distributions across the same landscape, but distinct habitat preferences or dispersal syndromes. At the next deepest node, pairs have less overlap, or are parapatric with each other, but in all cases, the general direction of adaptive radiation is from conditions that are less arid to those that are more arid.

Thus, the feasibility of stable, temporal niches at the micro-evolutionary scale sets in motion a cascade of evolutionary dynamics that can result in the accumulation of functional redundancy within a climate zone. More than just explaining the evolutionary origin of functional redundancy, this model predicts consequences for the presumed insurance function of redundancy in the context of species loss and modern climate change. First, lineages that have evolved in this way are expected to increase environmental competence with each additional species and, as a group, should have a greater capacity for maintaining ecological function under climate change than lineages prevented from adaptive radiation. That said, an alternative to speciation is the evolution of exceptional phenotypic plasticity or genetic diversity (e.g. *Arabidopsis thaliana*; Lasky *et al.* 2012). Nonetheless, insurmountable functional tradeoffs and inescapable hybridisation may constrain niche opportunities for most species. Second, species that evolved recently in sympatry are more likely to have largely overlapping fundamental niches, and thereby the greatest capacity to substitute for one another in space and time, should one species become extinct. Last, in older lineages, species loss has asymmetrical consequences on ecosystem function: the loss of the more derived, more tolerant species is expected to reduce the environmental competence of the lineage more than the loss of less derived, more competitive species. Thus, the conversion of natural forest into plantations of species composed of the most productive of congeners may come at the price of reduced climate adaptation.

8.5 CONCLUSIONS

The framework for the role of niche dynamics in ecosystem function and climate response presented here consists of a hierarchical structure involving two factors. First are functional niches, expressed in the traits of adult phenotypes and exhibiting a high degree of phylogenetic signal (Blomberg 2003, Losos 2008a, b). Second are recruitment niches, needed for the stable coexistence of functionally redundant, closely related species and supported by the greater responsiveness of seed and seedling traits to environmental variability (Figure 8.1). Our purpose has been to examine the implications of this framework for ecosystem productivity, stability, sensitivity to species loss and climate adaptation.

More succinctly, what is the role of functional redundancy in natural communities?

For the system of redundancy we have outlined here, we perceive no necessary or immediate impact on ecosystem productivity. Instead, we suggest that temporal niche processes have provided long-term functional stability through generating species that as a group inhabit a wider range of recruitment niches for a given plant functional trait (PFT). Rather than subdividing the ancestral niche, which could potentially increase the extinction risk to individual species, we suggest that speciation supported by temporal processes expands out from the ancestral recruitment niche based on novel or new combinations of reproductive traits. Published geographical distributions of a number of genera support this inference (cited above).

This mechanism of speciation could also support the persistence of ecosystem function in the face of contemporary climate change, at least for as long as fundamental functions of the adult phenotype remain viable. A broad portfolio of recruitment niches cannot rescue a group of species if it no longer has a viable adult phenotype, but it may decrease the chance that an entire PFT becomes locally extinct for failure to recruit. A broad recruitment portfolio may also increase the chance of successful geographical range shift, as some sympatrically evolved recruitment syndromes may be well suited for long-distance dispersal and establishment in novel terrain. Not all functionally redundant species would be expected to persist or to migrate successfully under contemporary climate change, least so species with the most highly specialised recruitment niches. However, our point is that directional evolution towards increased tolerance and generalism within a clade heightens the chance of that clade enduring climate change. This may in fact provide some explanation as to why supergenera exist and contribute disproportionally to redundancy in ecosystems. A genus that is genetically or ecologically predisposed to respond to temporal environmental variation by sympatric speciation may experience both an accelerated speciation rate and reduced extinction risk under climate change.

We believe the evolutionary process of temporal dynamics and congeneric proliferation to be widespread and important but, patently, not all plant genera have undergone this sort of adaptive radiation. Similarly, not all plant supergenera need to have arisen from temporal niche processes. The primary issue is that we have outlined a coexistence and speciation process that generates testable predictions of large-scale species distributions and local-scale ecology with direct relevance to ongoing and future ecosystem function.

ACKNOWLEDGEMENTS

GAF was partially supported by a grant from the US National Science Foundation (DEB-1120330).

REFERENCES

Axelrod, D. I. (1972). Edaphic aridity as a factor in angiosperm evolution. *American Naturalist* **106**, 311–320.

Axelrod, D. I. and Raven, P. H. (1985). Origins of the Cordilleran flora. *Journal of Biogeography* **12**, 21–47.

Balvanera, P., Pfisterer, A. B., Buchmann, N. *et al.* (2006). Quantifying the evidence for biodiversity effects on ecosystem functioning and services. *Ecology Letters* **9**, 1146–1156.

Barua, D., Butler, C., Tisdale, T. E. and Donohue, K. (2012). Natural variation in germination responses of *Arabidopsis* to seasonal cues and their associated physiological mechanisms. *Annals of Botany* **109**, 209–226.

Becerra, J. X. (2005). Timing the origin and expansion of the Mexican tropical dry forest. *Proceedings of the National Academy of Sciences, USA* **102**, 10919–10923.

Beckage, B. and Gross, L. J. (2006). Overyielding and species diversity: what should we expect? *New Phytologist* **172**, 140–148.

Blomberg, S. P. (2003). Testing for phylogenetic signal in comparative data: behavioral traits are more labile. *Evolution* **57**, 717–745.

Box, E. O. (1996). Plant functional types and climate at the global scale. *Journal of Vegetation Science* **7**, 309–320.

Burns, J. H. and Strauss, S. (2011). More closely related species are more ecologically similar in an experimental test. *Proceedings of the National Academy of Sciences, USA* **108**, 5302–5307.

Cardinale, B. J., Matulich, K. L., Hooper, D. U. *et al.* (2011). The functional role of producer diversity in ecosystems. *American Journal of Botany* **98**, 572–592.

Cardinale, B. J., Palmer, M. A. and Collins, S. L. (2002). Species diversity enhances ecosystem functioning through interspecific facilitation. *Nature* **415**, 426–429.

Cardinale, B. J., Srivastava, D. S., Duffy, J. E. *et al.* (2006). Effects of biodiversity on the functioning of trophic groups and ecosystems. *Nature* **443**, 989–992.

Cardinale, B. J., Wright, J. P., Cadotte, M. W. *et al.* (2007). Impacts of plant diversity on biomass production increase through time because of species complementarity. *Proceedings of the National Academy of Sciences, USA* **104**, 18123–18128.

Chapin, F. S., Sala, O. E., Burke, I. C. *et al.* (1998). Ecosystem consequences of changing biodiversity – Experimental evidence and a research agenda for the future. *Bioscience* **48**, 45–52.

Chesson, P. (1994). Multispecies competition in variable environments. *Theoretical Population Biology* **45**, 227–276.

Chesson, P., Pacala, S. and Neuhauser, C. (2001). Environmental niches and ecosystem functioning. In A. Kinzig, S. Pacala and D. Tilman (eds), *The Functional Consequences of Biodiversity*. Princeton, NJ: Princeton University Press, pp. 213–245.

Chesson, P. L. and Warner, R. R. (1981). Environmental variability promotes coexistence in lottery competitive-systems. *American Naturalist* **117**, 923–943.

Couvreur, T. L. P., Porter-Morgan, H., Wieringa, J. J. and Chatrou, L. W. (2011). Little ecological divergence associated with speciation in two African rain forest tree genera. *BMC Evolutionary Biology* **11**, 296.

Darwin, C. (1859). *On the Origin of Species by Means of Natural Selection, or the Preservation of Favoured Races in the Struggle for Life*. London: John Murray.

Dawson, T. E. (1993). Hydraulic lift and water-use by plants: implications for water-balance, performance and plant-plant interactions. *Oecologia* **95**, 565–574.

Dieckmann, U. and Doebeli, M. (1999). On the origin of species by sympatric speciation. *Nature* **400**, 354–357.

Doak, D. F., Bigger, D., Harding, E. K. *et al.* (1998). The statistical inevitability of stability-diversity relationships in community ecology. *American Naturalist* **151**, 264–276.

Doebeli, M. and Dieckmann, U. (2003). Speciation along environmental gradients. *Nature* **421**, 259–264.

Fornara, D. A. and Tilman, D. (2009). Ecological mechanisms associated with the positive diversity-productivity relationship in an N-limited grassland. *Ecology* **90**, 408–418.

Gilpin, M. E. and Justice, K. E. (1972). Reinterpretation of invalidation of principle of competitive exclusion. *Nature* **236**, 273.

Gonzalez, A. and Loreau, M. (2009). The causes and consequences of compensatory dynamics in ecological communities. *Annual Review of Ecology, Evolution and Systematics* **40**, 393–414.

Grime, J. P. (1994). The role of plasticity in exploiting environmental heterogeneity. In M. M. Caldwell and R. W. Pearcy (eds), *Exploitation of Environmental Heterogeneity by Plants*. San Diego, CA: Academic Press, pp. 1–19.

Hampe, A. and Petit, R. J. (2005). Conserving biodiversity under climate change: the rear edge matters. *Ecology Letters* **8**, 461–467.

Haxeltine, A. and Prentice, I. C. (1996). BIOME3: An equilibrium terrestrial biosphere model based on ecophysiological constraints, resource availability, and competition among plant functional types. *Global Biogeochemical Cycles* **10**, 693–709.

Hector, A., Bazeley-White, E., Loreau, M., Otway, S. and Schmid, B. (2002). Overyielding in grassland communities: testing the sampling effect hypothesis with replicated biodiversity experiments. *Ecology Letters* **5**, 502–511.

Hector, A., Hautier, Y., Saner, P. *et al.* (2010). General stabilizing effects of plant diversity on grassland productivity through population asynchrony and overyielding. *Ecology* **91**, 2213–2220.

Herben, T., Novakova, Z., Klimesova, J. and Hrouda, L. (2012). Species traits and plant performance: functional trade-offs in a large set of species in a botanical garden. *Journal of Ecology* **100**, 1522–1533.

Hillebrand, H., Bennett, D. M. and Cadotte M. W. (2008). Consequences of dominance: a review of evenness effects on local and regional ecosystem processes. *Ecology* **89**, 1510–1520.

Hooper, D. U., Adair, E. C., Cardinale, B. J. *et al.* (2012). A global synthesis reveals biodiversity loss as a major driver of ecosystem change. *Nature* **486**, 105–108.

Hooper, D. U., Chapin, F. S., Ewel, J. J. *et al.* (2005). Effects of biodiversity on ecosystem functioning: a consensus of current knowledge. *Ecological Monographs* **75**, 3–35.

Hughes, C. E., Pennington, R. T. and Antonelli, A. (2013). Neotropical plant evolution: assembling the big picture. *Botanical Journal of the Linnean Society* **171**, 1–18.

Huntington, T. G. (2006). Evidence for intensification of the global water cycle: review and synthesis. *Journal of Hydrology* **319**, 83–95.

Huston, M. A. (1997). Hidden treatments in ecological experiments: re-evaluating the ecosystem function of biodiversity. *Oecologia* **110**, 449–460.

IPCC (2007). *Contribution of Working Group I to the Fourth Assessment Report of the Intergovernmental Panel on Climate Change, 2007*. Cambridge, UK: Cambridge University Press.

Isbell, F. I., Polley, H. W. and Wilsey, B. J. (2009). Biodiversity, productivity and the temporal stability of productivity: patterns and processes. *Ecology Letters* **12**, 443–451.

Keller, I. and Seehausen, O. (2012). Thermal adaptation and ecological speciation. *Molecular Ecology* **21**, 782–799.

Kelly, C. K. and Bowler, M. G. (2002). Coexistence and relative abundance in forest tree species. *Nature* **417**, 437–440.

Kelly, C. K. and Bowler, M. G. (2005). A new application of storage dynamics: differential sensitivity, diffuse competition and temporal niches. *Ecology* **86**, 1012–1022.

Kelly, C. K., Bowler, M. G., Joy, J. B. and Williams, J. N. (2010). Fractional abundance and the ecology of community structure. *ArXiv* **1008.2527v1**.

Kelly, C. K., Bowler, M. G. Pybus, O. G. and Harvey, P. H. (2008). Phylogeny, niches and relative abundance in natural communities. *Ecology* **89**, 962–970.

Kelly, D. (1994). The evolutionary ecology of mast seeding. *Trends in Ecology and Evolution* **9**, 465–470.

Lavorel, S., Díaz, S. and Cornelissen, J. H. C. (2007). Plant functional types: are we getting any closer to the Holy Grail? In J. Canadell, D. Pataki and L. Pitelka (eds), *Terrestrial Ecosystems in a Changing World*. Berlin: Springer-Verlag, pp. 149–160.

Lasky, J. R., Des Marais, D. L., McKay, J. K. *et al.* (2012). Characterizing genomic variation of *Arabidopsis thaliana*: the roles of geography and climate. *Molecular Ecology* **21**, 5512–5529.

Loera, I., V. Sosa and Ickert-Bond, S. M. (2012). Diversification in North American arid lands: niche conservatism, divergence and expansion of habitat explain speciation in the genus *Ephedra*. *Molecular Phylogenetics and Evolution* **65**, 437–450.

Loreau, M. (1998). Biodiversity and ecosystem functioning: a mechanistic model. *Proceedings of the National Academy of Sciences, USA* **95**, 5632–5636.

Loreau, M. (2000). Biodiversity and ecosystem functioning: recent theoretical advances. *Oikos* **91**, 3–17.

Loreau, M. (2004). Does functional redundancy exist? *Oikos* **104**, 606–611.

Loreau, M., Mouquet, N. and Gonzalez, A. (2003). Biodiversity as spatial insurance in heterogeneous landscapes. *Proceedings of the National Academy of Sciences, USA* **100**, 12765–12770.

Losos, J. B. (2008a). Phylogenetic niche conservatism, phylogenetic signal and the relationship between phylogenetic relatedness and ecological similarity between species. *Ecology Letters* **11**, 995–1003.

Losos, J. B. (2008b). Rejoinder to Wiens (2008): phylogenetic niche conservatism, its occurrence and importance. *Ecology Letters* **11**, 1005–1007.

MacArthur, R. and Levins, R. (1967). Limiting similarity, convergence, and divergence of coexisting species. *American Naturalist* **101**, 377–385.

MacArthur, R. H. (1955). Fluctuations of animal populations, and a measure of community stability. *Ecology* **36**, 533–536.

Marquard, E., Weigelt, A., Temperton, V. M. *et al.* (2009). Plant species richness and functional composition drive overyielding in a six-year grassland experiment. *Ecology* **90**, 3290–3302.

May, R. M. (1975). Patterns of species abundance and diversity. In M. L. Cody and J. M. Diamond (eds), *Ecology and Evolution of Communities*. Cambridge, MA: Harvard University Press, pp. 81–120.

Mayr, E. (1942). *Systematics and the Origin of Species*. New York: Columbia University Press.

Menzel, A., Sparks, T. H., Estrella, N. *et al.* (2006). European phenological response to climate change matches the warming pattern. *Global Change Biology* **12**, 1969–1976.

Metcalf, S. E. and Nash, D. J. (2012). *Quaternary Environmental Change in the Tropics*. Chichester, UK: John Wiley and Sons.

Mouquet, N., Moore, J. L. and Loreau, M. (2002). Plant species richness and community productivity: why the mechanism that promotes coexistence matters. *Ecology Letters* **5**, 56–65.

Naeem, S. (1998). Species redundancy and ecosystem reliability. *Conservation Biology* **12**, 39–45.

Norberg, J., Swaney, D. P., Dushoff, J. *et al.* (2001). Phenotypic diversity and ecosystem functioning in changing environments: a theoretical framework. *Proceedings of the National Academy of Sciences, USA* **98**, 11376–11381.

Obeso, J. R. (2002). The costs of reproduction in plants. *New Phytologist* **155**, 321–348.

Odum, E. P. (1953). *Fundamentals of Ecology*. Philadelphia, PA: Saunders.

Philippi, T. and Seger, J. (1989). Hedging one's evolutionary bets, revisited. *Trends in Ecology and Evolution* **4**, 41–44.

Preston, F. W. (1948). The commonness and rarity of species. *Ecology* **29**, 254–283.

Queenborough, S. A., Burslem, D. F. R. P., Garwood, N. J. and Valencia, R. (2009). Taxonomic scale-dependence of habitat niche partitioning and biotic neighbourhood on survival of tropical tree seedlings. *Proceedings of the Royal Society B* **276**, 4197–4205.

Rymer, P. D., Manning, J. C., Goldblatt, P., Powell, M. P. and Savolainen, V. (2010). Evidence of recent and continuous speciation in a biodiversity hotspot: a population genetic approach in southern African gladioli (*Gladiolus;* Iridaceae). *Molecular Ecology* **19**, 4765–4782.

Savolainen, V., Anstett, M. C., Lexer, C. *et al.* (2006). Sympatric speciation in palms on an oceanic island. *Nature* **441**, 210–213.

Scheff, J. and Frierson, D. (2012). Twenty-first-century multimodel subtropical precipitation declines are mostly midlatitude shifts. *Journal of Climate* **25**, 4330–4347.

Schluter, D. (2001). Ecology and the origin of species. *Trends in Ecology and Evolution* **16**, 372–380.

Schwartz, M. W., Brigham, C. A., Hoeksema, J. D. *et al.* (2000). Linking biodiversity to ecosystem function: implications for conservation ecology. *Oecologia* **122**, 297–305.

Schwinning, S. and Parsons, A. J. (1996). Analysis of the coexistence mechanisms for grasses and legumes in grazing systems. *Journal of Ecology* **84**, 799–813.

Tilman, D. (1982). *Resource Competition and Community Structure*. Princeton, NJ: Princeton University Press.

Tilman, D. (1996). Biodiversity: population versus ecosystem stability. *Ecology* **77**, 350–363.

Tilman, D. (1999). The ecological consequences of changes in biodiversity: a search for general principles. *Ecology* **80**, 1455–1474.

Valente, L. M., Savolainen, V. and Vargas, P. (2010). Unparalleled rates of species diversification in Europe. *Proceedings of the Royal Society B* **277**, 1489–1496.

Vander Kloet, S. P. and Hill, N. M. (2000). *Bacca quo vadis*: regeneration niche differences among seven sympatric *Vaccinium* species on headlands of Newfoundland. *Seed Science Research* **10**, 89–97.

Vandermeer, J. H. (1992). *The Ecology of Intercropping*. Cambridge: Cambridge University Press.

Walker, B. H. (1992). Biodiversity and ecological redundancy. *Conservation Biology* **6**, 18–23.

Wardle, D. A. (1999). Is 'sampling effect' a problem for experiments investigating biodiversity-ecosystem function relationships? *Oikos* **87**, 403–407.

Warren, B. H., Bakker, F. T., Bellstedt, D. U. *et al.* (2011). Consistent phenological shifts in the making of a biodiversity hotspot: the Cape flora. *BMC Evolutionary Biology* **11**, 39.

Westoby, M. and Wright, I. J. (2006). Land-plant ecology on the basis of functional traits. *Trends in Ecology and Evolution* **21**, 261–268.

Williams, J. N. and Kelly, C. K. (2013). Deconstructing the signal: phylogenetic structure, elevation change and the implications for species coexistence. *Evolutionary Ecology Research* In press.

Woodward, F. I. and Kelly, C. K. (1997). Plant functional types: towards a definition by environmental constraints. In T. M. Smith, H. H. Shugart and F. I. Woodward (eds), *Plant Functional Types*. Cambridge, UK: Cambridge University Press, pp. 47–65.

Wright, I. J., Clifford, H. T., Kidson, R. *et al.* (2000). A survey of seed and seedling characters in 1744 Australian dicotyledon species: cross-species trait correlations and correlated trait-shifts within evolutionary lineages. *Biological Journal of the Linnean Society* **69**, 521–547.

Yachi, S. and Loreau, M. (1999). Biodiversity and ecosystem productivity in a fluctuating environment: The insurance hypothesis. *Proceedings of the National Academy of Sciences, USA* **96**, 1463–1468.

Yachi, S. and Loreau, M. (2007). Does complementary resource use enhance ecosystem functioning? A model of light competition in plant communities. *Ecology Letters* **10**, 54–62.

Zhang, Y., Chen, H. Y. H. and Reich, P. B. (2012). Forest productivity increases with evenness, species richness and trait variation: a global meta-analysis. *Journal of Ecology* **100**, 742–749.

Part II Application to specific questions

The second section of this book looks to addressing specific questions with the analytical tools of temporal processes. Temporal processes take on a wider framework here than in the previous section, including consideration of internally driven fluctuations. The extent to which species aggregated within a landscape are experiencing temporal variation is likely to become an important topic in the future, and we can look forward to fruitful development of these explorations as and when they are challenged by real data from real communities.

Yoh Iwasa and colleagues and Akiko Satake and colleagues provide a tandem approach to masting, the periodic synchrony of reproduction in forest trees. Iwasa *et al.* first deal with masting within species, after which Satake *et al.* address multispecies masting. These chapters depend upon elaboration of a very simple biological model for resource reserves, the idea being that if a tree depletes its resources severely in an episode of fruiting it will have no resources to spare for reproduction for several years thereafter. The studies offer a fascinating review of complicated time-dependent phenomena emerging from a model that is very simple, both biologically and mathematically. Although not directly related to the role of environmental fluctuation in promoting coexistence, there is an indirect link: the phenomenon of masting generates a fluctuating environment for seed predators, and fluctuating levels of infestation constitute a fluctuating environment for the species of insect which prey upon the larvae that feed upon the seeds.

Both of the remaining chapters use models of temporal dynamics in annual plants to frame their questions. The coexistence of species through temporal variation in environmental conditions takes its simplest form when the environment in which species coexist is homogeneous. Robin Snyder investigates the effect of both spatial and temporal variations on the coexistence of annual species. After a user-friendly discussion of her methods, Snyder describes her principal result: spatiotemporal variation of the environment is more effective at promoting coexistence than pure temporal variation, with the important caveat that this result holds for the same variance of temporal and spatiotemporal fluctuations in the environment. Different homes favour different species, and of course for a uniform environment the temporal variance will be the only active ingredient. Coexistence is best promoted by environmental variation when species prefer different environments and perhaps surprisingly a factor promoting coexistence is short-range dispersal of seeds with species aggregation.

Michael Bowler and his colleagues introduce a version of the seedbank model for competition between two ecologically similar species subject to varying levels of herbivory. The authors consider the situation where species are not only ecologically similar but also genetically closely related, so that gene flow or hybridisation may be possible, having in mind the problem of the spread of an insect resistant trait introduced for agribusiness into a wild community of related species. Bowler *et al.* show that if some disadvantage is attached to herbivore resistance then the wild variety may well persist. This model could be helpful in understanding a-posteriori aspects of the spread or otherwise of an insect resistant transgene, but the model is not restricted to agricultural use. These considerations might guide development of genetically modified crops that pose minimal threat to the larger environment, or usefully applied to better understanding the evolutionary path of herbivore resistance in natural systems.

YOH IWASA, AKIKO SATAKE AND YUUYA TACHIKI

9

Evolution of synchronised and intermittent reproduction (masting) of trees: key role of regeneration dynamics

9.1 INTRODUCTION

The variable and synchronous production of seeds by plant populations is called masting or mast seeding and is observed in diverse forests (Kelly 1994). Many flowers and fruits are produced one year (called a mast year) but little reproductive activity occurs during the several subsequent years until the next mast year (Herrera *et al.* 1998, Koening and Knops 1998, 2000, Koening *et al.* 1999). The variance in the reproductive activity of trees between years is large. It cannot be simply a result of environmental fluctuation in annual productivity (Tamura and Hiura 1998).

Many studies on masting have focused on adaptive significance (Kelly and Sork 2002). A popular hypothesis is the predator satiation theory – that is, seed predators starve during non-mast years, while they are unable to consume all the seeds during mast years (Janzen 1971, Silvertown 1980, Nilsson and Wästljung 1987, van Schaik *et al.* 1993). An alternative but not mutually exclusive hypothesis is pollination efficiency: in mast years, trees receive a lot of outcross pollen, which may improve fruiting success compared with reproduction in non-mast years (Nilsson and Wästljung 1987, Smith *et al.* 1990, van Schaik *et al.* 1993, Shibata *et al.* 1998, Kelly *et al.* 2001, Rees *et al.* 2002, Satake and Bjørnstad 2004).

However, these hypotheses on the evolutionary advantages of masting explain neither the mechanism causing masting nor how the adjustment of timing of reproduction over different individuals is

Temporal Dynamics and Ecological Process, ed. C. K. Kelly, M. G. Bowler, G. A. Fox. Published by Cambridge University Press.

achieved. To explain the mechanisms responsible for the intermittent reproduction of trees, the resource budget model was proposed by Isagi *et al.* (1997), and has been analysed in detail by Satake and Iwasa (2000, 2002a, 2002b). The model assumes that a tree gains a constant resource income every year from photosynthesis, and that the tree may not reproduce while the reserve level is below a threshold. Once the resource reserve exceeds the threshold, the tree flowers, and its ovules may be fertilised by outcrossed pollen. The investment in seeds and fruits produced from these ovules subsequently depletes the reserves of the tree. Depending on the degree of resource depletion, the reproductive activity of a single tree may be at a constant rate, or fluctuating periodically or chaotically (Isagi *et al.* 1997, Satake and Iwasa 2000, 2002a, 2002b).

A key parameter in the control of fluctuation is the resource depletion coefficient k. When k is smaller than or equal to 1, each tree reproduces every year and no intermittent reproduction occurs. In contrast, when k is greater than 1, each tree reproduces intermittently. In this case, the level of reserves of each tree shows chaotic fluctuation, with a positive Lyapunov coefficient (Satake and Iwasa 2000). This implies that trees with very similar reserve levels initially would become completely different after a small number of years – highlighting the difficulty of maintaining synchronised reproduction of trees without additional processes.

If the seed and fruit production of a tree is limited by the availability of outcrossed pollen supplied by other trees in the same forest, different tree individuals are coupled by pollen supply and show a synchronised reproductive pattern without an environmental cue, which is a mechanism called 'pollen coupling' by Satake and Iwasa (2000). Since the spatial range of pollen exchange is much shorter than the range of the whole forest, whether or not a short-range local interaction between trees is able to achieve synchronisation over the whole forest is an important theoretical question, which can be studied as a coupled map lattice (Satake and Iwasa 2002a). The effect of the environmental fluctuation common to trees far from each other would also be important in making trees synchronised, called the 'Moran effect' (Satake and Iwasa 2002b), which is discussed in Chapter 11.

Tachiki and Iwasa (2008) discussed the evolution of the resource depletion coefficient k. Tachiki and Iwasa considered N individuals that exchange their pollen with each other. We trace the level of resource reserve, flowering activity and seeding success of each individual through time. When we think of the evolution of parameter k, we consider that a mutant with a different value k' invaded the population dominated by the resident of phenotype k. The mutant may go extinct or increase in number and finally take over the whole population. If the number of individuals N is not very large, there is considerable stochasticity in the process of invasion and replacement of mutants. Even a mutant that has on average a higher reproductive success than the resident goes extinct with a large probability, and a mutant less fit than the resident can take over in a finite population. The situation is the same as the game in a finite population (Nowak *et al.* 2004), in which the advantage of a mutant over the resident is determined by whether its fixation probability is greater than that expected for a neutral mutant.

To address the evolution of k in a finite forest population model, Tachiki and Iwasa developed a new graphical method called finite pairwise invasibility plot (fPIP). This method extends the pairwise invasibility plot (PIP) for an infinitely large population (Metz *et al.* 1992, Kisdi and Meszéna 1995, Geritz *et al.* 1997, 1999) to a finite population.

In this chapter, we review the series of works on resource budget models, on the synchrony between trees, and also on the condition for the evolution of masting. To our great surprise, we found that whether masting can evolve is not controlled by the benefit of improved seed production and by their survivorship. The assumption of how trees are replaced by the following generations is important in determining whether masting evolves. For example, masting can never evolve if gaps (created by the death of canopy trees) are filled by seeds produced in the same year, even if it greatly improves the pollination efficiency and escape from seed predators. Instead, masting can evolve if seeds produced in different years compete with each other. We studied the case when seedlings can survive over several years after germination waiting for a lucky supply of tree fall gaps. Species that maintain a seedling bank can evolve masting, but those species without a seedling bank on the forest floor do not evolve masting. We also discuss the effect of specialist seed predators in the evolution of masting of trees.

9.2 MODEL

We consider the resource budget of an individual tree. From photosynthesis, a mature tree gains net production P_s each year, which is accumulated in the trunk or branches. Let $S(t)$ be the amount of energy reserve at the beginning of year t. If the sum $S(t) + P_s$ is below a critical level L_T, the tree does not reproduce, and it saves all the resource reserve for the following year. However if the sum exceeds L_T, the tree uses some resources for flowering. Satake and Iwasa (2000) assumed that the expenditure for flowering is proportional to the excess, $a(S(t) + P_s - L_T)$, in which a is a positive constant (Isagi *et al.* (1997) assumed $a = 1$). Flowering plants may be pollinated and set seeds and fruits. If the cost for fruits is proportional to the cost of flowers, it is expressed as $R_c a(S(t) + P_s - L_T)$, in which R_c is the ratio of fruiting cost to flowering cost. After reproduction, the energy reserve of the tree falls:

$$S(t+1) = \begin{cases} S(t) + P_s & \text{if} \quad S(t) + P_s \leq L_T \\ S(t) + P_s - a(R_c + 1)(S(t) + P_s - L_T) & \text{if} \quad S(t) + P_s > L_T \end{cases}. \tag{9.1}$$

By introducing a non-dimensionalised variable $Y(t) = (S(t) + P_s - L_T)/P_s$, Equation (9.1) is rewritten as

$$Y(t+1) = \begin{cases} Y(t) + 1 & \text{if} \quad Y(t) \leq 0 \\ -kY(t) + 1 & \text{if} \quad Y(t) > 0 \end{cases}, \tag{9.2}$$

in which $k = a(R_c + 1) - 1$. The parameter k is the depletion coefficient, which indicates the degree of resource depletion after a reproductive year, divided by the excess amount of energy reserve before the year. From Equation (9.2), $Y(t) \leq 1$ holds. $Y(t)$ is positive if and only if the tree invests some resource in reproductive activity in year t.

Figure 9.1 illustrates trajectories for three different values of k. As shown in Figure 9.1a, when k is less than 1 ($0 < k < 1$), $Y(t)$ quickly converges to the stable equilibrium $1/(k + 1)$. Then the tree reproduces every year at a constant rate. In contrast, when k is greater than 1, $Y(t)$ keeps fluctuating with a chaotic time series (Figures 9.1b and c).

For the model given by Equation (9.2), the derivative of the map is $F'(Y) = 1$ for $Y \leq 0$, but $F'(Y) = -k$ for $Y > 0$, and hence the Lyapunov exponent is positive for $k > 1$, implying that a small initial difference in the amount of reserve between individuals increases with time (Satake and Iwasa 2000). Hence, the perfect synchronisation cannot be maintained without additional processes to make trees synchronised.

Coupling of trees through pollen limitation

Fruiting efficiency may depend on the flowering activity of the other trees in the same forest, because pollination efficiency changes with the number of flowering plants (Nilsson and Wästljung 1987, Smith *et al.* 1990). Consider a tree that flowers in a year in which only a small fraction of the other trees flower. The tree fails to produce many fruits

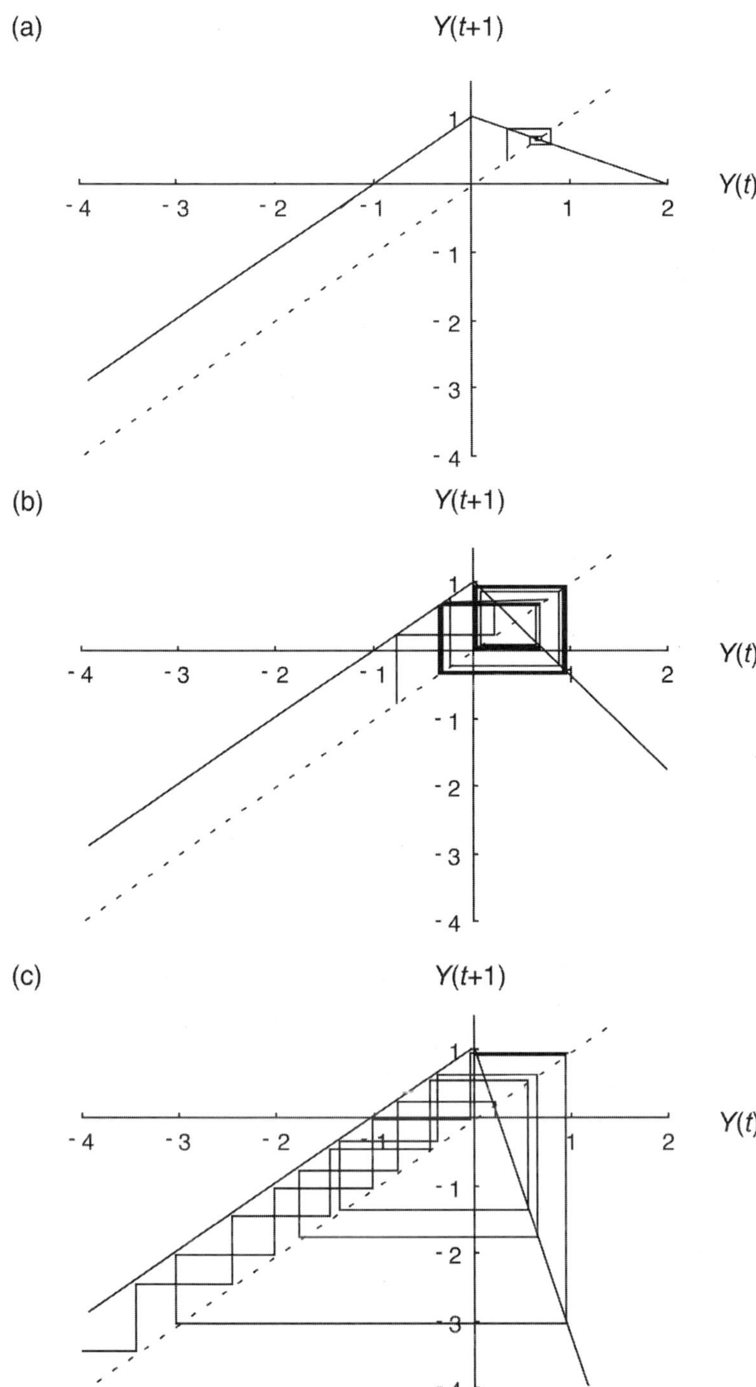

Figure 9.1 Return map and the trajectory of a tree without pollen limitation. Horizontal axis is for the normalised energy reserve of a tree

because of pollen limitation, and it does not experience heavy resource depletion. The tree continues to flower in the following years, until the year comes in which many other trees in the forest also flower at a high blooming intensity. Then they all experience a large fruit set and resource depletion, which gives a mechanism to make different individuals synchronised in the face of the chaotic tendency of each individual.

Consider a forest including N individuals with index $\mathbf{x}$, which indicates the location of the individual tree in the forest. To model the pollen limitation of reproduction, we replace k in Equation (9.2) by $kP_{\mathbf{x}}(t)$.

$$Y_{\mathbf{x}}(t+1) = \begin{cases} Y_{\mathbf{x}}(t)+1, & \text{if} \quad Y_{\mathbf{x}}(t) \leq 0 \\ -kP_{\mathbf{x}}(t)Y_{\mathbf{x}}(t)+1, & \text{if} \quad Y_{\mathbf{x}}(t) > 0 \end{cases}, \tag{9.3a}$$

in which $P_{\mathbf{x}}(t)$ is the pollination success for a tree located at $\mathbf{x}$ in year t. It is determined by the average flowering intensity of its neighbours:

$$P_{\mathbf{x}}(t) = \left(\frac{1}{z} \sum_{\mathbf{y} \in U_{\mathbf{x}}} [Y_{\mathbf{y}}(t)]_{+} \right)^{\beta}, \tag{9.3b}$$

where $[Y]_{+} = Y$ if $Y > 0$; $[Y]_{+} = 0$ if $Y \leq 0$. $U_{\mathbf{x}}$ in Equation (9.3b) stands for the set of neighbouring trees that are pollen donors for the focal tree at $\mathbf{x}$. z is the number of trees in $U_{\mathbf{x}}$, not including the focal tree.

$P_{\mathbf{x}}(t) = 1$ holds when all the neighbours reproduce at full intensity ($Y_{\mathbf{y}}(t) = 1$ for all $\mathbf{y}$ in $U_{\mathbf{x}}$). Small values of $P_{\mathbf{x}}(t)$ indicate pollen limitation of seed and fruit production. Parameter β in Equation (9.3b) controls the dependence of fruit production on outcross pollen availability. If β is close to zero, fruit production is almost independent of the reproductive activity of the other trees. Small β corresponds to either a high pollination efficiency or a high density of trees because a small fraction of flowering in the rest of the forest is sufficient to achieve a good fruiting success. In contrast, a large β implies a strong dependence of seed and fruit production on the reproductive activity of other trees in the forest. Hence β indicates pollen coupling strength.

Caption for figure 9.1 (cont.) $Y_{\mathbf{x}}(t)$; vertical axis is for value in the subsequent year $Y_{\mathbf{x}}(t+1)$. The solid line indicates the relationship given by Equation (9.3a) when the pollen availability is not limited ($P_{\mathbf{x}}(t) = 1$). **(a)** $k = 0.5$, stable equilibrium. **(b)** $k = 1.4$, chaotic fluctuation. **(c)** $k = 4.3$, chaotic fluctuation with longer intervals between mast years. After Satake and Iwasa (2000).

9.3 GLOBAL POLLEN COUPLING

First we consider the case in which the neighbourhood U_x is the whole forest. Equations (9.3a and b) are examples of coupled nonlinear dynamics in which each element in isolation has chaotic dynamics. It exhibits diverse dynamical behaviours that are qualitatively similar to coupled logistic maps (Kaneko 1990, Hastings 1993).

The model includes two parameters, k and β. The depletion coefficient k indicates the magnitude of chaos of each single tree, while the coupling strength β corresponds to the process causing synchronous reproduction. Figure 9.2 is a phase diagram, in which the horizontal and vertical axes are k and β, respectively. Satake and Iwasa (2000) classified five phases:

1. Annual reproduction: If $k < 1$, all the mature trees in the forest reproduce every year.
2. Desynchronised phase: If k is sufficiently large compared to β, any two trees in the forest reproduce in a different way and there is no synchronous reproduction. The amount of pollen produced in the forest is constant because there are always some trees that reproduce, but each individual tree shows intermittent reproduction.
3. and 4. Coherent phases: If β is large compared to k, all the trees in the forest show synchronised and fluctuating reproduction. This is further classified into (3) a coherent chaotic phase and (4) a coherent periodic phase (see below).
5. Clustering phase: Between the desynchronised phase and the coherent phase, there is a region in which trees are separated into several clusters, with trees of the same cluster reproducing in a synchronised manner.

In Figure 9.2, the coherent phase, in which all the trees in the forest engage in perfectly synchronised reproduction, is further classified as a coherent chaotic phase and a coherent periodic phase. Periodic reproduction occurs when parameter k is close to an integer. The system shows a stable periodic reproduction that is close to a cycle with period of $n + 1$ if parameter k is close to an integer n. The trajectory is composed of a mast year in which all the trees in the forest reproduce nearly at the maximum strength, followed by n years of no or very small reproduction. Although an isolated tree is always chaotic, if multiple trees are coupled, they can show periodic fluctuation.

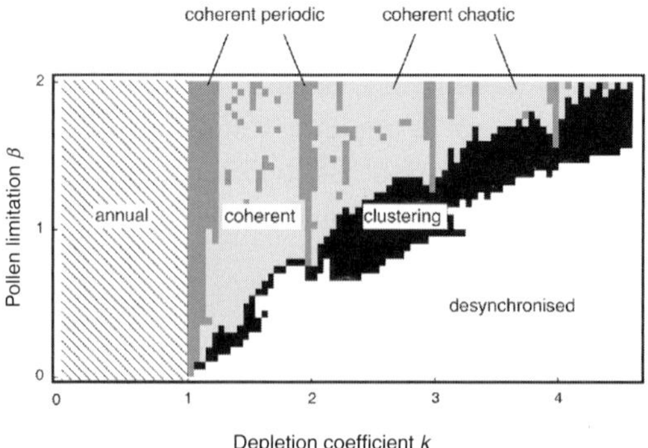

Figure 9.2 Phase plane of the global coupling model. Horizontal axis is depletion coefficient k, and vertical axis is coupling strength β. Five regions are classified. For each run, we calculated effective number of clusters based on $NC = 1/\sum_i p_i^2$, in which p_i is the fraction of trees belonging to the ith cluster. The region with $k > 1$ is classified according to the average of NC for 100 replicates: desynchronised phase ($N - 1 < \overline{NC} \leq N$) coherent phase ($1 \leq \overline{NC} < 2$), and clustering phase ($2 \leq \overline{NC} \leq N - 1$). Total number of individuals is $N = 100$. From Satake and Iwasa (2000).

In the coherent phase, all the trees are perfectly synchronised and have the same energy reserve levels. Along such a trajectory, all the N Lyapunov exponents are calculated (Satake and Iwasa 2000). For the synchronised trajectory to be locally stable, the Lyapunov exponents corresponding to the difference between trees must be negative. This condition gives the following inequality if the number of trees N is large:

$$\prod_{\substack{t \text{ s.t.} \\ Y(t)>0}} kP(t) < 1, \tag{9.4}$$

where the product is calculated for t with $Y(t) > 0$ along the trajectory, i.e. only for those years in which trees reproduce. Since all the trees have the same Y_i, we have dropped the subscript i. $P(t)$ is the pollen availability ($P(t) = Y(t)^\beta$), which is 1 when all the trees reproduce at full intensity, and it is less than 1 if pollen availability is limited. In a forest of trees with intermittent reproduction, k is larger than 1, and in fact was estimated to be about 5 for Japanese beeches (Isagi *et al.* 1997). If $P(t)$ is always close to 1, Equation (9.3b) cannot be satisfied. Equation (9.3b) is satisfied only if the trajectory includes sufficiently many years in which trees reproduce ($Y(t) > 0$) but pollen availability $P(t)$ is clearly smaller than 1 ($0 < P(t) << 1$). In those years, trees engage in some flowering but cannot set many fruits because of limited outcross pollen availability. A stable coherent trajectory must include a significant fraction of imperfect masting years.

9.4 LOCAL POLLEN COUPLING

The reproductive activity of trees shows synchronisation over a long distance (>100 km) (Koenig and Knops 1998) which is several orders of magnitude greater than the spatial range of direct pollen exchange, measured by genetic markers (Hamrick and Loveless 1989, Cruzan 1998, Dow and Ashley 1998). Satake and Iwasa (2002a) asked whether the local interaction of trees is able to create the reproductive synchronisation of a much larger spatial scale, and studied a coupled map lattice model in which trees exchange pollen only within local neighbourhoods.

Consider a forest composed of a number of trees arranged on lattice points of a two-dimensional square grid. Trees are indexed by $\mathbf{x} = (x_1, x_2)$, in which x_1 and x_2 are integers ($x_1, x_2 = 1, 2,\ldots, L$). Each tree accumulates photosynthate, reproduces, and experiences subsequent depletion of stored resources in a way given by Equation (9.3a). The pollen availability for a tree at $\mathbf{x}$ is determined by the average flowering intensity of its neighbours. $U_{\mathbf{x}}$ in Equation (9.3b) stands for the set of neighbouring trees that are pollen donors for the focal tree at $\mathbf{x}$. In the simplest case, $U_{\mathbf{x}}$ is composed of four nearest neighbours ($z = 4$).

To analyse the spatiotemporal pattern of reproductive activity of trees, we consider the fruit production activity (or seed crop size) of a tree at $\mathbf{x} = (x_1, x_2)$ in year t ($t = 1, 2,\ldots, T$), denoted by $\varphi_{\mathbf{x}}(t) = kP_{\mathbf{x}}(t)[Y_{\mathbf{x}}(t)]_+$. The mean seed crop of all the trees in the forest in year t is $\overline{\varphi}(t)$. The between-year variance V_{by} is the magnitude of fluctuation of the mean seed crop $\overline{\varphi}(t)$ over the years, and the within-year variance V_{wy} is the variance of seed crop size between trees in a single year, and then averaged over the years. The ratio of between-year variance to the total variance $V_{by}/(V_{by} + V_{wy})$ is an index of global synchrony. If the depletion coefficient k is smaller than 2, the model with nearest neighbour pollen exchange can show strong synchronisation. This implies that strong synchronisation of tree reproduction over the whole forest can be produced by short-range interactions.

Figure 9.3 illustrates the bifurcation diagram, which shows a clear difference in behaviour between $1 < k < 2$ and $k > 2$. For k smaller than 2, $\overline{\varphi}(t)$ visits two sets of values. The forest in a year is either at a high reproductive level or at a low reproductive level, alternately, showing strong synchrony. If k is larger than 2, $\overline{\varphi}(t)$ is centred around a single value (Figure 9.3), which implies that the mean reproductive activity does not change between years. In such a desynchronised forest,

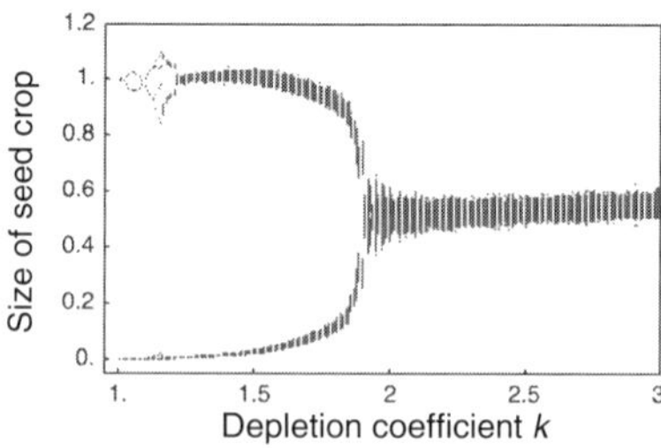

Figure 9.3 Bifurcation diagram. Starting from a random initial condition, we ran the model for 20 000 years, and then plotted 250 iterations. Horizontal axis is depletion coefficient k. Neighbourhood size is $D = 1$. Coupling strength is $\beta = 1.6$. From Satake and Iwasa (2002a).

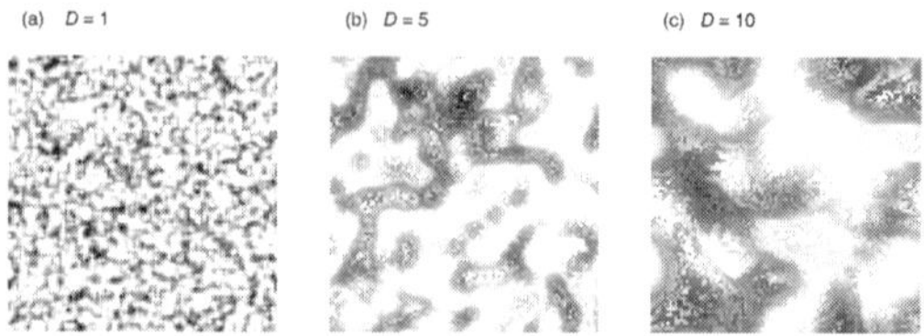

Figure 9.4 The spatial patterns generated by the model. The density plot indicates the seed output of individual trees, $\varphi_x(t)$. Dark squares are for high seed crops. Open squares are trees without seed crop ($\varphi_x(t) = 0$). The pollen dispersal ranges are **(a)** $D = 1$; **(b)** $D = 5$; and **(c)** $D = 10$. Parameters are: $k = 3.6$ and $\beta = 1.6$. From Satake and Iwasa (2002a).

the fraction of trees that bloom, the mean pollen availability and the total number of seeds produced in the forest are almost constant.

Even when a strong global synchrony occurs, the synchronisation between individuals in the forest is not perfect. There remains some spatial variation of $\varphi_x(t)$ between individuals. The spatial pattern is not uniform but has a characteristic heterogeneity, as illustrated in Figure 9.4. This makes a contrast with the globally coupled model studied in section 9.3, in which perfect synchronisation is likely to occur (Satake and Iwasa 2000). However, the spatial range of this heterogeneity is of rather short distance, with less than several times pollen dispersal range, according to the analysis of spatial covariance (Satake and Iwasa 2002a).

The availability of outcross pollen for a tree depends on the neighbours beyond the nearest four (Satake and Iwasa 2002a), in which pollen dispersal range U_x is a circle centred around the tree with radius D, which can be greater than 1. The overall spatial patterns look quite similar, and the conditions in which synchronisation occurs remain the same, but the spatial scale of the pattern is larger and becomes coarser grained as the size of the neighbourhood increases (Figure 9.4).

9.5 EVOLUTION OF MASTING

So far, we have been focusing on the spatiotemporal pattern of trees, and this can be analysed considering only the resource depletion of trees by reproduction. In the rest of the chapter, we discuss when the trees evolve to show masting or to show constant reproduction.

As we describe the resource dynamics in terms of the resource budget model, if pollen-coupling strength β is sufficiently large, and if the resource depletion coefficient k evolves to be significantly greater than 1, then synchronised and intermittent reproduction (masting) should be realised in the forest. In contrast, if k evolves to a value close to 1, each tree will reproduce every year, showing no sign of masting. The key parameter is the resource depletion coefficient k.

Tachiki and Iwasa (2008) focused on the evolution of the resource depletion coefficient k, assuming that β is kept a sufficiently large constant. Tachiki and Iwasa considered how a mutant type with different k invades and replaces the original type, and examined the conditions in which k evolves to be a value clearly larger than 1, and showing intermittent reproduction.

When we consider the evolution of tree traits, such as k, we must consider reproductive success of types of different k by counting the number of offspring of different types. We must also consider how the seeds produced in this way can survive, grow, finally reach the canopy and produce many seeds – we need to consider events in the whole life cycle, not just the resource depletion dynamics.

Life-cycle model

We assume that a forest is composed of N sites, each of which is either empty or filled by a single tree that might be immature or mature. To describe the condition of the i-th site, we introduce $\theta_i(t)$. If the i-th site is empty at the beginning of year t, $\theta_i(t) = 0$. If it is occupied by an immature tree, $\theta_i(t) = 1$. If it is occupied by a mature canopy tree, $\theta_i(t) = 2$.

Within each year, there are four phases. (1) In the growth phase, each tree photosynthesises, and accumulates resources in its reserve (tree trunk). (2) This is followed by the reproduction phase, in which trees with sufficient resources make flowers and fruits. (3) In the death phase, a small proportion of trees are randomly killed by disturbances, leaving an empty site. Although a tree can die at any time of the year, we proceed as if tree fall events occur only after the reproduction phase. (4) Finally, in the recruitment phase, each gap site is filled by an offspring started from a seed chosen at random. After a number of years in which this new individual remains immature, it reaches the canopy and starts reproduction. We explain each of these processes in turn below.

To consider the dynamics of flowering and fruiting, we adopt the resource budget model as explained in section 9.2.

Gap formation and gap filling

In a forest, only a very small fraction of seeds can survive to become canopy trees. This is especially difficult under a dense canopy, because it is very dark on the forest floor. However, when a canopy tree dies, an empty site or a 'gap' is created, where the local light availability is much improved. Thus, for a seed to establish successfully, it must be lucky enough to be in a gap site, and lucky enough to win the race among seedlings to fill the gap site.

To describe the process of regeneration of a forest, we consider a model similar to the Moran process in population genetics (Moran 1962). In the reproduction phase, each mature tree produces seeds. The quantity of seeds produced by a tree in any one year may depend on the quantity of accumulated resources invested in reproduction and the availability of outcross pollen. To be specific, the seeds produced by a tree in the i-th site in a year is $\varphi_i(t) = k_i P_i(t)[Y_i(t)]_+$, if $\theta_i(t) = 2$. Therefore, the total amount of seeds in the forest is

$$\varphi(t) = \sum_{\theta_i=2} \varphi_i(t), \tag{9.5}$$

where the sum is calculated only for sites occupied by (mature) canopy trees.

We first analyse the case in which all vacant sites will be filled by the seeds produced in the same year. This is plausible if some trees always produce seeds and the number of seeds produced by a tree is enormous. Under this simplifying assumption, the total number of trees N is kept constant, as for the Moran process (Moran 1962).

At the beginning of a year, the number of trees is N. Since the annual mortality of canopy trees is δ, the number of newly formed vacant sites $F_{\text{new}}(t)$ follows a binomial distribution $B(N,\delta)$. In the recruitment phase, each of these vacant sites is occupied by an offspring of the i-th tree with probability $\varphi_i(t)/\varphi(t)$. A new individual remains immature for τ years, during which site i is occupied by an immature tree, satisfying $\theta_i(t) = 1$. After τ years, the tree reaches the canopy and starts reproduction, satisfying $\theta_i(t) = 2$.

Reproductive success is achieved either as a female (by producing fruits) or as a male (by being a pollen donor). For the simplest genetic system to be compatible with this picture, we assume that the trees are sexual but haploid. Plants undergo sexual reproduction, and a seed has two sets of genes (diploid) derived from its parents. The zygote then experiences meiosis immediately, and the offspring tree is haploid. For each genetic locus, an offspring inherits an allele from either its mother or its father with a ratio of 50:50. For trees of a diploid genotype, the calculation becomes more complex due to the genetic dominance of alleles, but the evolutionary process is very similar.

New mutants are generated with a small probability (0.01 per generation) by random mutation at reproduction. Mutants have a trait k' slightly different from the parent k. The magnitude of difference from the parent $|k' - k|$ is 0.1. A novel mutant always starts from a single individual, and its descendants may increase or decrease during subsequent generations.

9.6 DIFFICULTY OF MASTING EVOLUTION

Since we have to trace the resource dynamics of each individual tree, the number of trees N is set at 100. This is plausible as the spatial range of pollen exchange between trees is of the order of 100 m or less (Dow and Ashley 1998), and hence N is not very large.

Figure 9.5a shows the result of numerical simulation of the evolutionary process. The model starts with a monomorphic population at the beginning of the simulation. The mean k in the population starts at $k = 5$. Initially, it moved to a smaller value with time but later became stabilised at a certain value. This final value of mean k is the evolutionary outcome. Figure 9.5b shows the evolutionary outcome – the number of seeds produced in the forest over 50 years. In trees evolved to reproduce every year, masting could not evolve.

Finite pairwise invasibility plot

The reproductive success of each tree depends on the behaviour of other individuals in the population, and the situation is the same as 'the game in a finite population' (Nowak *et al.* 2004, Taylor *et al.* 2004, Nowak 2006). In such a situation, calculation of marginal fitness is no longer a good predictor of the evolutionary outcome, but instead we need to calculate the fixation probability of a mutant that appeared in the population dominated by the resident. Specifically, consider a

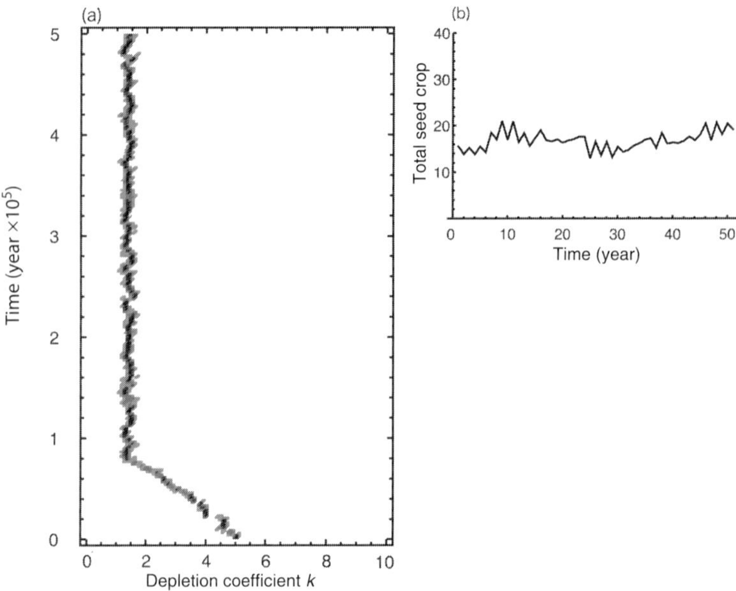

Figure 9.5 **(a)** Evolutionary trajectory. Horizontal axis represents the resource depletion coefficient k that evolves. Vertical axis represents time. Population starting from $k = 5$ moved and finally reached the evolutionarily stable value of k. **(b)** Total seed production in the forest for final 50 years of the evolutionary endpoint of the trajectory in (a). Trees in the forest reproduce every year. Parameters are: $N = 100$, $\delta = 0.04$, $\tau = 10$, $\beta = 1.75$. From Tachiki and Iwasa (2008).

population in which there is a single mutant with parameter k' and $N - 1$ resident individuals with parameter k. The run will end with either the extinction or fixation of the mutant. From a large number of replicates, we can calculate the fraction of runs with successful fixation. If the fraction of fixation runs exceeds that expected for the neutral case ($1/N$), we conclude that the mutant was favoured by natural selection (Taylor *et al.* 2004). If the fixation rate was less than that expected for the neutral mutants, we conclude that the resident was favoured by natural selection. In between these two cases, we conclude that the mutant was neutral. We distinguish these three cases using statistics.

We ran these simulations for all possible combinations of the resident's k and the mutant's k', and marked the result on a two-dimensional square. The graph is a finite population version of the PIP, which we abbreviate as fPIP. Figure 9.6a–c shows fPIPs with different values of parameter β (the other parts of Figure 9.6 will be

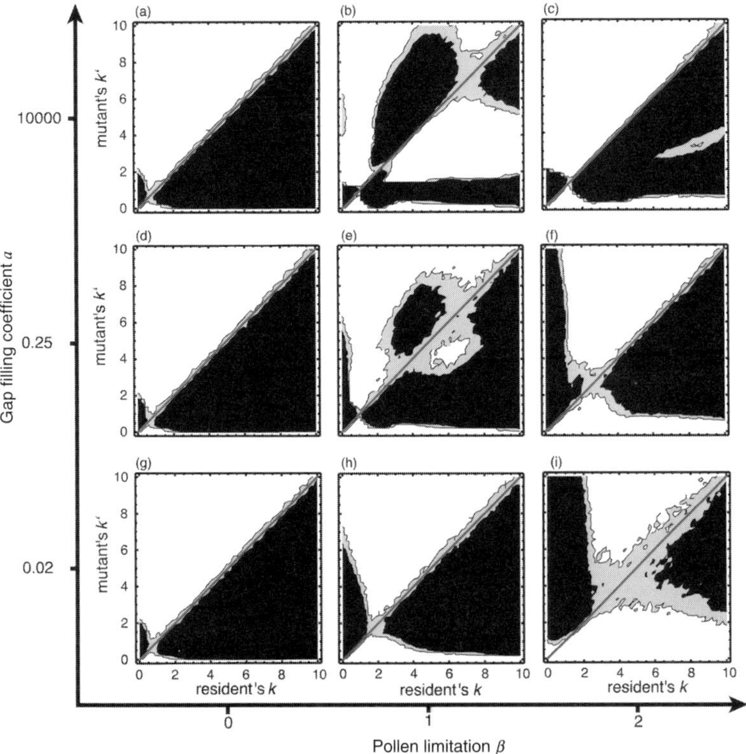

Figure 9.6 Finite pairwise invasibility plots for different combinations of parameters a and β. The vertical axis represents the mutant's resource depletion coefficient k', and the horizontal axis represents that of the resident's k, in the range $0 \leq k \leq 10$. In each panel, three regions are classified. The black region indicates that the mutant has a fixation probability significantly higher than the neutral expectation. The white region indicates that the mutant has a significantly lower fixation probability. Grey means that the mutant has the same fixation probability as in the neutral case. Masting behaviour can evolve only if the gap filling parameter a (section 9.7) is small and β is large, when the evolutionary attractor is significantly greater than 1. Parameters are: $N = 100$, $\delta = 0.04$, $\tau = 10$. **(a)** $a = 10\ 000$, $\beta = 0$, **(b)** $a = 10\ 000$, $\beta = 1$, **(c)** $a = 10\ 000$, $\beta = 2$, **(d)** $a = 0.25$, $\beta = 0$, **(e)** $a = 0.25$, $\beta = 1$, **(f)** $a = 0.25$, $\beta = 2$, **(g)** $a = 0.02$, $\beta = 0$, **(h)** $a = 0.02$, $\beta = 1$, **(i)** $a = 0.02$, $\beta = 2$. From Tachiki and Iwasa (2008).

explained later). In each panel, the horizontal axis is the resident's trait k, while the vertical axis is the mutant's trait k'. When the mutant had a fixation probability significantly higher than the neutral case, the corresponding site is shaded in black. In contrast, when the mutant had a fixation probability significantly lower than the neutral case, the corresponding site is left white. No significant difference from the neutral expectation was shown in the grey area along the diagonal line, where the mutant's k' is equal to the resident's k.

Using this graph, we can find evolutionary attractors to which the phenotype converges as a result of evolution. Since mutants are similar to the parent in phenotype, we can focus on the result near the diagonal line. If the area above the line is black and the area below the line is white, the mutant with a higher value of k' has a higher fixation probability than the resident and the mutant with a lower value of k' cannot invade. In such a case, we conclude that k should evolve to a higher value.

Figure 9.6c illustrates the case starting from a population with k smaller than $k^* = 1.4$. Mutants with k' slightly larger than the resident's k have a fixation probability greater than that of the neutral case (indicated in black). They can invade and subsequently replace the resident. Similarly, a resident population with k greater than k^* is likely to be invaded and replaced by the mutants with smaller k more than by the mutants with the larger k. Thus k^* is an evolutionary attractor because a population with different initial k will evolve gradually towards k^*. We expect that in the long run, k would approach $k^* \approx 1.4$, although the stochasticity may sometimes cause the fixation of a mutant that is not favoured by natural selection.

Also note that k as large as 1.4 would produce trees to show reproduction every year. In the analysis by Satake and Iwasa (2000), trees show annual reproduction for $k \leq 1.0$, and they show intermittent reproduction for $k > 1.0$. The difference of the results in the current model can be explained the presence of immature trees which do not contribute flowering, which makes $\frac{1}{N}\sum_{i=1}^{N} Y_i$ smaller than 1 even if all the mature trees flowers at the highest intensity. This would result in lower pollen availability in the current model than in Satake and Iwasa (2000), and hence trees tend to reproduce annually even for k greater than 1. Note that due to this effect of shortage of pollen caused by immature trees, trees show annual reproduction when k evolves to around 1.4, instead of exactly 1, in the following analyses.

Masting cannot evolve

Analyses similar to Figure 9.6a–c with more cases with intermediate β indicate that, when β is either very small ($\beta \approx 0$) or large ($\beta > 1.75$), there is only one evolutionary attractor satisfying $k^* \approx 1$. When β has an intermediate value, there are two evolutionary attractors: One attractor is $k^* \approx 1$, corresponding to trees reproducing annually. The other attractor corresponds to trees showing intermittent reproduction but without synchronisation (i.e. the desynchronised region in Satake and Iwasa (2000)). We examined the model in detail for a wide range of β, and concluded that irrespective of the choice of β, trees cannot evolve to the state in which they show intermittent and synchronised reproduction. The model always evolved either to annual reproduction (each tree reproduces every year) or to a desynchronised forest (each tree reproduces intermittently but not all trees are synchronised).

The lack of masting evolution observed in the evolutionary simulations is plausible if all the gaps formed in a year are filled within the same year. Since δ is a constant, gaps are formed at the same rate between years, and hence the opportunity for successful recruitment is independent of the total seed crop in the forest. By masting, trees may be able to produce a large amount of seeds owing to improved pollination efficiency. But the competition among those seeds to become a canopy tree is intense in mast years, because the probability of recruitment success for a single seed is inversely proportional to the amount of seeds produced in that year. In contrast, a few seeds that happen to be produced in non-mast years enjoy a high chance of becoming a canopy tree in the future. The benefit to a greater number of surviving seeds is cancelled by the intense competition among abundant surviving seeds in mast years. Hence, the reproductive success of a tree would be improved by producing seeds every year, rather than entering into masting reproduction.

9.7 CARRYOVER OF UNFILLED GAP SITES

This intuitive argument suggests that masting may be favoured by natural selection if not all the gaps are filled in a year. Suppose that there are a considerable number of empty sites (or gaps) that remain unfilled for several years. The carryover of unfilled sites would mitigate the competition among trees in a mast year. Then, the total reproductive success in a mast year with a big seed crop becomes greater than that in a non-mast year with a poor seed crop, providing some advantage to reproducing in mast years.

To examine this possibility, we considered the dynamics of the number of vacant sites (Tachiki and Iwasa 2008). In the recruitment phase, the number of seeds that establish in a site is drawn from a Poisson distribution with mean $a\varphi(t)$, in which $\varphi(t)$ is the total seed crop and a is the gap filling coefficient. A large a implies that all the gaps created in a year by the death of canopy trees are filled within the same year, as was assumed in section 9.6. In contrast, a small a means that there is considerable carryover of unfilled gap sites that are produced in previous years. The probability of one or more successful recruitment events occurring is $1 - e^{-a\varphi(t)}$. Then the probability that an offspring of the i-th tree is established in a particular vacant site is $(\varphi_i(t)/\varphi(t))\left(1 - e^{-a\varphi(t)}\right)$.

Let $F(t)$ be the number of vacant sites in the forest at the beginning of year t (i.e. the number of i satisfying $\theta_i(t) = 0$). Since the number of sites occupied by trees (including immature and mature ones) is $N - F(t)$, the number of newly formed vacant sites $F_{new}(t)$ follows a binominal distribution $B(N - F(t), \delta)$. The number of available gap sites in the recruitment phase is $F(t) + F_{new}(t)$. In the recruitment phase, some vacant sites are filled, but others remain unfilled, and are carried over to the following year $F(t + 1)$.

As pointed out before, if a is very large (Figure 9.6a–c), all gap sites are filled even if seed production in the forest is small. The resource depletion coefficient k always evolves to about 1, and the trees in the final population reproduce every year. Since the number of available vacant sites is nearly constant (to be exact, following the binomial distribution with mean $N\delta$), the strategy of reproducing every year is advantageous.

In contrast, masting can evolve when a is small and β is of the order of 1 or greater. Three examples are found in Figure 9.6f, h and i, where k evolves to values significantly larger than 1. With k greater than 1 and β relatively large, an intermittent and synchronised reproductive pattern is realised as a result of evolution.

For trees to achieve high reproductive success in mast years, a sufficient number of vacant sites must be available, which are supplied by the carryover from several previous nonmast years, as is the case for small a. In Figure 9.6i, the evolutionary attractor of k^* is an interval between 2 and 5. Within this interval of k, any pair of mutant and resident ks are nearly neutral to each other. Hence, the average phenotype in the population fluctuates due to random drift.

Figure 9.7a shows a direct simulation of the above situation. The simulation confirms the prediction. A population starting with $k = 1$ evolves towards an evolutionary attractor. Once the population reaches the evolutionary attractor, its phenotypic value of k fluctuates. A forest dominated by trees with k^* reproduces synchronously, the seed crops showing a chaotic time series (masting) (Figure 9.7b).

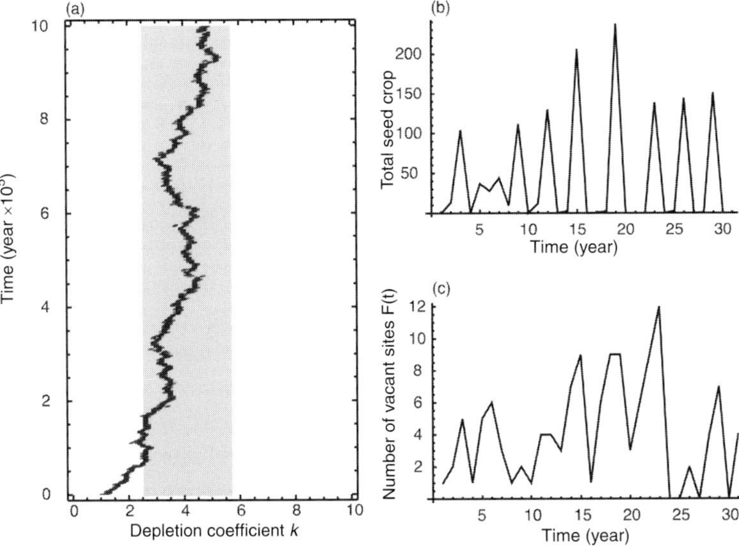

Figure 9.7 **(a)** Evolutionary trajectory when the gap filling coefficient a is small. The initial value is $k = 1$. The vertical axis represents time and the horizontal axis represents k. Once the population reaches the evolutionary attractor, k fluctuates within the evolutionary attractor, indicated by shading ($2.5 < k < 5.5$). This result corresponds to the fPIP in Figure 9.6i. **(b)** Total seed crop in the forest for the last 30 years of the evolutionary endpoint of the trajectory in (a). Trees in the forest reproduce intermittently and synchronously. **(c)** The number of gaps at the beginning of year t, $F(t)$. The number of available free sites in year t is the sum of $F(t)$ and newly created gaps $F_{new}(t)$. Vacant sites are carried over for several years. $F(t)$ is fluctuating. In years in which large amounts of seeds are produced synchronously, they can fill many vacant sites. This creates the reproductive advantage of masting. In all panels, parameters are: $N = 100$, $\delta = 0.04$, $\tau = 10$, $a = 0.02$ and $\beta = 2$. From Tachiki and Iwasa (2008).

We conclude that trees can evolve masting behaviour only through improving pollination efficiency if a considerable fraction of vacant sites created in a year remains unfilled for several years.

9.8 SEEDLING BANK

However, the carryover of unfilled gaps is unlikely to be a very important process in many forests. Gaps formed in a year tend to be filled by the seedlings from the seeds produced in the same year as well as those produced in previous years. The seedlings can survive for several years under dark canopy but cannot grow much because of the shortage of light. However, when a canopy tree dies, light on the forest floor improves immediately and seedlings start growing quickly. After competition among them, one individual is supposed to fill the canopy after some number of years.

Seedling bank is a terminology for the population of seedlings that is waiting for the opportunity of growing under canopy (Shibata and Nakashizuka 1995, Morin and Laprise 1997, Marks and Gardescu 1998, George and Bazzaz 1999). If a large number of seedlings produced in a year by masting can contribute not only to the acquisition of gaps formed in the same year but also gaps formed in subsequent years, then the total regeneration success by the seeds produced in a mast year would be greater than that by the seeds produced in nonmast years. Hence, some of the advantage to the parent trees of reproducing in mast years should remain after it is reduced by the intensive competition among offspring.

We have carried out the simulation. Every year new seedlings join the seedling bank. Some fraction of these seedlings can survive in the seedling bank for several years. When a canopy gap is formed, a single individual seedling would be selected randomly from the seedling bank, and after a certain number of years it reaches canopy and becomes a fully mature tree. In this setting, gaps are filled within the year they are created and there is no carryover of empty space, but seeds produced in different years must compete with each other for the acquisition of gap sites to fill.

Figure 9.8 illustrates the simulation results. Whether or not synchronised and intermittent reproduction of trees can evolve should critically depend on the survivorship curve of seedlings. In Figure 9.8a, only a small fraction of seedlings can contribute to the gap acquisition competition in the following year and after the second year all the seedlings are killed. The trees evolve to reproduce every year without any indication of masting – the effect of the seedling bank was not effective. In contrast, if a considerable fraction of seedlings can contribute to the gap acquisition for 2 or 3 years the resource depletion coefficient k evolves to about 4, as shown in the middle part of Figure 9.8b. The forest in the evolutionary endpoint shows clearly fluctuating seed crops, as illustrated in the bottom part of Figure 9.8b. In Figure 9.8c, a greater fraction of seedlings can contribute to gap acquisition for several years. The resource depletion coefficient k evolves to an even larger value ($k^* \approx 5$), and the trees in the forests show big mast years with intervals of two or three nonmast years (bottom part of

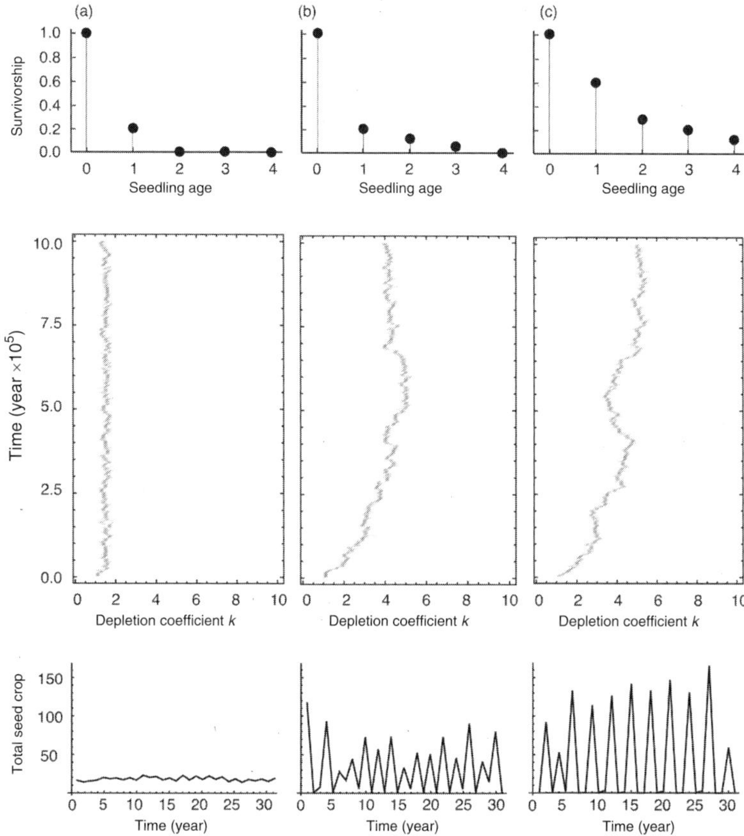

Figure 9.8 Effect of seedling bank on the evolution of depletion coefficient k and masting. Three different cases of seedling survivorship are shown. **Top**: fraction of seedlings that survive as a function of age. **Middle**: evolutionary trajectory of depletion coefficient k. **Bottom**: time series of the total seed production of the whole forest. **(a)** indicates the case in which seedlings cannot survive more than one year. k evolved to about 1.5 and the trees evolved to have annual reproduction. Masting did not evolve. **(b)** Indicates the case in which seedlings can survive for several years. The depletion coefficient k evolved to about 4 and the forest shows synchronised and intermittent reproduction (masting) with the interval about 2 years. **(c)** Is the case with even higher survivorship of seedlings. The depletion coefficient k evolved to about 5, and the forest evolves to show clear masting with intervals of about 3 years. Note that in these models there is no seed predator. The benefit of joining mast years comes from the improved pollination success only. Other parameters are: $N = 100$, $\delta = 0.04$, $\tau = 10$ and $\beta = 2$.

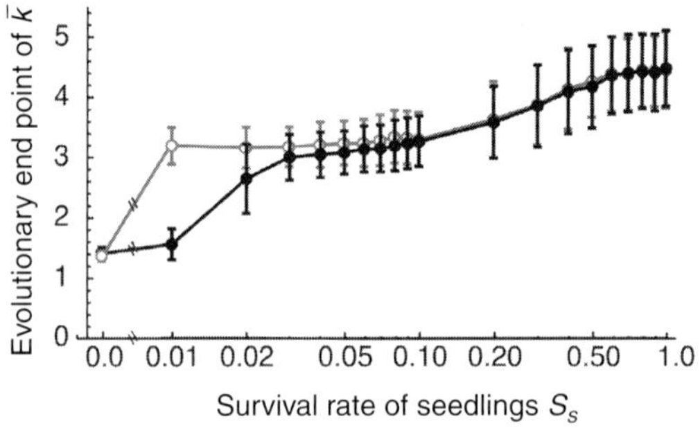

Figure 9.9 Depletion coefficient k at the evolutionary endpoint. Horizontal axis is for s_s, the annual survivorship of seedlings. Solid circles are the mean values in the absence of predators, and open circles are those in the presence of specialist seed predators. Bars indicate standard deviations. In the absence of predators (solid circles), k evolved to a higher value for a greater seedling survivorship. When k is less than 1.5, the trees show annual reproduction. When k is greater than 1.5, trees show intermittent reproduction. In the presence of predators (open circles), the k evolved to a higher value than in the absence of predators (solid circles), especially when the seedling survivorship is small but positive ($0.01 \leq s_s \leq 0.05$). Then masting evolved in the presence of predators but not in the absence of predators. In contrast both for no seedling bank ($s_s = 0$) and for large seedling survivorship ($s_s \geq 0.1$), there is no effect of specialist predators – the depletion coefficient evolved to the same value. Other parameters are: $N = 100$, $\delta = 0.04$, $\tau = 10$, $\beta = 2$, $a_p = 10$, $b = 20$ and $c = 5$.

Figure 9.8c). Comparison of these three cases demonstrates the importance of seedling survivorship to the evolution of masting of trees.

We also studied how the resource depletion coefficient k at the evolutionary endpoint changes with the survivorship of seedlings. We assumed that the probability of newly produced seeds surviving t years is s_s^t ($t = 0, 1, 2, 3, \ldots$), if they stay in the forest floor as seedlings waiting for lucky gap openings. The solid circles in Figure 9.9 illustrate the results of the mean value of k in the end of evolutionary simulations. The horizontal axis is the annual survivorship of seedlings s_s. We ran the simulation for 500 000 years and then sampled the population mean value of k. We replicated this procedure 500 times and examined the difference of the evolutionary outcome between presence and absence of predators. There is a fluctuation of k due to the finite population size. Solid circles indicate the mean and bars indicate the standard deviations (those with open circles will be explained below).

We can see that the value of k increases with seedling survivorship. As seedlings have higher survivorship, the seedling bank becomes more important, the trees evolve to have a larger depletion coefficient k, and show intermittent and synchronised reproduction. This clearly demonstrates the importance of the seedling bank in masting evolution. Here the benefit trees receive by masting is a higher pollination success only.

9.9 SPECIALIST SEED PREDATOR

Predator satiation has been regarded as the most important source of selective advantage for trees engaging in masting (Kelly 1994). There has been experimental and observational support for this hypothesis (Kelly and Sork 2002).

To test the predator satiation theory, we consider an insect predator that specialises on the seeds of the focal tree species. We consider a model that was developed for the apple fruit moth in Norway by Satake *et al.* (2004), for a pre-dispersal seed predator of rowan. Insects lay eggs in the fruit and larvae eat seed. Larvae spend the pupal stage in the seeds and, after diapause, they emerge in the same season as they are laid by their mother. From one seed in which an apple fruit moth has laid an egg, a single adult moth can hatch out.

The population dynamics of apple fruit moth can be described in terms of the Nicholson–Bailey model for host–parasitoid systems. Let $M(t)$ be the number of adult moths in year t. Let $\varphi(t)$ be the amount of seeds produced in year t. Suppose that the number of seeds attacked by a single adult moth is a function of the seed crop given as $f(\varphi(t))$. Then the mean fraction of seeds eaten is $f(\varphi(t))M(t)/\varphi(t)$. If predators attack seeds randomly and if all seeds are exposed equally to predators, the number of seeds attacked by a single individual predator follows a Poisson distribution with this mean value. Thus, the fraction of fruits attacked one or more times is $1 - \exp[-f(\varphi(t))M(t)/\varphi(t)]$. The number of adult moths in the next year, $M(t+1)$, will be

$$M(t+1) = c\varphi(t)\left[1 - \exp\left(-\frac{f(\varphi(t))M(t)}{\varphi(t)}\right)\right], \tag{9.6}$$

where c is a constant, which is the product of the number of seeds per unit amount of seed crop and the survivorship of a larva. We assume that functional response $f(\varphi(t))$ is given by a type II function:

$$f(\varphi) = \frac{a_p\varphi}{1 + b\varphi}, \tag{9.7}$$

which increases with $\varphi(t)$ but saturates to maximum attack rate a_p/b for very large $\varphi(t)$. Then the fraction of the fruits attacked decreases with the number of seeds $\varphi(t)$. It means that a larger seed production reduces the chance of being eaten for each individual seed, which is called 'functional satiation'. In addition, the fraction of attacked fruits will be low when the abundance of predators is low, which is called 'numerical satiation'.

The number of seeds that escape predators for tree i is given by

$$\tilde{\varphi}_i(t) = \varphi_i(t)\exp(-f(\varphi(t))M(t)/\varphi(t)). \tag{9.8}$$

Seeds germinate and enter the seedling bank and wait until the next gap opens by tree fall of canopy trees. The seed survivorship depending on the specialised seed predator is given by Equation (9.8), which can be combined with the evolutionary model including tree life cycles explained in sections 9.5–9.6. On the other hand, Equation (9.6) gives the population dynamics of a specialist seed predator. Then we have a system model for the evolution of masting of trees in the presence of a specialist seed predator.

We assume here that all seeds are equally likely to be attacked. Individual trees that produce a large number of seeds and those producing only a few seeds both experience the same seed mortality in a given year.

Since we are considering a perfectly specialised seed predator, if there is a year without seed production, the predator should go extinct. After extinction, we considered the recovery of the predator population at a low rate of migration from outside of the area ($M(t) = 0.5$) in the simulations.

Evolutionary outcomes in the presence of seed predators

If there is no seedling bank (i.e. seedlings cannot survive over the years), there is no possibility for intermittent reproduction to evolve even in the presence of specialist seed predators (results not shown). This is exactly the same as that in the absence of specialist predators. Figure 9.10 illustrates the evolutionary results for different survivorship of seedlings. Parameters related to predation (a_p, b and c in Equations (9.6–9.7) are fixed.

Figure 9.10a shows the result in the presence of specialist seed predators, but other parameters are the same as Figure 9.8a without predators. The contribution of the seedling bank is small (see top part of Figure 9.10a), and the evolved value of k is about 1.5 (middle part of Figure 9.10a). Masting did not evolve, which is the same as the case without specialist predators (Figure 9.8a).

Figure 9.10b shows results for a mild level of seedling survivorship (see top). The forest evolves to show more pronounced

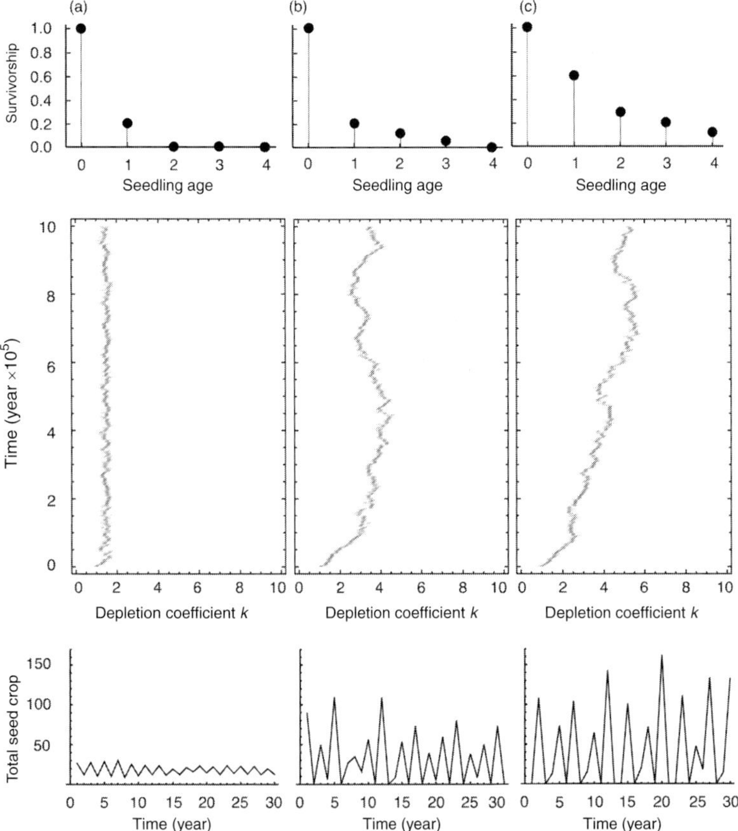

Figure 9.10 Effect of seedling bank on the evolution of k and of masting in the presence of a specialised seed predator (compare with Figure 9.8). Three cases of seedling survivorship correspond to the three cases in Figure 9.8. (a) Low survivorship of seedlings. k evolved to about 1.5, and masting did not evolve. (b) Indicates the case in which seedlings can survive for several years. The depletion coefficient k evolved to fluctuate between 3 and 4.5, and the forest showed synchronised and intermittent reproduction (masting) with intervals of about 2 years. (c) Even higher survivorship of seedlings. The depletion coefficient k evolved to fluctuate between 4 and 5.5, and the forest showed clear masting, similar to the results in Figure 9.8c. In these models trees in mast enjoy a higher fitness benefit both from a higher pollination efficiency and predator satiation. Other parameters are: $N = 100$, $\delta = 0.04$, $\tau = 10$, $\beta = 2$, $a_P = 10$, $b = 20$ and $c = 5$.

intermittent reproduction. Figure 9.10c shows the case with even higher survivorship of seedlings in the presence of specialist seed predators. We also observed that, if seedlings have a sufficiently high survivorship, the evolved trees may show masting with long intervals. In such a case, however, the presence and absence of seed predators did not affect the evolutionary results much (see Figures 9.8b and c).

In Figures 9.8 and 9.10, the endpoint of a single evolutionary trajectory is shown for each set of parameters. However, the value of k fluctuates considerably due to the stochasticity caused by the finite number of individuals (see middle parts of Figures 9.8 and 9.10). To know the range of evolutionary outcomes, we need to run a large number of replicates for the same set of parameters. Bars with open circles in Figure 9.9 illustrate the depletion coefficient k to which populations evolve when there are specialist seed predators. We can compare these results with those when there are no seed predators, as indicated by bars with solid circles in Figure 9.9. The horizontal axis is the annual survivorship of seedlings s_s.

When no seedling bank is formed ($s_s = 0$), masting evolved neither in the presence nor in the absence of predators. In both cases, the resource depletion coefficient k evolved to about 1.5, showing annual reproduction of every mature tree. This implies that, without a seedling bank, predators cannot make masting evolve.

When survivorship of seedlings is small but positive ($0.01 < s_s < 0.05$), the depletion coefficient k evolved to a value significantly larger in the presence of seed predators than in the absence of seed predators. For $s_s = 0.01$, synchronised and intermittent reproduction of trees evolved in the presence of specialist seed predators ($k^* = 3.3$), while annual reproduction of every tree evolved in the absence of specialist seed predators ($k = 1.5$).

However, when the survivorship of seedlings is very large (say $s_s > 0.1$), the difference between the presence and absence of seed predators disappears again – the level of the resource depletion coefficient k at the endpoint of evolution was the same in the absence and presence of specialist predators.

9.10 DISCUSSION

We analysed in detail coupled map models for pollen-coupled trees proposed by Isagi *et al.* (1997), in which the dynamics of the energy reserve of each tree are explicitly considered. When coupling is global, depending on the strength of coupling by pollen limitation relative to the magnitude of energy reserve depletion after reproduction, the whole forest may show diverse dynamic behaviours, such as perfectly synchronised periodic reproduction, synchronised reproduction with a chaotic time series, clustering phenomena and chaotic reproduction without synchronisation.

In masting, many trees reproduce synchronously but the synchrony is often not perfect (Sharp and Sprague 1967, Koenig *et al.* 1994). Even for synchronised reproduction, the time series of mast years is not exactly periodic (Sork *et al.* 1993, Koenig *et al.* 1994), which is regarded as deviations from ideally synchronised periodic reproduction caused by external environmental fluctuations (Sork *et al.* 1993) or by individual variability in production rate (Yamauchi 1996). The pollen-coupled tree model proposes an alternative view – each tree has an inherent tendency to fluctuate chaotically, and the synchronisation and periodicity are secondary characteristics formed through pollen coupling. The observed imperfect periodicity might be a natural outcome of energy reserve dynamics, because the synchronised reproductive level follows a chaotic fluctuation. In addition, individuals in a real forest are heterogeneous with respect to annual productivity, reproductive threshold and pollen sensitivity, which may make imperfect synchronisation and imperfect periodicity more likely.

Although pollen dispersal over 500 m has been reported from the study using genetic markers (e.g. Hamrick and Loveless 1989), it is unlikely that trees that are separated more than 100 km can directly exchange pollen. The study of the coupled map lattice shows that fluctuating reproduction of trees synchronised over the whole forest can be achieved by local pollen coupling, but the period of fluctuation cannot be longer than two (Satake and Iwasa 2002a), which is not consistent with the observed longer interval between mast of beech and other trees. An important factor for the synchronisation of trees in masting over a long distance, sometimes over 100 km, is the common climatic fluctuation that is experienced by distant trees (Koenig and Knops 1998). However, the common environmental fluctuation alone is not able to achieve a highly synchronised forest because of the chaotic nature of individual trees (Satake and Iwasa 2002b). In Chapter 11, we discuss the importance of common environmental fluctuation (Moran effect; Moran 1953) and of pollen coupling as processes to generate synchronised reproduction among forest trees.

Importance of seedling banks

We examined the conditions needed for trees to evolve to show masting. We adopted the globally coupled map model for tree reproduction (Satake and Iwasa 2000) and discussed the evolution of the resource depletion coefficient k of a tree, which regulates resource investment to fruits given a number of fertilised eggs. The model considers pollen coupling as a mechanism for synchronisation. We explicitly modelled the whole life cycle of the tree, including the dynamics of tree fall and recruitment (or gap dynamics), which are often neglected when the evolutionary advantage of masting is discussed. Our analysis demonstrated the importance of the recruitment dynamics in the evolution of masting – masting cannot evolve if all the gaps created in a year are filled within the same year, but can evolve if a considerable proportion of vacant sites formed by tree falls remain unfilled over multiple years. In addition, outcross pollen limitation needs to be sufficiently strong for the evolution of masting. Somewhat surprisingly, it can evolve without considering seed predators, which are often supposed to provide the most important evolutionary benefit to masting.

To analyse the evolutionary game in a finite population, Tachiki and Iwasa (2008) developed the finite PIP (fPIP). This graphic tool had been used previously by Kamo *et al.* (2007) and Proulx and Day (2001), although they did not discuss the statistical significance of the evolutionary advantage of mutants. Since stochasticity is unavoidable in the evolution of a finite population, an appropriate statistical test is needed to discuss the evolution properly.

If many trees in a forest show very high fruiting success in a mast year, and there is almost no seed production over the following several years (Figure 9.7b), the abundance of gap sites would be reduced after the mast year and increase slowly over subsequent nonmast years (Figure 9.7c). Hence, the recruitment success of a tree depends strongly on the availability of vacant sites, which is affected by the time series of the fruiting success of all the trees in the forest. In addition, as outcross pollen availability for a mutant is controlled by the flowering pattern of the resident, the temporal pattern of a mutant's seed production over the years would be determined by the behaviour of the resident trees in the forest in a complex manner.

The Moran process and lottery model have often been used for the dynamics and the evolution of sessile organisms (Chesson and Warner 1981, Kisdi and Meszéna 1995, Muko and Iwasa 2000, Kisdi and Geritz 2003, Higgins *et al.* 2008, Schoolmaster 2008), in which a vacant site produced by the death of an individual is filled immediately. Our results suggest that this assumption may not always be appropriate. The consideration of unoccupied space has been emphasised in the ecology of marine benthic species (Iwasa and Roughgarden 1986, Roughgarden and Iwasa 1986). When we consider the dynamics or evolution of sessile organisms, we should take account of the mechanisms by which vacant sites are created and filled.

In addition, masting cannot evolve if all the vacant sites are filled in the recruitment phase every year. Seeds produced in a mast year can enjoy high survivorship due to the predator satiation effect, but they face very intensive competition for vacant sites that are required for their establishment. If the same number of vacant sites were needed for recruitment in different years, the advantage of having many surviving seeds for a tree joining the synchronised reproduction would be cancelled by the intensive competition among seeds. We confirmed that masting cannot evolve even in the presence of seed predators if all vacant sites are filled every year.

Role of seed predator satiation

The predator satiation theory (Janzen 1971, Silvertown 1980, Nilsson and Wästljung 1987) is an important and plausible hypothesis for the evolutionary advantage of masting. There is much experimental evidence supporting the advantage of seed survivorship (Kobro *et al.* 2003).

We considered the population dynamics of a moth that is a parasitoid or predator feeding on seeds. If all the vacant sites are filled by offspring from the seeds produced in the same year, masting cannot evolve – even if predator satiation greatly improves the survivorship of seeds in mast years in addition to the improved availability of outcross pollen.

However, if seedlings can survive more than 1 year, the dynamics differ from the case without seed predators. This effect is most pronounced when the seedling survivorship is rather low (Figure 9.9).

Interestingly, when the survivorship of seedlings is sufficiently high, the difference between the presence and the absence of specialist predators disappears – in both cases the strongly synchronised and intermittent reproduction of trees evolves (Figures 9.8c, 9.10c). The value of the depletion coefficient at the evolutionary endpoint was also very similar, as shown in Figure 9.9 when the seedling survivorship is large, $s_s > 0.1$.

From the modelling study in this chapter, we can conclude that both seedling banks and specialised seed predators contribute to the evolution of masting in trees.

Since moths have one generation per year, a single nonmast year is able to deplete the predator population very effectively. Predators of a different type, such as large mammals, might have an impact on the masting evolution of trees different from that of the specialised invertebrate seed predators studied in this paper. It will be an important theme of future theoretical study to analyse the effect of seed predators that differ in dispersal ability, alternative food sources and population growth potential.

Gap adapted species such as maples (e.g. *Acer rubrum*) and riparian trees or trees adapted to highly disturbed areas such as willows tend to reproduce every year. These species produce numerous small seeds that are adapted to dispersal and consequently seedlings from them have small survivorship in overwintering, in comparison to tree species with acorns. Considering the results of this chapter, this lack of extensive seedling banks as well as the lack of specific seed predators might explain their annual reproduction. However, sugar maple (*Acer saccharum*) in North American temperate forests shows clearly synchronised and intermittent reproduction. This can be explained by the fact that these sugar maples form a seedling bank in which seedlings can survive for many years (Marks and Gardescu 1998), which is consistent with the results of our analysis. *Acer saccharum* has large seeds (or a high seed weight:wing length ratio), and it is shade tolerant and slow growing. Its masting behaviour fits with similar behaviour of other larger seeded trees, rather than with other maple species (e.g. *Acer rubrum*) (Peter Marks personal communication).

ACKNOWLEDGEMENTS

This work was supported in part by a Grant-in-Aid from the Japan Society for the Promotion of Science to Y.I. another to S.A, and also supported by a PRESTO program of Japan Science Technology Agency to AS. We thank the following people for their very useful comments: P. Abrams, T. Akita, O. N. Bjørnstad, U. Dieckmann, M. Doebeli, M. Hiramatsu, T. Hiura, M. Ikegami, T. Ikegami, K. Isagi, N. Kachi, M. Kamo, K. Kaneko, W. D. Koenig, T. Kohyama, H. Kudo, S. A. Levin, P.L. Marks, T. Masaki, Y. Morishita, T. Nakashizuka, D. Rand, A. Sasaki, N. Shigesada, K. Shimizu and H. Tanaka.

REFERENCES

Chesson, P. L. and Warner, R. R. (1981). Environmental variability promotes coexistence in lottery competitive systems. *American Naturalist* **117**, 923–943.

Cruzan, M. B. (1998). Genetic markers in plant evolutionary ecology. *Ecology* **79**, 400–412.

Dow, B. D. and Ashley, M. V. (1998). High levels of gene flow in bur oak revealed by paternity analysis using microsatellites. *Journal of Heredity* **89**, 62–70.

George, L. O. and Bazzaz, F. A. (1999). The fern understory as an ecological filter: emergence and establishment of canopy-tree seedlings. *Ecology* **80**, 833–845.

Geritz, S. A. H., Meijden, E. V. D. and Metz, J. A. J. (1999). Evolutionary dynamics of seed size and seedling competitive ability. *Theoretical Population Biology* **55**, 324–343.

Geritz, S. A. H., Metz, J. A. J., Kisdi, É. and Meszéna, G. (1997). Dynamics of adaptation and evolutionary branching. *Physical Review Letters* **78**, 2024–2027.

Hamrick, J. L. and Loveless, M. D. (1989). The genetic structure of tropical tree populations: associations with reproductive biology. In J. H. Bock and Y. B. Linhart (eds), *The Evolutionary Ecology of Plants*. Boulder, CO: Westview Press, pp. 129–149.

Hastings, A. (1993). Complex interactions between dispersal and dynamics – lessons from coupled logistic equations. *Ecology* **74**, 1362–1372.

Herrera, C. M., Jordano, P., Guitian, J. and Traveset, A. (1998). Annual variability in seed production by woody plants and the masting concept: reassessment of principles and relationship to pollination and seed dispersal. *American Naturalist* **152**, 576–594.

Higgins, S. I., Flores, O. and Schurr, F. M. (2008). Cost of persistence and the spread of competing seeders and sprouters. *Journal of Ecology* **96**, 679–686.

Isagi, Y., Sugimura, K., Sumida, A. and Ito, H. (1997). How does masting happen and synchronise? *Journal of Theoretical Biology* **187**, 231–239.

Iwasa, Y. and Roughgarden, J. 1986. Interspecific competition among metapopulations with space-limited subpopulations. *Theoretical Population Biology* **30**, 194–214.

Janzen, D. H. (1971). Seed predation by animals. *Annual Review of Ecology and Systematics* **2**, 465–492.

Kamo, M., Sasaki, A. and Boots, M. (2007). The role of trade-off shapes in the evolution of parasites in spatial host populations: an approximate analytical approach. *Journal of Theoretical Biology* **244**, 588–596.

Kaneko, K. (1990). Clustering, coding, switching, hierarchical ordering, and control in a network of chaotic elements. *Physica D* **41**, 137–172.

Kelly, D. (1994). The evolutionary ecology of mast seeding. *Trends in Ecology and Evolution* **9**, 465–470.

Kelly, D., Hart, D. E. and Allen, R. B. (2001). Evaluating the wind-pollination benefits of mast seeding. *Ecology* **82**, 117–126.

Kelly, D. and Sork, V. L. (2002). Mast seeding in perennial plants: why, how, where? *Annual Review of Ecology and Systematics* **33**, 427–447.

Kisdi, É. and Geritz, S. A. H. (2003). On the coexistence of perennial plants by the competition–colonization trade-off. *American Naturalist* **161**, 350–354.

Kisdi, É. and Meszéna, G. (1995). Life histories with lottery competition in a stochastic environment: ESSs which do not prevail. *Theoretical Population Biology* **47**, 191–211.

Kobro, S., Søreide, L., Djønne, E. *et al.* (2003). Masting of rowan *Sorbus aucuparia* L. and consequences for the apple fruit moth *Argyresthia conjugella* Zeller. *Population Ecology* **45**, 25–30.

Koening, W. D. and Knops, J. M. H. (1998). Scale of mast-seeding and tree-ring growth. *Nature* **396**, 225–226.

Koening, W. D. and Knops, J. M. H. (2000). Patterns of annual seed production by northern hemisphere trees: a global perspective. *American Naturalist* **155**, 59–69.

Koening, W. D., Knops, J. M. H., Carmen, W. J. and Stanback, M. T. (1999). Spatial dynamics in the absence of dispersal: acorn production by oaks in central coastal California. *Ecography* **22**, 499–506.

Koenig, W. D., Mumme, R. L., Carmen, W. J. and Stanback, M. T. (1994). Acorn production by oaks in central coastal California: variation within and among years. *Ecology* **75**, 99–109.

Marks, P. L. and Gardescu, S. (1998). A case study of sugar maple (*Acer saccharum*) as a forest seedling bank species. *Journal of the Torrey Botanical Society* **125**, 287–296.

Metz, J. A. J., Nisbet, R. M. and Geritz, S. A. H. (1992). How should we define 'fitness' for general ecological scenarios? *Trends in Ecology and Evolution* **7**, 198–202.

Moran, P. A. P. (1953). The statistical analysis of the Canadian lynx cycle. II. Synchronisation and meteorology. *Australian Journal of Ecology* **1**, 291–298.

Moran, P.A.P. (1962). *Stochastic Process in Evolutionary Theory*. Oxford: Clarendon Press.

Morin, H. and Laprise, D. (1997). Seedling bank dynamics in boreal balsam fir forests. *Canadian Journal of Forest Research* **27**, 1442–1451.

Muko, S. and Iwasa, Y. (2000). Species coexistence by permanent spatial heterogeneity in a lottery model. *Theoretical Population Biology* **57**, 273–284.

Nilsson, S. G. and Wästljung, U. (1987). Seed predation and cross-pollination in mast-seeding beech (*Fagus sylvatica*) patches. *Ecology* **68**, 260–265.

Nowak, M. A. (2006). *Evolutionary Dynamics: Exploring the Equations of Life*. Cambridge, MA: Harvard University Press.

Nowak, M. A., Sasaki, A., Taylor, C. and Fudenberg, D. (2004). Emergence of cooperation and evolutionary stability in finite populations. *Nature* **428**, 646–650.

Proulx, S. R. and Day, T. (2001). What can invasion analyses tell us about evolution under stochasticity in finite populations? *Selection: Molecules, Genes, and Memes* **2**, 1–15.

Rees, M., Kelly, D. and Bjørnstad, O. N. (2002). Snow tussocks, chaos, and the evolution of mast seeding. *American Naturalist* **160**, 44–59.

Roughgarden, J. and Iwasa, Y. (1986). Dynamics of a metapopulation with space-limited subpopulations. *Theoretical Population Biology* **29**, 235–261.

Satake, A. and Bjørnstad, O. N. (2004). Spatial dynamics of specialist seed predators on synchronised and intermittent seed production of host plants. *American Naturalist* **163**, 591–605.

Satake, A. and Iwasa, Y. (2000). Pollen coupling of forest trees: forming synchronised and periodic reproduction out of chaos. *Journal of Theoretical Biology* **203**, 63–84.

Satake, A. and Iwasa, Y. (2002a). Spatially limited pollen exchange and a long-range synchronisation of trees. *Ecology* **83**, 993–1005.

Satake, A. and Iwasa, Y. (2002b). The synchronised and intermittent reproduction of forest trees is mediated by the Moran effect, only in association with pollen coupling. *Journal of Ecology* **90**, 830–838.

Satake, A., Bjørnstad, O. N. and Kobro, S. (2004). Masting and trophic cascades: interplay between rowan trees, apple fruit moth, and their parasitoid in Southern Norway. *Oikos* **104**, 540–550.

Schoolmaster, D. R., Jr. (2008). Recruitment limitation modifies the net effects of shared enemies on competitively inferior plants. *Journal of Ecology* **96**, 114–121.

Sharp, W. M. and Sprague, V. G. (1967). Flowering and fruiting in white oaks. Pistillate flowering, acorn development, weather, and yields. *Ecology* **48**, 243–251.

Shibata, M. and Nakashizuka, T. (1995). Seed and seedling demography of four co-occurring *Carpinus* species in a temperate deciduous forest. *Ecology* **76**, 1099–1108.

Shibata, M., Tanaka, H. and Nakashizuka, T. (1998). Causes and consequences of mast seed production of four co-occurring *Carpinus* species in Japan. *Ecology* **79**, 54–64.

Silvertown, J. W. (1980). The evolutionary ecology of mast seeding in trees. *Biological Journal of the Linnean Society* **14**, 235–250.

Smith, C. C., Hamrick, J. L. and Kramer, C. L. (1990). The advantage of mast years for wind pollination. *American Naturalist* **136**, 154–166.

Sork, V. L., Bramble, J. and Sexton, O. (1993). Ecology of mast-fruiting in three species of North American deciduous oaks. *Ecology* **74**, 528–541.

Tachiki, Y. and Iwasa, Y. (2008). Role of gap dynamics in the evolution of masting of trees. *Evolutionary Ecology Research* **10**, 893–905.

Tamura, S. and Hiura, T. (1998). Proximate factors affecting fruit set and seed mass of *Styrax obassia* in a masting year. *Écoscience* **5**, 100–107.

Taylor, C., Fudenberg, D., Sasaki, A. and Nowak, M. A. (2004). Evolutionary game dynamics in finite populations. *Bulletin of Mathematical Biology* **66**, 1621–1644.

van Schaik, J. W., Terborgh, J. W. and Wright, S. J. (1993). The phenology of tropical forests: adaptive significance and consequences for primary consumers. *Annual Review of Ecology and Systematics* **24**, 353–377.

Yamauchi, A. (1996). Theory of mast reproduction in plants – storage size dependent strategy. *Evolution* **50**, 1795–1807.

ROBIN E. SNYDER

10

Spatiotemporal variation can promote coexistence more strongly than temporal variation

10.1 INTRODUCTION

Spatiotemporal variation in environmental conditions is ubiquitous and can help or hinder species coexistence. It is not always clear, however, when we should expect environmental variation to promote coexistence nor what forms of variation can do so most effectively. This chapter seeks to answer two questions. First, when does spatiotemporal variation most promote coexistence and why? It turns out that spatiotemporal variation promotes coexistence most effectively via spatial mechanisms and this leads us to the second question: is spatiotemporal variation more or less able to promote coexistence than purely temporal variation and why? For those who are interested, this chapter also gathers together mathematical calculations which have been scattered throughout the appendices of several papers (Snyder 2006, 2007, 2008). Here I state in one place how to get from environmental autocorrelation functions to spatiotemporal population distributions to a calculation of how much variation increases or decreases an invader's long-run growth rate.

I explore these issues using a model of competing annual plants that is presented in the next section. The model discussion is followed by an overview of the mathematical methods, which should be accessible to everyone, and a more detailed presentation of the mathematics, for those who may wish to use these methods themselves. Next I present the results of a numerical experiment in which I systematically vary each species' dormancy, seed dispersal distance and competition distances, and determine how much environmental variation promotes

Temporal Dynamics and Ecological Process, ed. C. K. Kelly, M. G. Bowler, G. A. Fox.

coexistence for each combination of traits. I find that regardless of the scale of spatial or temporal autocorrelation in environmental conditions, spatiotemporal variation most promotes coexistence when species have traits that encourage population aggregation and species segregation. While low dormancy and short-range between-species competition contribute to these tendencies, short-range seed dispersal is essential. I also find that spatiotemporal variation has a much higher capacity to promote coexistence than temporal variation does, and I discuss reasons for this. The chapter ends with a discussion section.

10.2 ANNUAL PLANT MODEL

The results in this chapter are derived from a model of competing annual plants. The number of species j seeds at location x at time t is $n_j(x,t)$. The seeds germinate with species-specific probability g_j and if they fail to germinate, survive until the following year with probability s_j. The number of dormant seeds at $(x,t+1)$ is thus $s_j(1-g_j)n_j(x,t)$.

Fecundity depends on local environmental conditions, which vary in space and time: in the absence of competition, an adult plant would produce $F_j(x,t)$ seeds. In this model, all of the effects of variable environmental conditions are mediated by fecundity, and for conciseness, I will refer to $F_j(x,t)$ as 'the environment', even though it is technically a response to the environment. Fecundity is exponentially correlated in space and time:

$$\mathrm{Cov}(F_j(x,t),F_k(x+x',t+t'))_{x,t} = \cos(\theta)V\exp(-x'/\xi)\exp(-t'/\tau), \tag{10.1}$$

where θ expresses the degree to which species j and k prefer the same environment ($\theta = 0$: species prefer identical environments, $\theta = \pi$: species prefer opposite environments) and V is the variance of environmental fluctuations. This means that conditions in nearby locations are similar, as are conditions in nearby times, but conditions here and now tell us little about conditions far away or a long time into the future. More specifically, conditions are roughly similar out to a distance ξ away and a time τ into the future: ξ is the spatial correlation length and τ is the temporal correlation length.

Seed production is reduced by competition: the number of seeds produced at (x,t) is $F_j(x,t)/C_j(x,t)$, where $C_j(x,t)$ is the competition experienced by a species j plant at (x,t). The number of seeds produced at (x,t) is therefore $F_j(x,t)g_jn_j(x,t)/C_j(x,t)$, where $g_jn_j(x,t)$ is the number of adults at (x,t).

Competition depends on a weighted average of local seedling density. The weights are given by a function called a competition kernel, which weights closer competitors more heavily. The two species are not equal competitors, however. The competitive effect of species k is multiplied by a factor γ_{jk} which accounts for species k's competitive ability against species j. Thus, the competitive effect of species k on species j is $\gamma_{jk}\int_{-\infty}^{\infty} U_{jk}(x-y)g_k n_k(y)dy$, where U_{jk} is the competition kernel for the effects of species k on species j. Finally, we must sum over the effects from all competitors, so that the total competition experienced by species j at (x,t) is $C_j(x,t) = \sum_k \gamma_{jk}\int_{-\infty}^{\infty} U_{jk}(x-y)g_k n_k(y)\, dy$.

Once seeds are produced, they disperse a distance z from their parent with probability $k_j(z)$. Thus, the full dynamics are given by

$$n_j(x,t+1) = \int_{-\infty}^{\infty} \begin{pmatrix} \text{Probability of} \\ \text{dispersing from} \\ y \text{ to } x \end{pmatrix} \begin{pmatrix} \text{number of} \\ \text{seeds produced} \\ \text{at } y \end{pmatrix} dy + \begin{pmatrix} \text{dormant} \\ \text{seeds} \end{pmatrix}$$

$$= \int_{-\infty}^{\infty} k_j(x-y)\frac{F_j(y,t)g_j n_j(y,t)}{C_j(y,t)}dy + s_j(1-g_j)n_j(x,t), \tag{10.2}$$

where

$$C_j(y,t) = \sum_k \gamma_{jk}\int_{-\infty}^{\infty} U_{jk}(z-y)g_k n_k(z,t)\, dz. \tag{10.3}$$

In this chapter I consider two annuals, one highly fecund but competitively inferior ($F_1 = 200$, $\gamma_{11} = \gamma_{21} = 0.9$), and one less fecund but competitively superior ($F_2 = 180$, $\gamma_{22} = \gamma_{12} = 1$). The seeds of both germinate at rate 0.07. I assume Laplacian dispersal kernels, with mean dispersal distance a_j: $k_j(z) = \exp(-|z|/a_j)/(2a_j)$. This is the dispersal kernel one would expect if seeds were passively dispersed by wind and fell out of the airstream with a constant probability per unit distance. Competition kernels take the same form, with characteristic competitive distance b_{jk}: $U_{jk}(z) = \exp(-|z|/b_{jk})/(2b_{jk})$. I consider multiple combinations of mean dispersal distance, competitive distance and seed survival probabilities, as discussed in section 10.4.

10.3 HOW TO CALCULATE THE LONG-RUN GROWTH RATE

Overview

I use the standard 'mutual invasibility' criterion for species coexistence. This states that species coexist if both species can reinvade the system if reduced to low density. The hypothetically reinvading species is termed the invader and its competitor, which is at the density it would achieve in the absence of the invader, is termed the resident. Determining whether two species will coexist thus requires us to calculate the invader's growth rate.

What is the appropriate growth rate? First, we presumably wish to consider coexistence over some region, and so we should be considering the invader's regional growth rate, defined as the ratio of this year's regional population to last year's regional population. Equivalently, we can define the regional growth rate as the ratio of the spatially averaged populations in successive years. Second, we wish to consider an invader's success over the long term. When growth fluctuates from year to year, a population's ultimate fate depends on the time average of the logarithm of the yearly growth rate (Lewontin and Cohen 1969). Thus, when growth varies in both time and space, coexistence depends on the invader's long-run regional growth rate: the time average of the logarithm of the regional growth rate.

To calculate the invader's long-run regional growth rate, I find an approximate expression which is valid when both the environmental variation and the resulting population variation are small. Invader growth depends on the resident and invader population distributions. These can be calculated in turn by breaking the environmental variation into components at single spatial and temporal scales, determining how the population responds to environmental variation at those scales, and reassembling. (The population responses are additive if the variation is small.) Let me expand on each step of this process.

The perturbative expression for the long-run growth rate is covered in Snyder (2008) and in the following section. Assuming that both the environmental variation and the resulting population variation are small relative to their means, we Taylor expand the expression for the local growth rate to second order in both environmental and population variation. This level of approximation allows us to keep some of the effect of nonlinearity while still permitting the Fourier analysis discussed below (i.e. the population responses remain additive). Having obtained a tractable expression for the local growth rate, we take the spatial average, then Taylor expand its logarithm to the same level of approximation and take the temporal average. This gives us an expression for the long-run regional growth rate. For those less concerned about the mathematics, the main point to remember is that this approach assumes that environmental variation and the resulting population variation are small. This approach is, therefore, more suitable for analysing the effects of ordinary year-to-year variation than for catastrophic disturbances.

The second-order approximation for long-run regional growth rate requires first-order approximations for the regional and invader population distributions. These we obtain using Fourier analysis. Fourier analysis expresses arbitrary patterns of environmental variation as sums of sinusoids at different spatial and temporal frequencies, much as a complex audio signal may be broken into sinusoidal signals at various frequencies for transmission along a cable. (The frequency of a sinusoid is proportional to 1 over its spatial or temporal period.) Once we have broken the environmental variation into its components, we can use a response function to determine the population response to each. We know that some spatial and temporal scales are biologically relevant while others are not: extremely small-scale or fast variation may produce no variation at the population level, while populations may be highly sensitive to variation at other scales, depending on species' traits. The response function tells us how much the population responds to variation at a given spatial and temporal scale. We obtain the population variation by multiplying each environmental variation component by the response function, evaluated at the appropriate spatial and temporal frequency. A large value of the response function means that the population responds strongly to variation at that spatial and temporal frequency while a small value means that the population is insensitive to variation at these scales. To extend the earlier metaphor, a response function is effectively a filter. Just as the electronic filters in a stereo can be used to amplify the bass or damp the treble in a complex audio signal, so the response function shows how population biology amplifies or damps environmental variation at different frequencies. Response functions are discussed in greater depth in Snyder (2007). Examples of how to use response functions to calculate the resident and invader population distributions can be found in Snyder (2006, 2007).

Using Fourier analysis also allows us to consider complex forms of environmental variation. For example, this chapter uses stochastic variation that is exponentially correlated in space and time (see section 10.2). This allows us to specify how predictable the environment is and how far a seed must travel in order to be likely to experience a novel environment.

How do we find the long-run growth rate for a correlated, stochastic environment? The second-order expression for the regional long-run growth rate is written in terms of variances and covariances involving population densities and the resident and invader environments. Using the Wiener–Khinchin theorem (e.g. Nisbet and Gurney 1982), we can take the Fourier transform of these variances and covariances and convert them into products of Fourier transforms of population densities and environments. We recall, however, that the Fourier transform of a population density can be expressed as a response function times the Fourier transform of the environment. At this point, the long-run growth rate is expressed in terms of products of Fourier transforms of the resident and invader environments. These can in turn be found from the correlation structure of the environment: the environmental fluctuations experienced by species j and k are adequately described by their spatiotemporal covariance, $\text{Cov}(F_j(x,t), F_k(x + x', t + t'))_{x,t}$, and the Wiener–Khinchin theorem relates variances and covariances to products of Fourier transforms. For this model, the environmental covariance function $\text{Cov}(F_j(x,t), F_k(x + x', t + t'))_{x,t}$ is written in terms of spatial and temporal correlation lengths, ξ and τ, (Equation (10.1)) and so in the end, the long-run invader growth rate can be written solely as a function of ξ and τ. Figure 10.1 provides a schematic guide and worked examples can be found in Snyder (2006, 2008).

Details

This section is for those who would like more details about how the long-term regional growth rate is calculated but who might find the original mathematical treatments in Snyder (2006, 2007, 2008) daunting. Let us begin by finding a mathematical definition for the long-run regional growth rate.

Long-term regional invader growth rate

The regional growth rate, $\tilde{\lambda}$, is the factor by which the spatially averaged population grows from one year to the next. That is, it is the ratio of next year's spatially averaged population to this year's spatially averaged population: $\tilde{\lambda}(t) = \langle n \rangle_x (t + 1)/\langle n \rangle_x (t)$, where $\langle \cdot \rangle_x$ denotes a spatial average. In a temporally varying environment, a population's fate depends on the time average of the logarithm of the growth rate, and so in a spatiotemporally varying environment, its fate depends on the time average of the logarithm of the regional growth rate: $\bar{r}_j = \langle \ln \tilde{\lambda}_j \rangle_t$, where $\bar{r}_j$ denotes the long-run growth rate of species j and $\langle \cdot \rangle_t$ represents the time average.

How do we find the regional growth rate, $\tilde{\lambda}_j$? Let us consider local growth. Before dispersal, the number of seeds at x next year is equal to the local growth rate times the number of seeds at x this year:

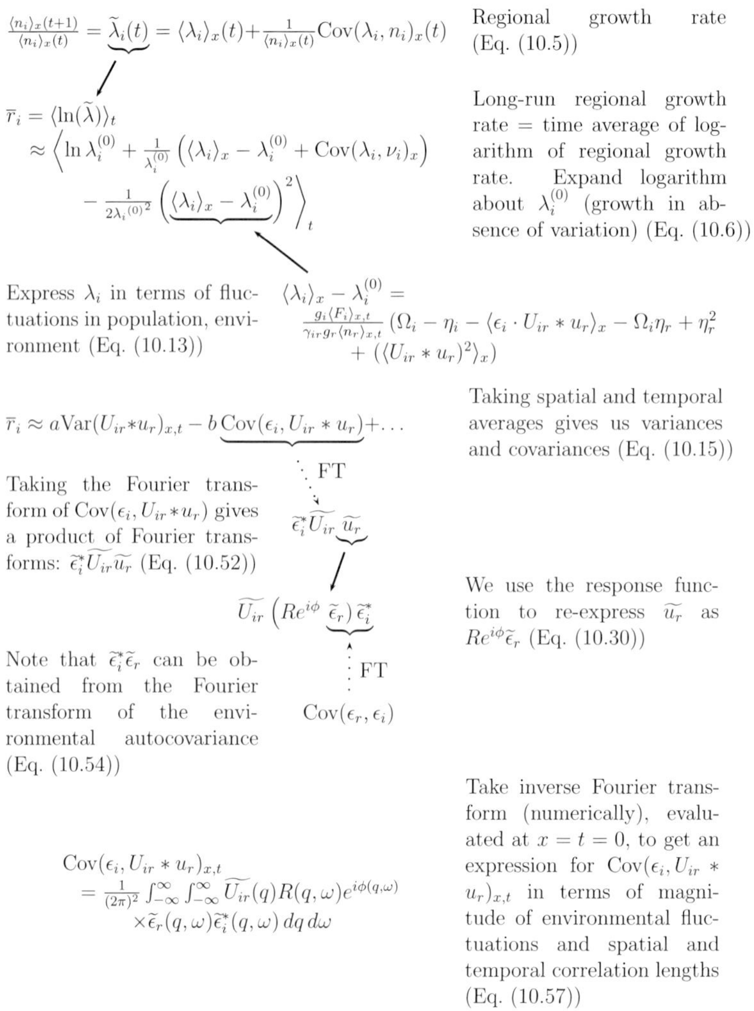

Figure 10.1 Flowchart for finding long-run invader growth from environmental autocovariance.

$n_j(x,t+1) = \lambda_j(x,t)n_j(x,t)$, where for this model, $\lambda_j = g_jF_j(x,t)/C_j(x,t) + s_j(1-g_j)$. Taking the spatial average of the equation for local growth, we get

$$\langle n_j\rangle_x(t+1) = \langle\lambda_j n_j\rangle_x(t). \tag{10.4}$$

We want an equation of the form $\langle n_j\rangle_x(t+1) = (\text{something})\langle n_j\rangle_x(t)$, where the 'something' will be identified as $\widetilde{\lambda}_j$. Remembering that $\mathrm{Cov}(a,b) = \langle ab\rangle - \langle a\rangle\langle b\rangle$, we rewrite the right-hand side of

Equation (10.4) as $\langle \lambda_j \rangle_x \langle n_j \rangle_x + \mathrm{Cov}(\lambda_j, n_j)_x$, where the subscript x is used with the covariance to remind us that this is a spatial covariance. We can bring out a factor of $\langle n_j \rangle_x$ to rewrite the covariance as $\langle n_j \rangle_x \mathrm{Cov}(\lambda_j, n_j / \langle n_j \rangle_x)_x$. Thus,

$$\langle n_j \rangle_x (t+1) = \underbrace{\left(\langle \lambda_j \rangle_x (t) + \mathrm{Cov}(\lambda_j, \nu_j)_x (t) \right)}_{\widetilde{\lambda}_j(t)} \langle n_j \rangle_x (t), \tag{10.5}$$

where $\nu_j(x,t) = n_j(x,t) / \langle n_j \rangle_x (t)$. We see that the regional growth rate, $\widetilde{\lambda}$, is equal to the spatially averaged growth rate, $\langle \lambda_j \rangle_x$, plus a term that gives extra weight to locations where the population is larger than average, $\mathrm{Cov}(\lambda_j, \nu_j)_x$.

At this point, $\bar{r}_j$ equals $\left\langle \ln\left(\langle \lambda_j \rangle_x + \mathrm{Cov}(\lambda_j, \nu_j)_x \right) \right\rangle_t$. To make any further progress, we must approximate. We Taylor expand the logarithm about $\lambda_j^{(0)}$ to obtain the second-order approximation

$$\bar{r}_j \approx \left\langle \ln \lambda_j^{(0)} + \frac{1}{\lambda_j^{(0)}} \left(\langle \lambda_j \rangle_x - \lambda_j^{(0)} + \mathrm{Cov}(\lambda_j, \nu_j)_x \right) - \frac{1}{2{\lambda_j^{(0)}}^2} \left(\langle \lambda_j \rangle_x - \lambda_j^{(0)} \right)^2 \right\rangle_t, \tag{10.6}$$

where each of the terms should be taken to $O(\sigma^2)$.[1] Note that $\mathrm{Cov}(\lambda_j, \nu_j)_{x,t}$ is $O(\sigma^2)$ and so the rest of the quadratic term, $\left(\mathrm{Cov}(\lambda_j, \nu_j)_{x,t} \right)^2 + 2\left(\langle \lambda_j \rangle_x - \lambda_j^{(0)} \right) \mathrm{Cov}(\lambda_j, \nu_j)_{x,t}$, is of too high an order to contribute.

At this point we need second-order expansions of $\left(\langle \lambda_j \rangle_x - \lambda_j^{(0)} \right)$ and $\left(\langle \lambda_j \rangle_x - \lambda_j^{(0)} \right)^2$, which means that we need to consider λ_j as a function of population (n_j) and the environment, which for this model is fecundity (F_j). Let

$$\langle n_j \rangle_x (t) = \langle n_j \rangle_{x,t} (1 + \eta_j(t)) \tag{10.7}$$

$$\langle F_j \rangle_x (t) = \langle F_j \rangle_{x,t} (1 + \Omega_j(t)), \tag{10.8}$$

where $\langle \eta_j \rangle_t = \langle \Omega_j \rangle_t = 0$, so that $\eta_j(t)$ and $\Omega_j(t)$ represent the temporal fluctuations of the spatially averaged population and environment

[1] The technical definition of $O(\sigma^n)$ is that if $g(x)$ is $O(\sigma^n)$, then $g(x)$ decreases with σ and $\left| \frac{g(x)}{\sigma^n} \right|$ can be made less than or equal to some positive constant K for σ small enough. On a more practical note, σ is some measure of smallness that we use for bookkeeping purposes. Taylor expansions approximate functions with power series. In order for an approximation based on Taylor expansions to be valid, we must keep all terms up to a given order. The expression for $\bar{r}_j$ is a second-order expansion and therefore we must keep all constant, $O(\sigma)$ and $O(\sigma^2)$ terms (but no higher).

relative to their spatiotemporal means. If $\eta_j(t)$ or $\Omega_j(t)$ is positive, then the spatially averaged population or environment is above the spatiotemporal average at time t, and if $\eta_j(t)$ or $\Omega_j(t)$ is negative, the spatial average is below the spatiotemporal average. Furthermore, let

$$n_j(x,t) = \langle n_j \rangle_x (1 + u_j(x,t)) \tag{10.9}$$

$$F_j(x,t) = \langle F_j \rangle_x (1 + \varepsilon_j(x,t)), \tag{10.10}$$

where $\langle u_j \rangle_x = \langle \varepsilon_j \rangle_x = 0$. Putting these together, we arrive at

$$\begin{aligned} n_j(x,t) &= \langle n_j \rangle_{x,t} (1 + u_j(x,t) + \eta_j(t) + u_j(x,t)\eta_j(t)) \\ F_j(x,t) &= \langle F_j \rangle_{x,t} (1 + \varepsilon_j(x,t) + \Omega_j(t) + \varepsilon_j(x,t)\Omega_j(t)). \end{aligned} \tag{10.11}$$

We can substitute these expressions for n_j and F_j into the expression for λ_j to arrive at

$$\lambda_j = \frac{g_j \langle F_j \rangle_{x,t} (1 + \varepsilon_j + \Omega_j + \varepsilon_j \Omega_j)}{\gamma_{jr} g_r \langle n_r \rangle_{x,t} (1 + \eta_r + U_{jr} * u_r + \eta_r (U_{jr} * u_r))} + s_j(1 - g_j). \tag{10.12}$$

Note that only the resident species contributes to competition: I make the usual assumption that the invader's density is too small to contribute much to the competition experienced by either species. Taylor expanding to second order in u, ε, η and Ω, which are all $O(\sigma)$, and taking the spatial average, we find that

$$\begin{aligned} \langle \lambda_j \rangle_x - \lambda_j^{(0)} = \frac{g_j \langle F_j \rangle_{x,t}}{\gamma_{jr} g_r \langle n_r \rangle_{x,t}} \Big(&\Omega_j - \eta_r - \langle \varepsilon_j \cdot U_{jr} * u_r \rangle_x - \Omega_j \eta_r + \eta_r^2 \\ &+ \langle (U_{jr} * u_r)^2 \rangle_x \Big) + O(\sigma^4), \end{aligned} \tag{10.13}$$

where $\lambda_j^{(0)} = \frac{g_j \langle F_j \rangle_{x,t}}{\gamma_{jr} g_r \langle n_r \rangle_{x,t}} + s_j(1 - g_j)$. This is where the variances and covariances come from. For example, $\mathrm{Cov}(\varepsilon_j, U_{jr} * u_r)_x = \langle \varepsilon_j \cdot U_{jr} * u_r \rangle_x - \langle \varepsilon_j \rangle_x \langle U_{jr} * u_r \rangle_x$, but the spatial averages of ε_j and $U_{jr} * u_r$ are zero. Thus,

$$\begin{aligned} \langle \lambda_j \rangle_x - \lambda_j^{(0)} = \frac{g_j \langle F_j \rangle_{x,t}}{\gamma_{jr} g_r \langle n_r \rangle_{x,t}} \Big(&\Omega_j - \eta_r - \mathrm{Cov}(\varepsilon_j, U_{jr} * u_r)_x - \Omega_j \eta_r + \eta_r^2 \\ &+ \mathrm{Var}(U_{jr} * u_r)_x \Big) + O(\sigma^4). \end{aligned} \tag{10.14}$$

Performing a similar expansion of $\left(\langle \lambda_j \rangle_x - \lambda_j^{(0)} \right)^2$ to $O(\sigma^2)$ and substituting back into Equation (10.6), we find

$$\bar{r}_j \approx \ln(\lambda_j^{(0)})$$
$$+\frac{1}{\lambda_j^{(0)}}\frac{g_j\langle F_j\rangle_{x,t}}{\gamma_{jr}g_r\langle n_r\rangle_{x,t}}\left[-\mathrm{Cov}(\varepsilon_j,U_{jr}*u_r)_{x,t}-\mathrm{Cov}(\Omega_j,\eta_r)_t+\mathrm{Var}(\eta_r)_t+\mathrm{Var}(U_{jr}*u_r)_{x,t}\right]$$
$$+\frac{1}{\lambda_j^{(0)}}\mathrm{Cov}(\lambda_j,\nu_j)_{x,t}-\frac{1}{2\lambda_j^{(0)2}}\left(\frac{g_j\langle F_j\rangle_{x,t}}{\gamma_{jr}g_r\langle n_r\rangle_{x,t}}\right)^2\left(\mathrm{Var}(\Omega_j)_t+\mathrm{Var}(\eta_r)_t-2\mathrm{Cov}(\Omega_j,\eta_r)_t\right). \tag{10.15}$$

Turning our attention to the $\mathrm{Cov}(\lambda_j,\nu_j)_{x,t}$ term, we note that further expansions yield

$$\mathrm{Cov}(\lambda_j,\nu_j)_{x,t}=\frac{g_j\langle F_j\rangle_{x,t}}{\gamma_{jr}g_r\langle n_r\rangle_{x,t}}\left(\mathrm{Cov}(\varepsilon_j,u_j)_{x,t}-\mathrm{Cov}(U_{jr}*u_r,u_j)_{x,t}\right) \tag{10.16}$$

to $O(\sigma^2)$. Observe that each covariance or variance term, including $\mathrm{Cov}(\lambda_j,\nu_j)_{x,t}$, is $O(\sigma^2)$, for each involves an average over a product of two of our $O(\sigma)$ perturbations: u_j, ε_j, η_j and Ω_j. We will see that each covariance and variance can be written in terms of environmental covariances (involving ε_j and Ω_j), and so each term is proportional to V, the variance of the environmental fluctuations.

Finally, we can calculate $\langle n_r\rangle_{x,t}$ by taking a spatiotemporal average of the local dynamics: $\langle n_r\rangle_{x,t}=\langle\lambda_r n_r\rangle_{x,t}$. All of the $O(\sigma)$ terms vanish, so that to $O(\sigma)$,

$$\langle n_r\rangle_{x,t}=\frac{F_r}{\gamma_{rr}(1-s_r(1-g_r))}+O(\sigma^2). \tag{10.17}$$

Equations (10.15) and (10.16) give us an expression for the invader's long-run regional growth rate $\bar{r}_i$ in terms of variances and covariances involving the resident and invader environments and population densities. That growth should depend on the distributions of residents and invaders makes intuitive sense: the invader should grow more rapidly if it is concentrated in good environments or in places where there is little competition from residents. This means, however, that in order to find $\bar{r}_i$, we need to calculate the resident and invader population distributions. We need only find $O(\sigma)$ approximations, since the covariance of two $O(\sigma)$ quantities is $O(\sigma^2)$, which is the level of approximation needed for $\bar{r}_i$.

Resident and invader population distributions

Let us begin by finding an expression for u_r. We can convert Equation (10.2) into an equation for $v_r(x,t)$ by dividing both sides by $\langle n_r \rangle_x (t+1)$ and substituting $\widetilde{\lambda}_r \langle n_r \rangle_x (t)$ for $\langle n_r \rangle_x (t+1)$ on the right-hand side. Thus,

$$v_r(t+1) = \frac{1}{\widetilde{\lambda}_r} \int_{-\infty}^{\infty} k_r(x-y) \frac{F_r(y,t) g_r v_r(y,t)}{C_r(y,t)} dy + \frac{1}{\widetilde{\lambda}_r} s_r(1-g_r) v_r(x,t). \tag{10.18}$$

To $O(\sigma)$, which, again, is all we need, $\widetilde{\lambda}_r$ can be replaced by $\lambda_r^{(0)}$. Substituting in our perturbative expressions for n_r and F_r (Equation (10.11)) and Taylor expanding to first order, we find that the purely temporal terms (Ω_r, η_j) cancel out and we are left with

$$u_r(x,t+1) = D_1(k_r * (u_r + \varepsilon_r - U_{rr} * u_r))(x,t) + D_2 u_r(x,t), \tag{10.19}$$

where

$$D_1 = \frac{\left(\dfrac{\langle F_r \rangle_{x,t}}{\gamma_{rr} \langle n_r \rangle_{x,t}} \right)}{\left(\dfrac{\langle F_r \rangle_{x,t}}{\gamma_{rr} \langle n_r \rangle_{x,t}} + s_r(1-g_r) \right)} \tag{10.20}$$

$$D_2 = \frac{s_r(1-g_r)}{\left(\dfrac{\langle F_r \rangle_{x,t}}{\gamma_{rr} \langle n_r \rangle_{x,t}} + s_r(1-g_r) \right)} \tag{10.21}$$

and where * represents a convolution: $(f * g)(x) = \int_{-\infty}^{\infty} f(x-y) g(y) dy$.

We would like to solve Equation (10.19) for u_r but the u_rs on the right-hand side are stuck in an integral. Equation (10.19) essentially says that the population at x is equal to a weighted average of the local seed production, with closer sites contributing more because they are the ones most likely to have seeds disperse to x. Here, Fourier analysis comes to the rescue. The Fourier transform of a convolution is the product of the Fourier transforms of the functions being convolved: $\widetilde{f*g} = \widetilde{f}\widetilde{g}$. Thus, taking the spatial Fourier transform,

$$\begin{aligned} \widetilde{u_r}(q,t+1) &= D_1 \widetilde{k_r}(q)(\widetilde{u_r}(q,t) + \widetilde{\varepsilon_r}(q,t) - \widetilde{U_{rr}}(q)\widetilde{u_r}(q,t)) + D_2 \widetilde{u_r}(q,t) \\ &= B_{u_r}(q)\widetilde{u_r}(q,t) + B_{\varepsilon_r}(q)\widetilde{\varepsilon_r}(q,t), \end{aligned} \tag{10.22}$$

where q is the spatial frequency and where

$$B_{u_r}(q) = D_1\widetilde{k_r}(q)(1 - \widetilde{U_{rr}}(q)) + D_2 \tag{10.23}$$

$$B_{\varepsilon_r}(q) = D_1\widetilde{k_r}(q). \tag{10.24}$$

The Fourier transform $\widetilde{u_r}(q,t)$ represents how much variation there is at frequency q (spatial period $2\pi/q$) at time t. By switching to the frequency domain, we can find an expression for $\widetilde{u_r}(q,t)$ and then take the inverse transform to regain $u_r(x,t)$.

Equation (10.22) expresses the spatial Fourier transform of this year's population in terms of the spatial Fourier transforms of last year's population and environment. Visually, we can represent this as a tree (Figure 10.2). $\widetilde{u_r}(q,t)$, which is at the root of the tree, depends on $\widetilde{\varepsilon_r}(q,t-1)$ and $\widetilde{u_r}(q,t-1)$, which in turn depend on $\widetilde{\varepsilon_r}(q,t-2)$ and $\widetilde{u_r}(q,t-2)$, and so forth. The contribution of the environment at time $t-k$ to $\widetilde{u_r}(q,t)$ is given by the path from $\widetilde{\varepsilon_r}(q,t-k)$ to $\widetilde{u_r}(q,t)$, where each link represents a multiplicative factor. For example, the path from $\widetilde{\varepsilon_r}(q,t-2)$ takes us through $B_{\varepsilon_r}(q)$, then $B_{u_r}(q)$, so that this path contributes $B_{\epsilon_r}(q)B_{u_r}(q)$. More generally, the path from $\widetilde{\varepsilon_r}(q,t-k)$ takes us through B_{ε_r} and then through $k - 1$ B_{u_r} links, for a contribution of $B_{u_r}^{k-1}B_{\varepsilon_r}$. Assuming that the transients have passed, so that we can ignore the path from $\widetilde{u_r}(q,0)$, we have

$$\widetilde{u_r}(q,t) = \sum_{j=0}^{t-1} B_{u_r}^{t-1-j}(q)B_{\varepsilon_r}(q)\widetilde{\varepsilon_r}(q,t=j). \tag{10.25}$$

We have successfully isolated $\widetilde{u_r}$ but writing it in terms of a sum is awkward. However, we can rewrite Equation (10.25) in the form of a discrete temporal convolution:

$$\widetilde{u_r}(q,t) = \sum_{j=0}^{\infty} \widetilde{M}(q,t-j)\widetilde{\varepsilon_r}(q,j), \tag{10.26}$$

where

$$\widetilde{M}(q,n) = \begin{cases} B_{u_r}^{n-1}(q)B_{\varepsilon_r}(q) & n > 0 \\ 0 & n \leq 0 \end{cases}, \tag{10.27}$$

which in turn allows us to get rid of the convolution by taking another Fourier transform (here a discrete temporal Fourier transform). Thus,

$$\widetilde{u_r}(q,\omega) = \widetilde{M}(q,\omega)\widetilde{\varepsilon_r}(q,\omega), \tag{10.28}$$

where ω is the temporal frequency and

$$\widetilde{M}(q,\omega) = \frac{B_{\varepsilon_r}(q)(e^{-i\omega} - B_{u_r}(q))}{1 + B_{u_r}^2(q) - 2B_{u_r}(q)\cos\omega}. \tag{10.29}$$

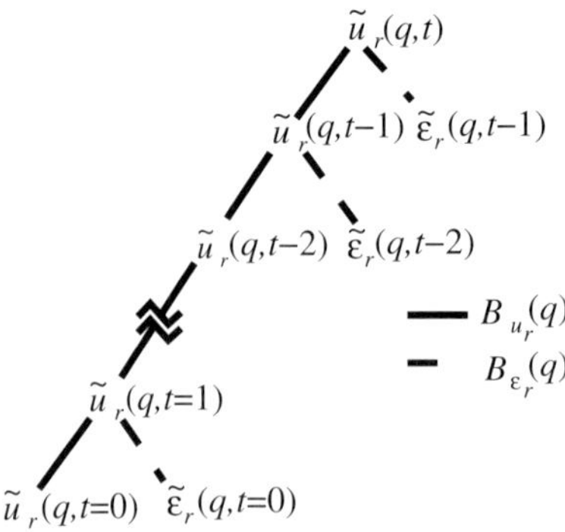

Figure 10.2 Schematic view of the dependence of this year's population (Fourier transformed in space) on the populations and environments of previous years. Solid lines denote multiplication by $B_{ur}(q)$ while dashed lines denote multiplication by $B_{\varepsilon_r}(q)$.

Here $\widetilde{M}(q,\omega)$ is a response function, also called a transfer function. As described in the overview, the response function damps or amplifies environmental variation ($\widetilde{\varepsilon_r}$) to yield the resulting population variation ($\widetilde{u_r}$). We can make our expression for $\widetilde{u_r}$ somewhat more intuitive (really!) if we express the response function in polar coordinates:

$$\widetilde{u_r}(q,\omega) = R(q,\omega)e^{i\phi(q,\omega)}\widetilde{\varepsilon_r}(q,\omega), \tag{10.30}$$

where

$$R(q,\omega) = [\mathrm{Re}(\widetilde{M}(q,\omega))^2 + \mathrm{Im}(\widetilde{M}(q,\omega))^2]^{1/2} = \frac{B_{\varepsilon_r}(q)}{\sqrt{1 + B_{u_r}^2(q) - 2B_{u_r}(q)\cos\omega}} \tag{10.31}$$

$$\phi(q,\omega) = \tan^{-1}\left(\frac{\mathrm{Im}(\widetilde{M}(q,\omega))}{\mathrm{Re}(\widetilde{M}(q,\omega))}\right) = \tan^{-1}\left(\frac{-\sin\omega}{\cos\omega - B_{u_r}(q)}\right) \tag{10.32}$$

and $\tan^{-1}$ ranges from $-\pi$ to π.

The modulus $R(q,\omega)$ specifies the amplitude of the population's response to variation at spatial frequency q and temporal frequency ω. As q and ω become large, $R(q,\omega)$ normally becomes small, reflecting the fact that populations cannot respond to very rapid or very small-scale environmental variation. The phase angle $\phi(q,\omega)$ tells us how long the delay is between variation in the environment and the corresponding variation in population. If ϕ is zero, the population is perfectly in sync with the environment (no delay). If $\phi = -\pi$, the population is precisely out of phase ('180° out of phase') with the environment.

Now that we have found an expression for u_r, let us complete our description of the resident population distribution by finding an expression for η_r. $\eta_r(t)$ represents the scaled fluctuations in the spatially averaged population, $\langle n_r \rangle_x(t)$, and so we begin by finding an equation for the dynamics of $\langle n_r \rangle_x$, good to $O(\sigma)$. Substituting $\langle n_r \rangle_{x,t}(1 + u_r(x,t) + \eta_r(t))$ for $n_r(x,t)$ and $\langle F_r \rangle_{x,t}(1 + \varepsilon_r(x,t) + \Omega_r(t))$ for $F_r(x,t)$ in Equations (10.2) and (10.3) and Taylor expanding to first order, we find

$$\begin{aligned} &\langle n_r \rangle_{x,t}(1 + u_r(x,t+1) + \eta_r(t+1)) = \\ &\frac{g_r \langle F_r \rangle_{x,t} \langle n_r \rangle_{x,t}}{\gamma_{rr} g_r \langle n_r \rangle_{x,t}} [1 + \varepsilon_r(x,t) + \Omega_r(t) + u_r(x,t) + \eta_r(t) - (U_{rr} * u_r)(x,t) - \eta_r(t)] \\ &\qquad + s_r(1 - g_r)\langle n_r \rangle_{x,t}[1 + u_r(x,t) + \eta_r(t)]. \end{aligned} \tag{10.33}$$

When we take the spatial average, all of the ε_r and u_r terms vanish (because $\langle \varepsilon_r \rangle_x = \langle u_r \rangle_x = 0$) and so

$$\eta_r(t+1) = B_\eta \eta_r(t) + B_\Omega \Omega_r(t), \tag{10.34}$$

where

$$B_\eta = s_r(1 - g_r) \tag{10.35}$$

$$B_\Omega = \frac{\langle F_r \rangle_{x,t}}{\gamma_{rr} \langle n_r \rangle_{x,t}}. \tag{10.36}$$

Once again, this year's population fluctuation depends on last year's population fluctuation and environmental fluctuation, and once again, we can write this as a convolution in time and use a temporal Fourier transform. We find that

$$\tilde{\eta}_r(\omega) = H(\omega) e^{i\chi(\omega)}, \tag{10.37}$$

where

$$H(\omega) = \frac{B_\Omega}{\sqrt{1 + B_\eta^2 - 2B_\eta \cos\omega}} \tag{10.38}$$

$$\chi(\omega) = \tan^{-1}\left(\frac{-\sin\omega}{\cos\omega - B_\eta}\right). \tag{10.39}$$

Now that we have an expression for the resident population distribution, we can find the invader population distribution. We begin with the invader equivalent of Equation (10.18) (substitute i for r). Remembering that the competition experienced by the invader depends only on the resident population distribution, we again Taylor expand to first order and arrive at

$$u_i(x,t+1) = d_1 k_i * (u_i + \varepsilon_i - U_{ir} * u_r)(x,t) + d_2 u_i(x,t), \tag{10.40}$$

where

$$d_1 = \frac{\left(\frac{g_i \langle F_i \rangle_{x,t}}{\gamma_{ir} g_r \langle n_r \rangle_{x,t}}\right)}{\left(\frac{g_i \langle F_i \rangle_{x,t}}{\gamma_{ir} g_r \langle n_r \rangle_{x,t}} + s_i(1-g_i)\right)} \tag{10.41}$$

$$d_2 = \frac{s_i(1-g_i)}{\left(\frac{g_i \langle F_i \rangle_{x,t}}{\gamma_{ir} g_r \langle n_r \rangle_{x,t}} + s_i(1-g_i)\right)}. \tag{10.42}$$

Once again, we take the spatial Fourier transform, and this time, when we express $\widetilde{u_i}(q,t)$ as a tree, we find that $\widetilde{u_i}(q,t)$ depends not only on all of the previous invader environments ($\widetilde{\varepsilon_i}$) and populations ($\widetilde{u_i}$) but also on all of the resident distributions ($\widetilde{u_r}$). Thus,

$$\begin{aligned}\widetilde{u_i}(q,t) &= \sum_{j=0}^{t-1} b_{u_i}^{t-1-j}(q)\left[b_{\varepsilon_i}(q)\widetilde{\varepsilon_i}(q,t=j) + b_{u_r}(q)\widetilde{u_r}(q,t=j)\right],\\ &= \sum_{j=0}^{t-1} b_{u_i}^{t-1-j}(q) b_{\varepsilon_i}(q)\left[\widetilde{\varepsilon_i}(q,t=j) + \frac{b_{u_r}(q)}{b_{\varepsilon_i}(q)}\widetilde{u_r}(q,t=j)\right],\end{aligned} \tag{10.43}$$

where

$$b_{u_i}(q) = d_1 \widetilde{k_i}(q) + d_2 \tag{10.44}$$

$$b_{\varepsilon_i}(q) = d_1 \widetilde{k_i}(q) \tag{10.45}$$

$$b_{u_r}(q) = -d_1 \widetilde{k_i}(q) \widetilde{U_{ir}}(q). \tag{10.46}$$

As before, we express this as a temporal convolution and take the temporal Fourier transform, giving us

$$\widetilde{u_i}(q,\omega) = G(q,\omega) e^{i\psi(q,\omega)}\left[\widetilde{\varepsilon_i}(q,\omega) + \frac{b_{u_r}(q)}{b_{\varepsilon_i}(q)}\widetilde{u_r}(q,\omega)\right], \tag{10.47}$$

where

$$G(q,\omega) = \frac{B_{\varepsilon_i}(q)}{\sqrt{1 + B_{u_i}^2(q) - 2B_{u_i}(q)\cos\omega}} \tag{10.48}$$

$$\psi(q,\omega) = \tan^{-1}\left(\frac{-\sin\omega}{\cos\omega - B_{u_i}(q)}\right). \tag{10.49}$$

But of course we have an expression for the resident population distribution, $\widetilde{u_r}$. Substituting $R(q,\omega)\exp(i\phi(q,\omega))\widetilde{\varepsilon}_r(q,\omega)$ for $\widetilde{u_r}(q,\omega)$, we arrive at our final expression for the invader distribution:

$$\widetilde{u_i}(q,\omega) = G(q,\omega)e^{i\psi(q,\omega)}\left[\widetilde{\varepsilon}_i(q,\omega) - \widetilde{U_{ir}}(q)R(q,\omega)e^{i\phi(q,\omega)}\widetilde{\varepsilon}_r(q,\omega)\right], \tag{10.50}$$

where we have replaced $b_{u_r}(q)/b_{\epsilon_i}(q)$ with $-\widetilde{U_{ir}}(q)$.

We are now finished. The invader population enters the expression for $\bar{r}_i$ only through $\mathrm{Cov}(\lambda_i, v_i)_{x,t} = \mathrm{Cov}(\lambda_i, u_i)_{x,t}$. Thus, we need not find $\eta_i(t)$. (This is fortunate since the invader population is normally growing or shrinking and so $\langle n_i \rangle_x$ cannot be expressed in terms of small fluctuations about a constant.)

From environmental autocovariance to long-run invader growth

We have derived an expression for the long-run regional growth rate in terms of variances and covariances of the fluctuations in environments and populations (Equation (10.15)) and have derived expressions for the population fluctuations in terms of the environmental fluctuations (Equations (10.30), (10.37), (10.50)). How do we put them together to get from environmental covariance functions to a growth rate? For an example, let us consider the first term in the equation for $\bar{r}_j$.

The first term in Equation (10.15) is proportional to $\mathrm{Cov}(\varepsilon_j, U_{jr} * u_r)_{x,t}$. Our strategy will be to take the Fourier transform of $\mathrm{Cov}(\varepsilon_j, U_{jr} * u_r)_{x,t}$, re-express this in terms of the environmental covariance functions, and then take the inverse Fourier transform to get back to the original covariance. Note that although $\bar{r}_j$ depends only on covariances at zero lag, a spatiotemporal covariance is in general a function of spatial and temporal lag $(\mathrm{Cov}(f,g)_{x,t}(x',t') = \langle (f(x+x',t+t') - \langle f \rangle_{x,t})(g(x,t) \quad \langle g \rangle_{x,t}) \rangle_{x,t})$, and so it is possible to find its Fourier transform. The Wiener–Khinchin theorem states that the Fourier transform of $\mathrm{Cov}(f,g)_{x,t}$ is given by

$$\mathcal{F}[\mathrm{Cov}(f,g)_{x,t}] = \lim_{N\to\infty} \frac{\widetilde{f}^{*(N)}(q,\omega)\widetilde{g}^{(N)}(q,\omega)}{N^2}, \tag{10.51}$$

where superscript * denotes the complex conjugate and $\widetilde{g}^{(N)}(q,\omega)$ equals the Fourier transform of $g(x,t)$ in the limit as N approaches infinity: $\widetilde{g}^{(N)}(q,\omega) = \sum_{x,t=-N/2}^{N/2} g(x,t)\exp(-i(qx+\omega t))$. For our covariance,

$f(x,t) = \varepsilon_j(x,t)$ and $g(x,t) = (U_{jr} * u_r)(x,t)$. Recall that Fourier transforms turn convolutions into products, so the Fourier transform of $U_{jr} * u_r$ is $\widetilde{U_{jr}}(q)\widetilde{u_r}(q,\omega)$. The Fourier transform of $\mathrm{Cov}(\varepsilon_j, U_{jr} * u_r)_{x,t}$ is thus

$$\mathfrak{F}[\mathrm{Cov}(\varepsilon_j, U_{jr} * u_r)_{x,t}] = \lim_{N\to\infty} \frac{\widetilde{\varepsilon_j}^{*(N)}(q,\omega)\widetilde{U_{jr}}^{(N)}(q)\widetilde{u_r}^{(N)}(q,\omega)}{N^2}. \tag{10.52}$$

We can put this expression entirely in terms of environmental fluctuations by substituting $R(q,\omega)e^{i\phi(q,\omega)}\widetilde{\varepsilon_r}^{(N)}(q,\omega)$ for $\widetilde{u_r}^{(N)}(q,\omega)$, so that

$$\mathfrak{F}[\mathrm{Cov}(\varepsilon_j, U_{jr} * u_r)_{x,t}] = \widetilde{U_{jr}}(q)R(q,\omega)e^{i\phi(q,\omega)} \lim_{N\to\infty} \frac{\widetilde{\varepsilon_j}^{*(N)}(q,\omega)\widetilde{\varepsilon_r}^{(N)}(q,\omega)}{N^2}. \tag{10.53}$$

At this point we are nearly able to calculate $\mathrm{Cov}(\varepsilon_j, U_{jr} * u_r)_{x,t}$: we know $U_{jr}(q)$, $R(q,\omega)$ and $\phi(q,\omega)$ and lack only an expression for $\lim_{N\to\infty}\widetilde{\varepsilon_j}^{*(N)}(q,\omega)\widetilde{\varepsilon_r}^{(N)}(q,\omega)/N^2$. Using the Wiener–Khinchin theorem again, we recognise $\lim_{N\to\infty} \widetilde{\varepsilon_j}^{*(N)}(q,\omega)\widetilde{\varepsilon_r}^{(N)}(q,\omega)/N^2$ as the Fourier transform of $\mathrm{Cov}(\varepsilon_j, \varepsilon_r)_{x,t}$:

$$\begin{aligned}
\lim_{N\to\infty} \frac{\widetilde{\varepsilon_j}^{*(N)}(q,\omega)\widetilde{\varepsilon_r}^{(N)}(q,\omega)}{N^2} &= \sum_{t'=-\infty}^{\infty}\int_{-\infty}^{\infty} \mathrm{Cov}(\varepsilon_j,\varepsilon_r)_{x,t}(x',t')e^{-i(qx'+\omega t')}dx' = \\
\cos(\theta)V \sum_{t'=-\infty}^{\infty}\int_{-\infty}^{\infty} &\exp(-\frac{|x'|}{\xi})\exp(-\frac{|t'|}{\tau})e^{-i(qx'+\omega t')}dx' \\
= \cos(\theta)V\left(\frac{2\xi}{\xi^2+q^2}\right)&\left(\frac{1-e^{-2/\tau}}{1-2e^{-1/\tau}\cos(\omega)+e^{-2/\tau}}\right),
\end{aligned} \tag{10.54}$$

assuming that space is continuous and time is discrete. We have thus expressed the Fourier transform of $\mathrm{Cov}(\varepsilon_j, U_{jr} * u_r)_{x,t}$ in terms of an environmental covariance function, $\mathrm{Cov}(\varepsilon_j, \varepsilon_r)$. All that remains is to take the inverse Fourier transform :[2]

[2] For a domain size of N, the inverse discrete Fourier transform of $\widetilde{f}(q)$ is given by

$$\frac{1}{N}\sum_{x=-N/2}^{N/2-1} \widetilde{f}(q)e^{iqx}, \tag{10.55}$$

but in the limit as N approaches infinity, this can be rewritten as

$$\frac{1}{2\pi}\int_{-\pi}^{\pi} \widetilde{f}(q)e^{iqx}dq. \tag{10.56}$$

$$\mathrm{Cov}(\varepsilon_j, U_{jr} * u_r)_{x,t} = \frac{1}{(2\pi)^2}\int_{-\pi}^{\pi}\int_{-\infty}^{\infty} \widetilde{U_{jr}}(q)R(q,\omega)e^{i\phi(q,\omega)}S_{\varepsilon_j\varepsilon_r}(q,\omega)\,dq\,d\omega, \tag{10.57}$$

where $S_{\varepsilon_j\varepsilon_r}(q,\omega) = \lim_{N\to\infty}\widetilde{\varepsilon_j}^{*(N)}(q,\omega)\widetilde{\varepsilon_r}^{(N)}(q,\omega)/N^2$ (S for 'spectral density'). This integral will normally need to be performed numerically.

We can use the same technique to evaluate the variances in the equation for long-run growth. Consider $\mathrm{Var}(\eta_r)_t$. This can be rewritten as $\mathrm{Cov}(\eta_r, \eta_r)_t$, with Fourier transform $\lim_{N\to\infty}\widetilde{\eta_r}^{*(N)}(\omega)\widetilde{\eta_r}^{(N)}(\omega)/N$.

We use response functions to rewrite this as $H(\omega)e^{-i\chi(\omega)}H(\omega)e^{i\chi(\omega)}\lim_{N\to\infty}\widetilde{\Omega_r}^{*(N)}(\omega)\widetilde{\Omega_r}^{(N)}(\omega)/N$. We note that $\lim_{N\to\infty}\widetilde{\Omega_r}^{*(N)}(\omega)\widetilde{\Omega_r}^{(N)}(\omega)/N$ is the Fourier transform of $\mathrm{Cov}(\omega_r, \omega_r)_t$:

$$\begin{aligned}\lim_{N\to\infty}\frac{\widetilde{\Omega_r}^{*(N)}(\omega)\widetilde{\Omega_r}^{(N)}(\omega)}{N} &= \sum_{t'=-\infty}^{\infty}\mathrm{Cov}(\omega_r,\omega_r)_t(t')e^{-i\omega t'} \\ &= \sum_{t'=-\infty}^{\infty} Ve^{-|t'|/\tau}e^{-i\omega t'} = V\frac{1-e^{-2/\tau}}{1-2e^{-1/\tau}\cos(\omega)+e^{-2/\tau}} \equiv S_{\Omega_r\Omega_r}(\omega).\end{aligned} \tag{10.58}$$

Thus,

$$\mathrm{Var}(\eta_r)_t = \frac{1}{2\pi}\int_{-\pi}^{\pi} H^2(\omega)S_{\Omega_r\Omega_r}(\omega)d\omega. \tag{10.59}$$

Again, this integral will normally need to be taken numerically.

We can follow the same procedure with all of the other terms in the expression for $\bar{r}_i$, so that long-run invader growth becomes a function of the spatial and temporal correlation lengths (ξ and τ).

10.4 A NUMERICAL EXPERIMENT

In order to test the effectiveness of spatiotemporal and temporal variation in promoting coexistence, I performed a large factorial experiment. For each species, seed survival rate was allowed to be either low (0.1) or high (0.9), representing mean seed lifetimes of 0.12 years and 9 years. Mean dispersal distance was allowed to be short (0.5), medium (2), or long (10). I take the unit of space to be the radius of an average plant, so that with the Laplacian kernels used here, a dispersal distance of 0.5 means that 86% of seeds land under their parent, while with a distance of 2, 39% land under their parent and with a distance of 10, only 10% land under their parent. Alternatively, we can say that dispersal distances of 0.5, 2 and 10 represent dispersal within one correlation length (almost all offspring experience similar conditions), between one or two correlation lengths (some offspring experience similar conditions and others do not), or essentially global (offspring experience random conditions). Competition distances were also allowed to be short, medium or long. Species preferred either identical environments (both find the same locations/times favourable) or opposite environments (the locations/times most favourable to one are least favourable to the other). Note, however, that even if both species have the same parameters and prefer the same environments, they are still distinguishable, as one is a superior competitor while the other has higher fecundity.

I also consider the case where between- and within-species competition distances do not vary independently. It is possible for the two to be independent if there are multiple sources of competition: for example, if a species-specific pathogen or pest produces long-range competition within a species while resource limitation produces short-range competition between species. However, if plants are simply competing for resources, then the spatial scale of competition is determined by the distances over which they can gather those resources, which is in turn governed by the spatial extent of their above- and below-ground growth. In this case, the scales of both within- and between-species competition are determined by the identities of the species involved. Thus, in addition to unconstrained competition scales, I also consider the case where the between-species competition distance is the average of the within-species competition distances for the two species involved. I refer to this as constrained competition.

Fecundity is exponentially correlated in space and time, so that for spatiotemporal variation, $\mathrm{Cov}(F(x,t),\ F(x+x',t+t'))_{x,t} = V\exp(-x'/\xi)$ $\exp(-t'/\tau)$, while for temporal variation, $\mathrm{Cov}(F(t),F(t+t'))_t = V\exp(-t'/\tau)$, where V is the variance, ξ is the spatial correlation length and τ is the temporal correlation length. Note that both forms of variation have the same variance. The degree to which environmental variation can affect coexistence depends on how much variation is present, i.e. on the variance. Setting the variances equal thus puts spatiotemporal and temporal variation on an equal footing.

Variation promotes coexistence by increasing an invader's long-run regional growth rate, $\bar{r}_i$. While it is possible to get coexistence by increasing $\bar{r}_i$ for the inferior competitor and decreasing $\bar{r}_i$ for the superior competitor, so that $\bar{r}_i > 0$ for both, this is a rather fragile sort of coexistence. Rather, let us focus on situations where variation increases $\bar{r}_i$ for both species as invader. I measure the ability of variation to promote coexistence as the geometric mean of the increases in long-run

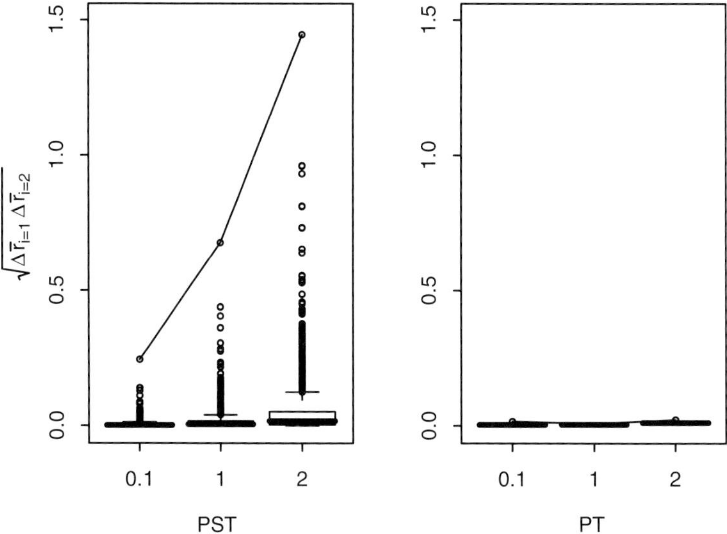

Figure 10.3 Boxplots of the degree to which variation promotes coexistence, as measured by the geometric mean of the increases in long-run growth for both species as invader $\left(\sqrt{\Delta\bar{r}_{i=1}\Delta\bar{r}_{i=2}},\ \Delta\bar{r}_i = (\bar{r}_i \text{ with variation}) - (\bar{r}_i \text{ without variation}),\right.$ $\left.\text{for } \Delta r_{i=1} \text{ and } \Delta r_{i=2} > 0 \text{ only}\right)$. For both spatiotemporal and temporal variation, boxplots are given for spatial and/or temporal correlation lengths of 0.1, 1 and 2. Each boxplot summarises the results of a numerical experiment measuring long-run invader growth $\overline{r_i}$ for a variety of parameter combinations, as described in section 10.4. The result for each parameter combination is calculated using $\overline{r_i}$s averaged over 50 realisations. As an aid to the eye, lines connect the maximum values of $\sqrt{\Delta\bar{r}_{i=1}\Delta\bar{r}_{i=2}}$ for each correlation length. Because the approximation presented in 'Details' in section 10.3 overestimates growth for very short dispersal distances, $\overline{r_i}$ has been calculated using simulations of the full nonlinear model (Equation (10.2)).

invader growth rate: $\sqrt{\Delta\bar{r}_{i=1}\Delta\bar{r}_{i=2}}$, where $\Delta\bar{r}_i = (\bar{r}_i$ with variation$)-$ $(\bar{r}_i$ without variation$)$ and where $\sqrt{\Delta\bar{r}_{i=1}\Delta\bar{r}_{i=2}}$ is defined when $\Delta\bar{r}_{i=1}$ and $\Delta\bar{r}_{i=2}$ are both positive.

The mathematical approach presented in the previous section is useful for developing an intuitive understanding of how environmental variation can promote coexistence and is vital for evolutionary studies, where $\bar{r}_i$ becomes invader fitness (e.g. Snyder 2006). However, it tends to overestimate long-run invader growth for very short dispersal distances. I have therefore checked all results by simulating the full model

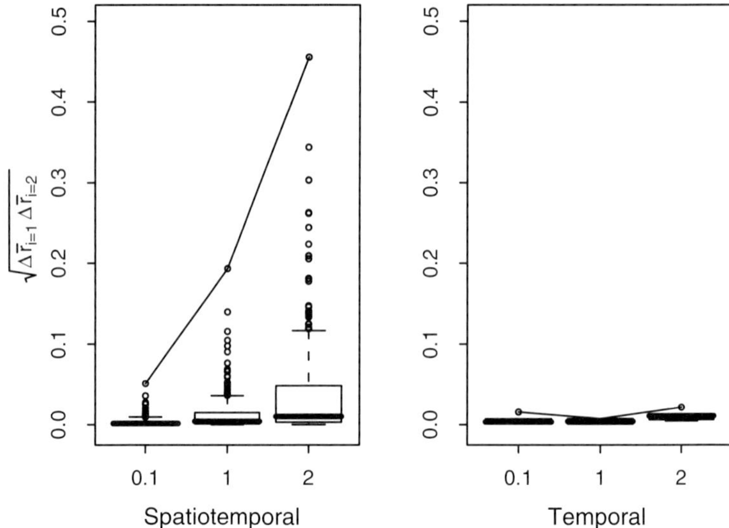

Figure 10.4 Boxplots of the degree to which variation promotes coexistence $\sqrt{\Delta\bar{r}_{i=1}\Delta\bar{r}_{i=2}}$ with constrained competition distances. The between-species competition distance is constrained to equal the average of the within-species competition distances of the two species involved. All else is as in Figure 10.3.

equations, without approximations, and the results presented in Figures 10.3–10.5 come from the numerical simulations.

10.5 RESULTS

Spatiotemporal variation has a higher capacity to promote coexistence than temporal variation (i.e. $\sqrt{\Delta\bar{r}_{i=1}\Delta\bar{r}_{i=2}}$ reaches larger values) for all spatial and temporal correlation lengths, regardless of whether the species prefer similar or different habitats, regardless of whether competition is constrained or unconstrained (Figures 10.3 and 10.4). Before we ask why spatiotemporal variation can promote coexistence so much more effectively, let us first ask what traits maximise the potential for coexistence for both spatiotemporal and temporal variation.

When competition is unconstrained, the same set of traits is responsible for spatiotemporal variation's success in promoting coexistence in all circumstances, a set of traits that I have referred to as 'tracking' traits (Snyder 2008). Tracking traits consist of short-range between-species competition, long-range within-species competition,

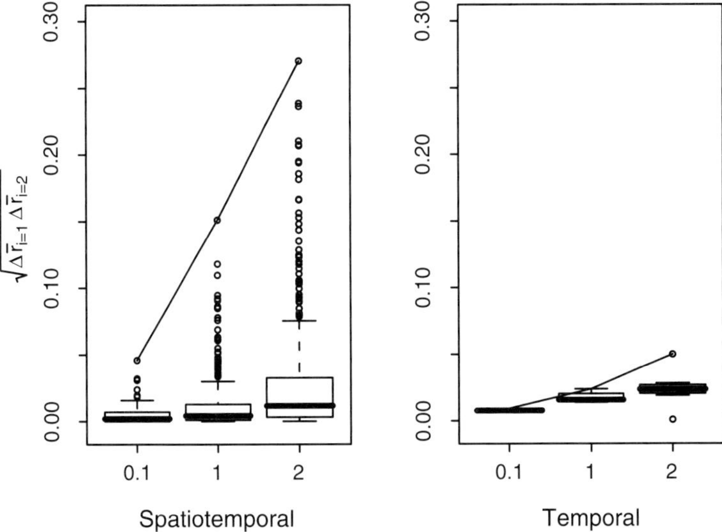

Figure 10.5 Boxplots of the degree to which variation promotes coexistence $\sqrt{\Delta\bar{r}_{i=1}\Delta\bar{r}_{i=2}}$ with invader competition included and constrained competition distances. Here we include the competition imposed by the invader (normally assumed to be zero), where a species is considered an invader as long as its spatially averaged density is under 10% of the resident's spatiotemporally averaged density. As in Figure 10.4, the between-species competition distance is constrained to equal the average of the within-species competition distances of the two species involved. All else is as in Figure 10.3.

short-lived seeds, and short-range dispersal – the shorter the better. Short-range dispersal causes both the resident and invader populations to become highly aggregated: seeds land near their parents and so the population accumulates in those areas where parents produce the most offspring. For the resident, these areas will be the places where the environment is most favourable to it. The invader, on the other hand, will become aggregated in the open space between resident clusters. As described in Roughgarden (1974) and Gurney and Nisbet (1976), long-range within-species competition allows tightly clustered residents to experience an average of crowded (within-cluster) and uncrowded (out-of-cluster) conditions, relieving competitive pressure and allowing the residents to become ever more tightly aggregated. Short-range between-species competition ensures that invaders will not be unduly affected by nearby clusters of residents. Finally, short-lived seeds mean that the system has a short 'memory' and can respond quickly to changing conditions. Because germination is not predictive – a constant fraction of seeds germinate in every location every year – the current population of adults reflects seed production patterns over the average seed lifetime.

If competition is constrained and the two species prefer different environments (i.e. $\theta = \pi$ in Equation (10.1)), the traits that most promote coexistence form a modified tracking strategy: dispersal remains short range and seeds remain short lived but now all competition is short range. If the two species prefer the same environment, then the traits which most promote coexistence depend on the spatial and temporal correlation length, and they promote coexistence only weakly – about as strongly as temporal variation promotes coexistence when species prefer different environments.

Intermediate levels of seed survival (and hence intermediate levels of dormancy) allow temporal variation to best promote coexistence when species prefer different environments. When they prefer the same environments, they cannot coexist.

Why is temporal variation so much less effective at promoting coexistence? Are populations unable to become aggregated in favourable times in the same way that they become aggregated in favourable places? This isn't quite it. What makes spatiotemporal (and spatial) variation so powerful is that it causes the invader's long-run growth rate to depend on the invader's aggregation as well as the resident's aggregation, and the invader, having a small population, experiences little self-limitation and thus is able to become very aggregated indeed.[3] Invader growth in a temporally variable environment, on the other hand, depends only on resident aggregation, which is limited by resident self-competition. The reason that invader growth depends on invader aggregation in a spatially heterogeneous environment is that the regional growth rate, $\widetilde{\lambda}_i$, is equal to the spatially averaged growth rate, $\langle \lambda_i \rangle_x$, plus a term that gives extra weight to locations with higher than average densities of invaders, $\mathrm{Cov}(\lambda_i, \nu_i)_x$ (Equation (10.5)). The long-run growth rate is then the time average of the logarithm of the regional growth rate. If the environment varies only in time, however, there are no locally dense locations to give extra weight to. The long-run growth rate is just the time average of the logarithm of the instantaneous growth rate, which is the same everywhere.

[3] Alert readers will note that this represents a change from my understanding in Snyder (2008).

In the maths presented here and in the simulations used to generate Figures 10.3 and 10.4, we make the traditional assumption that invaders do not compete with themselves at all. We therefore overestimate the invader's ability to aggregate and overestimate the invader long-run growth rate under spatiotemporal variation. It is reasonable to ask whether the relative success of spatiotemporal variation in promoting coexistence is due to this simplification. I have re-run the simulations including competition from invaders, defining a species as an invader as long as its spatially averaged population is under 10% of the resident's spatiotemporally averaged population. I find that while including invader competition does reduce the degree to which spatiotemporal variation is able to increase invader long-run growth rates, spatiotemporal variation still promotes coexistence more strongly than temporal variation, even when competition is constrained (Figure 10.5).

Differences in whether invader aggregation contributes to coexistence explain most but not all of the differences in the ability of spatiotemporal and temporal variation to promote coexistence. In a temporally varying environment, population aggregation contributes to invader growth via $\frac{g_i \langle F_i \rangle_{x,t}}{\gamma_{ir} g_r \langle n_r \rangle_{x,t}} \mathrm{Cov}(\Omega_i, \eta_r)_t$ – this is the covariance between environment and competition that is at the heart of Chesson's temporal storage effect (Chesson 1994) – while in a spatiotemporally varying environment, population aggregation contributes via $\mathrm{Cov}(\lambda_i, \nu_i)_{x,t}$ and, to a lesser extent, via $\frac{g_i \langle F_i \rangle_{x,t}}{\gamma_{ir} g_r \langle n_r \rangle_{x,t}} \mathrm{Cov}(\varepsilon_i, U_{ir} * u_r)_{x,t}$ – the heart of Chesson's growth–density covariance and spatial storage effect (Chesson 2000). If we consider sinusoidal variation at a single spatial and temporal frequency, it is possible to find situations where temporal aggregation is stronger than spatiotemporal aggregation – $-\frac{g_i \langle F_i \rangle_{x,t}}{\gamma_{ir} g_r \langle n_r \rangle_{x,t}} \mathrm{Cov}(\Omega_i, \eta_r)_t > \mathrm{Cov}(\lambda_i, \nu_i)_{x,t}$ – and yet temporal aggregation still has less of an effect on invader growth than spatiotemporal aggregation. The reason for this has to do with the fact that reproduction is pulsed.

While temporal variation can increase long-run invader growth (e.g. if residents and invaders prefer different environmental conditions), it simultaneously depresses long-run growth if reproduction is pulsed (Lewontin and Cohen 1969). Growth in discrete time is a multiplicative process in which low-growth years hurt more than high-growth years help. Mathematically, we see this in Equation (10.15): the terms that arise from temporal variation, $\mathrm{Cov}(\Omega_i, \eta_r)_t$ and $\mathrm{Var}(\Omega_i)_t$, appear two times with opposing signs. The contribution of, for example, $\frac{-1}{\lambda_j^{(0)}} \frac{g_i \langle F_i \rangle_{x,t}}{\gamma_{ir} g_r \langle n_r \rangle_{x,t}} \mathrm{Cov}(\Omega_i, \eta_r)_t$ is partially cancelled out by $\frac{1}{2\lambda_j^{(0)^2}} \left(\frac{g_i \langle F_i \rangle_{x,t}}{\gamma_{ir} g_r \langle n_r \rangle_{x,t}} \right)^2 \mathrm{Cov}(\Omega_i, \eta_r)_t$. The second term arises from the expansion of the logarithm in the equation $\bar{r}_i = \langle \ln \widetilde{\lambda}_i \rangle_t$.

Spatiotemporal variation, on the other hand, produces little temporal variation in the regional growth rate $\widetilde{\lambda}_i$. The regional growth rate is $\mathrm{Cov}(\lambda_i, \nu_i)_x$ plus the spatial average of the local growth rate, $\langle \lambda_i(x,t) \rangle_x$, and $\langle \lambda_i(x,t) \rangle_x$ consists of terms like $\mathrm{Cov}(\varepsilon, U_{ir} * u_r)_x$ and $\mathrm{Var}(U_{ir} * u_r)_x$. None of these spatial covariances or variances varies much in time (in the language of section 10.3, they are $O(\sigma^2)$), and so spatiotemporal variation does not depress the long-run growth rate to the same degree that temporal variation does. Thus it is that, even when $-\frac{g_i \langle F_i \rangle_{x,t}}{\gamma_{ir} g_r \langle n_r \rangle_{x,t}} \mathrm{Cov}(\Omega_i, \eta_r)_t > \mathrm{Cov}(\lambda_i, \nu_i)_{x,t}$, the full contribution of temporal segregation to long-run growth can be less than that of spatiotemporal segregation.

10.6 DISCUSSION

I find that spatiotemporal variation has a higher capacity to promote coexistence than temporal variation and that it most promotes coexistence when it enables species to become spatially segregated. Spatial segregation requires short-range seed dispersal and is stronger if between-species competition is short range while within-species competition is long range. However, if competition distances are determined by the growth habits of the individuals involved, so that between- and within-species competition distances cannot vary independently, it is still possible for species to become spatially segregated if they prefer different environmental conditions and all competition is short range.

The primary reason that spatiotemporal variation can promote coexistence more strongly than temporal variation is that in a spatially heterogeneous environment, invader aggregation in high-growth areas can increase long-run invader growth. Because the invader has a small population, it experiences much less self-limitation than the resident, even in locally dense areas, and so can become highly aggregated. In contrast, invader aggregation (in time or space) does not affect long-run invader growth in the presence of temporal variation. When reproduction is pulsed (i.e. discrete-time models are appropriate), the benefits of temporal variation are further limited because while an invader may take advantage of favourable periods, increasing its long-run growth, temporal variation also depresses the long-run growth rate. This, however, is a secondary effect.

Although spatiotemporal variation has a higher capacity to promote coexistence, it is important to remember that it does not promote coexistence more strongly than temporal variation in all circumstances. What happens in a given situation depends on the traits of the species involved, and these evolve to maximise individual fitness, not the likelihood of coexistence. For example, temporal variation in germination fractions can promote coexistence if competing species have their highest germination fractions in different years, but because a variable germination fraction produces a fitness cost, variable germination may not co-evolve (Snyder and Adler 2011).

It is interesting to note that in this system coexistence is always most effectively promoted by short-range dispersal, not long-range dispersal. In a spatiotemporally variable environment, we often think of coexistence as occurring via a competition–colonisation tradeoff, in which the inferior competitor persists by being better at colonising newly available habitat (Levins and Culver 1971). There is a competition–colonisation tradeoff in this model in one sense, in that the species which inflicts less competition has a higher fecundity. However, one might expect coexistence to be strongest when the inferior competitor's seeds disperse farther than those of the superior competitor, and this is not the case. Evidently, the inferior competitor gains more by staying out of the superior competitor's way than it does by trying to be faster at exploiting newly favourable locations. It is not clear why this is so. It may be that longer-range dispersal does not, in fact, enable a species to colonise newly favourable locations more quickly in this model. Longer-range dispersal leads to a more uniform population distribution, which means that the longer disperser has a larger population in newly favourable locations, but it does not get the location to itself: in the absence of demographic stochasticity, both species have some population everywhere.

In short, the ability of low-density species to become highly aggregated enables spatiotemporal variation to promote coexistence more strongly than temporal variation. Selective pressures may produce traits which discourage aggregation – after all, there are many plants with winged or plumed seeds – but where aggregation is possible, spatial heterogeneity can provide an important means for coexistence.

REFERENCES

Chesson, P. (1994). Multispecies competition in variable environments. *Theoretical Population Biology* **45**, 227–276.

Chesson, P. (2000). General theory of competitive coexistence in spatially-varying environments. *Theoretical Population Biology* **58**, 211–237.

Gurney, W.S.C. and Nisbet, R.M. (1976). Spatial pattern and the mechanism of population regulation. *Journal of Theoretical Biology* **59**, 361–370.

Levins, R. and Culver, D. (1971). Regional coexistence of species and competition between rare species. *Proceedings of the National Academy of Sciences, USA* **68**, 1246–1248.

Lewontin, R. and Cohen, D. (1969). On population growth in a randomly varying environment. *Proceedings of the National Academy of Sciences, USA*, **62**, 1056–1060.

Nisbet, R.M. and Gurney, W.S.C. (1982). *Modelling Fluctuating Populations.* New York: John Wiley and Sons, reprinted by The Blackburn Press.

Roughgarden, J. (1974). Population dynamics in a spatially varying environment: How population size "tracks" spatial variation in carrying capacity. *The American Naturalist* **108**, 649–664.

Snyder, R.E. (2006). Multiple risk reduction mechanisms: Can dormancy substitute for dispersal? *Ecology Letters* **9**, 1106–1114.

Snyder, R.E. (2007). Spatiotemporal population distributions and their implications for species coexistence in a variable environment. *Theoretical Population Biology* **72**, 7–20.

Snyder, R.E. (2008). When does environmental variation most influence species coexistence? *Theoretical Ecology* **1**, 129–139.

Snyder, R.E. and Adler, P.B. (2011). Can the storage effect evolve? *The American Naturalist* **178**, E76–84.

AKIKO SATAKE, YUUYA TACHIKI AND YOH IWASA

11

Roles of pollinator attraction and environmental fluctuation in inducing flowering synchrony

11.1 INTRODUCTION

Masting, or mast seeding, is the synchronous seed production in certain years by a population of plants (Kelly 1994). The reproductive activity of plants in forests often fluctuates considerably between years, and flowering and fruit production are synchronised over long distances (Koenig and Knops 1998, 2000, Koenig *et al.* 1999).

Mast seeding has clear disadvantages such as higher density-dependent mortality of seedlings and lost opportunities for reproduction (Waller 1979). In addition, field observations suggest that seed production fluctuates more than the level that can be explained by climate-mediated variation in resource availability (Büsgen and Münch 1929). Two major questions arise from ecological studies on masting: Why is masting evolutionarily favoured regardless of apparent disadvantages? How do plants allocate resources to realise variable flowering efforts in a way synchronised over different individuals?

A family of resource budget models that have recently been studied provides answers to both questions: on the physiological mechanism by which plants may successfully generate such intermittent and synchronous reproduction (Isagi *et al.* 1997, Satake and Iwasa 2000, 2002a, b) and on the conditions for which masting is adaptive (Rees *et al.* 2002, Tachiki and Iwasa 2008). Resource budget models assume that plants accumulate resources every year and set flowers and fruits at a rate limited by pollen availability when the stored resources exceed a reproductive threshold level. The model predicts that individual plants flower

Temporal Dynamics and Ecological Process, ed. C. K. Kelly, M. G. Bowler, G. A. Fox.
Published by Cambridge University Press.

intermittently when their resources are depleted after heavy flowering and fruiting, and that synchrony emerges in self-organised fashion by coupling through the need to receive outcross pollen from other plants (Satake and Iwasa 2000).

In contrast, a traditional hypothesis attributes the causes of masting to simultaneous induction by environmental cues such as low temperature and drought. Intermittent and synchronised flowering might be triggered by rare environmental conditions shared by multiple plant species, a phenomenon called ‘general flowering’. There is a positive relation between environmental cues and the likelihood of floral initiation in Dipterocarpaceae (Ashton *et al.* 1988, Sakai *et al.* 2006). However, recent observations show that environmental cues are not sufficient for general flowering because some individuals do not flower even when they encounter favourable environmental cues (Sakai *et al.* 2006). This suggests a potential interplay between resource availability and environmental cues in floral initiation: in order for a plant to mast, requirements of both environmental cues and resource availability must be satisfied.

In this chapter, we review a series of theoretical studies of masting by focusing on the interplay of resource dynamics and environmental fluctuation. First, we propose a new hypothesis, ‘pollinator coupling’, to explain the mechanism of interspecific synchrony in flowering. In Southeast Asia, more than 80 Dipterocarpaceae and other species flower synchronously and supra-annually (Ashton *et al.* 1988, Sakai *et al.* 1999, Curran and Leighton 2000). Such flowering synchronised between different species is likely to be caused by environmental cues such as low temperature and drought. An alternative explanation in addition to environmental cueing may be pollinator attraction. If different plant species share generalist pollinators, interspecific synchrony in flowering may facilitate pollinator visits (Thomson 1981, 1982, Roy 1994), enhancing seed production (e.g. Pellmyr 1986, Laverty 1992, Roy 1996, Moeller 2004, Ghazoul 2006). We develop a mathematical model that incorporates pollinator behaviour, and theoretically explore the conditions in which synchrony in flowering among different plant species is likely to occur.

Second, we investigate the role of common environmental fluctuations in inducing synchronised flowering. Empirical studies have confirmed the impact of correlated climatic fluctuations at landscape level on spatial synchrony in reproduction (Koenig and Knops 2000, Rees *et al.* 2002, Schauber *et al.* 2002) – spatially correlated fluctuation of climatic variables, such as solar radiation, temperature and precipitation level that influence floral development, pollination, seed ripening and resource acquisition may produce synchronised reproduction at a regional scale. This is an example of the Moran effect (Moran 1953, Royama 1992) – population synchrony is caused by spatially correlated climatic forces. We explore the impact of correlated environmental fluctuation in inducing synchronised flowering within a single plant species.

Finally, we study intraspecific geographic variation in the temporal pattern of mast seeding, which is a reasonably common but poorly understood phenomenon. Rowan, *Sorbus aucuparia* L., exhibits large-scale geographic variation in masting across southern Norway. Along the west coast trees mast in alternate years, while in the east the mast crops are every 3 years (Kobro *et al.* 2003, Satake *et al.* 2004). Such geographic variation may be induced by local adaptation to local ecological conditions or heterogeneity in productivity. We seek to elucidate the theoretical plausibility of these hypotheses based on the resource budget model (Satake and Bjørnstad 2008).

11.2 SYNCHRONISED FLOWERING AND POLLINATOR COUPLING

Plant species in diverse taxonomic groups reproduce in synchrony, but with large interannual variation in reproductive effort (Kelly 1994, Kelly and Sork 2002). One of the well-known examples is 'general flowering' in Southeast Asia: more than 80 species of Dipterocarpaceae and related groups flower synchronously and supra-annually, resulting in heavy fruit fall that carpets the forest floor with germinating seeds (Ashton *et al.* 1988, Sakai *et al.* 1999, Curran and Leighton 2000).

Synchronisation of flowering between species is likely to be caused by environmental cues. Temperature, an increase in solar radiation and water stress are considered potential triggers of flowering. For example, based on an 11-year record of flowering in Pasoh Forest Reserve in Malaysia, Ashton *et al.* (1988) suggested that flowering by a suite of dipterocarp species might be cued by a series of cool nights associated with El Niño. From 10 years of observations of flowering events in Lambar National Hills Park in Malaysia, Sakai *et al.* (2006) showed that dry spells always preceded flowering, leading to the conclusion that water stress could be the most plausible trigger of flowering in dipterocarp species. Yasuda *et al.* (1999) and Numata *et al.* (2003) pointed out a correlation between low temperature and water stress, suggesting a difficulty in distinguishing these two hypotheses.

In self-incompatible species using animal pollination, an alternative but not mutually exclusive explanation may be the need of pollinator attraction. Plant species within many communities have substantially overlapping flowering periods and share generalist pollinators. If different plant species share generalist pollinators, interspecific synchrony in flowering may facilitate pollinator visits (Thomson 1981, 1982, Roy 1994), enhancing seed production (e.g. Pellmyr 1986, Laverty 1992, Roy 1996, Moeller 2004, Ghazoul 2006). But these co-flowering species may compete for pollinator visits (e.g. Rathcke 1988). Even if patches with co-flowering species attract more pollinators, heterospecific pollen transfer may increase (Waser and Fugate 1986), clogging stigmas, or conspecific pollen may be lost to surrounding heterospecifics (Campbell and Motten 1985). Thus, whether plant species sharing generalist pollinators aggregate or segregate their flowering would depend critically on the relative importance of facilitation and competition of pollination within plant–pollinator networks (Rathcke 1983).

In this section, we develop a model of facilitation and competition of pollination, and explore when synchronised flowering among different species is realised. A basic model for plant–pollinator interaction builds on a resource budget model studied by Satake and Iwasa (2000). A family of resource budget models have recently been studied, and they provide a physiological mechanism by which plants may successfully generate intermittent and synchronous reproduction (Isagi *et al.* 1997, Satake and Iwasa 2000, 2002a, b, Crone *et al.* 2005). However, these resource budget models only explain synchrony within a single species, and no effects of pollinators have been examined. In a new model, we explicitly incorporate pollinator behaviour in a resource budget model.

Resource budget model for multiple plant species

In Iwasa *et al.* (this volume) we explained the basic structure and the behaviour of the resource budget model. Here we focus on the case with two plant species indexed by $l = 1$ and 2. The forest is composed of these two species and the numbers of individuals of species 1 and 2 are N_1 and N_2, respectively. Their spatial structure is not important because trees in a forest are assumed to behave independently of their location. We trace the resource budget of each individual.

Let $S_i^l(t)$ be the level of stored resource of the ith individual of species l. The plant accumulates resources, P_S, by photosynthesis in a year. If stored resources exceed a reproductive threshold L_T, the tree sets flowers using the amount of resources proportional to the excess,

$S_i^l(t) + P_S - L_T$. Then the flowers are pollinated with rate $P_i^l(t)$, and the plant sets fruits. The energetic costs of fruit development following flower fertilisation are determined by k^l : the ratio of fruiting cost to flowering cost that may differ between species in general.

With these assumptions, the energy dynamics of the ith individual of plant species l are written as

$$S_i^l(t+1) = \begin{cases} S_i^l(t) + P_S & \text{if } S_i^l(t) + P_S \le L_T \\ S_i^l(t) + P_S - \left(1 + P^l{}_i(t)k^l\right)\left(S^l{}_i(t) + P_S - L_T\right) & \text{otherwise} \end{cases} \tag{11.1}$$

In this model, a plant having a larger k^l experiences a more intensive resource depletion with the same amount of flowers and outcross pollen availability, resulting in a lower level of resource reserve after the reproduction event. The above equation is equivalent to Equation (9.1) with $a = 1$.

We therefore call k^l the 'depletion coefficient' (Satake and Iwasa 2000). In order to simplify Equation (11.1), we introduce a non-dimensional variable, $Y_i^l(t) = \left(S_i^l(t) + P_S - L_T\right)/P_S$. Using $Y_i^l(t)$, we rewrite Equation (11.1) as

$$Y_i^l(t+1) = \begin{cases} Y_i^l(t) + 1 & \text{if } Y_i^l(t) \le 0 \\ -k^i P_i^l(t) Y_i^l(t) + 1 & \text{otherwise} \end{cases}, \tag{11.2}$$

where $P_i^l(t)$ is the pollination rate of the ith individual of species l in year t.

The flowering intensity of the ith individual is

$$F_i^l(t) = P_S[Y_i^l(t)]_+/c^l. \tag{11.3}$$

$[Y_i^l(t)]_+ = Y_i^l(t)$ if $Y_i^l(t) > 0$, otherwise it is 0. c^l is the unit cost of producing a single flower. The total number of flowers produced by species l is $F^l(t) = \sum_{i=1}^{N} F_i^l(t)$. All previous studies of the resource budget model assume the pollination rate, $P_i^l(t)$, is a simple function of flower density produced by other trees of the same species, which corresponds to wind pollination. In this chapter, we assume that plants are animal pollinated, and we extend the model by incorporating processes causing facilitation or inhibition of pollination success by different tree species.

Facilitation and competition of pollination across plant species

Many empirical studies have demonstrated evidence of pollination facilitation. The number of pollinator visits to a local patch increases as the number of flowers in the patch increases. Here we model this pollination facilitation by assuming that pollinator visits to a patch increase as the number of flowers in the patch increases, but saturate at a certain level because the total number of pollinators is limited.

We consider the simplest case in which there are two plant species (species 1 and 2) and one pollinator. Plant species 1 and 2 share the single pollinator. To model the pollination facilitation between these two plant species, we assume that the rate of pollinator visitation to the patch in year t is a Hill function:

$$N(t) = \frac{N_{\max}\left(F^1(t) + F^2(t)\right)^n}{\theta^n + \left(F^1(t) + F^2(t)\right)^n} \tag{11.4}$$

where $N_{\max}$ is the maximum visitation rate to the patch, θ is the number of flowers at which the pollinator visitation rate is half its maximum value, and n is a Hill constant. The larger n becomes, the more closely $N(t)$ approaches the all-or-none switch from alternative flowering patches to the focal patch. The magnitude of θ is influenced by two factors: flowering intensity in alternative patches and abundance of pollinators. Pollinators visit the focal patch when flowering resources are more abundant in the focal patch than those in alternative patches. Therefore θ becomes large as the flowering intensity in alternative patches increases, meaning that higher intensity of flowering is necessary to attract pollinators to visit the patch. Even if alternative patches are more attractive for pollinators, they will visit the less attractive patches when pollinators are abundant because competition becomes severe in the attractive patches. Thus θ decreases as the abundance of pollinators increases, meaning that lower flowering intensity is sufficient for pollinator attraction. For simplicity, we assume that the alternative patches always provide flowering resources at a constant level, and pollinator population size stays constant in the following analysis.

On the other hand, plant species 1 and 2 also compete with each other for pollination services. Here, we apply a modelling approach similar to that proposed by Feldman *et al.* (2004) and make the assumptions that: (1) pollinators do not discriminate between the plant species, and randomly visit flowers in the patch; and (2) a pollinator can pollinate the plant only if its last visit was to the same plant species. Given these assumptions, the pollinator visitation rate to a single flower is $N(t)/\left(F^1(t) + F^2(t)\right)$. The probability that the last visit was to a different individual of species l is $\left(F^l(t) - F_i^l(t)\right)/\left(F^1(t) + F^2(t)\right)$. Since we assume self-incompatible plants, the available pollen for the ith individual of species l is $F^l(t) - F_i^l(t)$. Thus, the expected number of successful visits to pollinate a single flower produced by the ith individual of species l in year t is

$$\phi_i^l(t) = N(t)\frac{F^l(t) - F_i^l(t)}{\left(F^1(t) + F^2(t)\right)^2} = \frac{N_{\max}\left(F^l(t) - F_i^l(t)\right)\left(F^1(t) + F^2(t)\right)^{n-2}}{\theta^n + \left(F^1(t) + F^2(t)\right)^n}. \tag{11.5}$$

We assume that the number of successful pollinator visits experienced by a flower of the ith individual of plant species l follows a Poisson distribution with mean $\phi_i^l(t)$. Thus, the pollination rate per year is

$$P_i^l(t) = 1 - e^{-\phi_i^l(t)}, \tag{11.6}$$

the probability that a flower of the ith individual of plant species l receives one or more successful pollinator visits. The pollination rate increases as $F^l(t)$ increases, but after it attains the peak, it decreases due to pollinator limitation (Figure 11.1). The pollination rate is also influenced by the number of flowers produced by the other species. Flowering of the other species enhances the pollination rate of one species if flowering intensity of the species is relatively small, while it degrades the pollination rate if flowering intensity of both species is large due to competition for pollination services (Figure 11.1). Thus, Equation (11.6) describes both facilitation and competition of plant species with respect to pollinator attraction. The Hill constant, n, in Equations (11.4) and (11.5), is a parameter that controls this balance. In the following, we analyse the conditions that lead to intraspecies and interspecies synchrony in flowering.

Synchrony within species and pollinator coupling

We first investigate the situation where only one species is included. Plants in a forest may reproduce synchronously, entirely asynchronously, or form clusters of trees in which members of the same cluster have long-term synchrony but different clusters flower asynchronously. Clusters are defined as follows: individuals i and j belong to the same cluster if the cumulative sum of the deviation between two resource dynamics is less than ε (i.e. $\sum_{t=0}^{T} |Y_i(t) - Y_j(t)| < \varepsilon$), where ε is a very small number. Note that spatial structure is not considered in this analysis because global pollen dispersal is assumed. Satake and Iwasa (2002a) extended this assumption explicitly to incorporate short-distance pollen

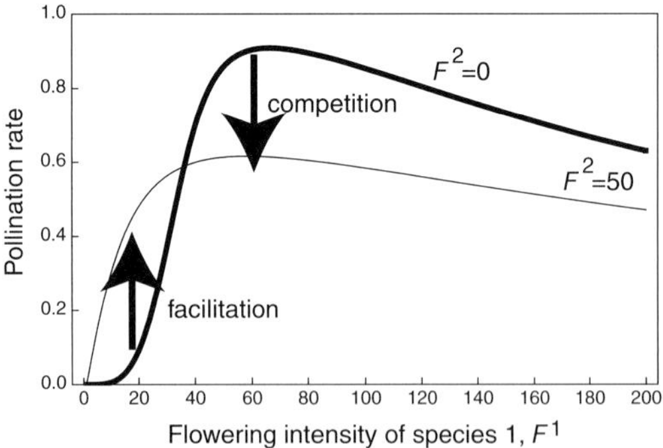

Figure 11.1 Pollination rates for different flowering intensities of species 1. Flowering intensity for species 2 facilitates pollination of species 1, and then at higher flowering intensities for species 1, it competitively suppresses pollination. Pollination rate is defined by Equation (11.6) in the text. Thick line: Flowering intensity of species 2 is 0. Thin line: Flowering intensity of species 2 is 50. Parameters are $N_{max} = 200$, $\theta = 50$ and $n = 5.0$.

dispersal. In order to quantify the level of synchrony, we calculate the effective number of clusters (*NC*), defined as $NC = 1/\sum_i p_i^2$, where p_i is the fraction of plants belonging to the *i*th cluster. When all plants are synchronised in flowering, $NC = 1$, while *NC* is close to the number of trees in the forest when all plants reproduce asynchronously. We repeated numerical calculation of the model 100 times with random initial conditions, and calculated the average $\overline{NC}$.

Based on the effective number of clusters, we classify four different phases: (1) complete synchrony ($1 \leq \overline{NC} < 2$), (2) clustering ($2 \leq \overline{NC} \leq N - 1$), (3) desynchrony ($N - 1 < \overline{NC} \leq N$) and (4) annual reproduction where all individuals reproduce every year. Results of the analysis illustrate that complete synchrony in reproduction occurs when the Hill constant *n* is large (Figure 11.2). When *k* is increased beyond 1, the transition from desynchrony to clustering occurs, and then from clustering to synchrony, as the Hill constant *n* increases. These transitions are caused by the change in pollination rate, and depend strongly on the magnitude of the Hill constant. When the Hill constant *n* is small, the pollination rate increases rapidly with flowering intensity, resulting in a high pollination rate even under low flowering intensity (e.g. pollination rate increases to 80% even when flowering intensity is as low as 20) (Figure 11.3a: thick line). This results in asynchronous flowering.

In contrast, when the Hill constant is large, high flowering intensity is necessary to attain reasonably large pollination success because of

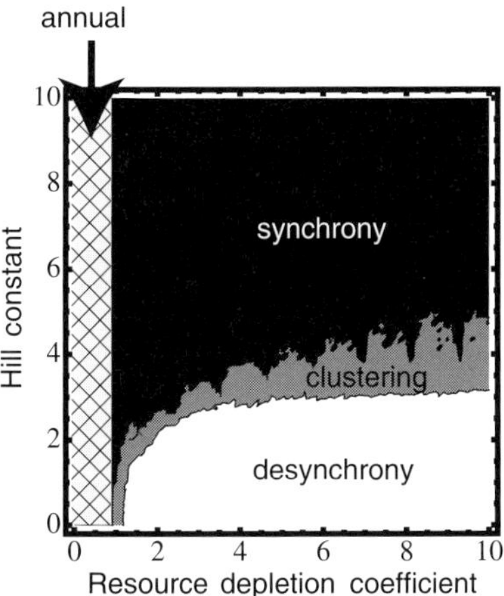

Figure 11.2 Phase diagram for one species. We classified four regions as explained in the text. Total number of individuals was 100. Other parameters are: $N_{max} = 200$ and $\theta = 50$.

the switching behaviour of pollinators (Figure 11.3a: thin line). Pollination success remains low under low flowering intensity, and increases as flowering intensity increases at an accelerating rate. Under such a situation, a plant that flowers in a manner not synchronised with other individuals cannot set many fruits due to pollen limitation. Thus, it does not experience resource depletion, and flowers again in the following year. Such an individual sets many fruits only when many other plants flower, resulting in synchrony in resource and reproductive cycles among different individuals that share the pollinator (Figure 11.3b). When pollen availability is always high as in the case of a small Hill constant, synchrony does not occur (Figure 11.3c). This mechanism of synchrony is qualitatively the same as the pollen-coupling hypothesis proposed by Satake and Iwasa (2000) where wind-pollinated species are considered (this volume).

If plant density is low in the focal patch, flowering together with other plants is necessary to attract pollinators and attain a high pollination rate, resulting in synchronised flowering, as exemplified by the effective cluster size of 1 (Figure 11.4). In contrast, if plant density is high in the focal patch in comparison to alternative flower sources, synchronised reproduction in the entire forest causes a slight reduction of the pollination rate due to competition for limited pollinator resources, but the pollination rate is still large enough to produce an intermediate number of fruits (Figure 11.3a). In addition, pollen limitation rarely happens in a forest with high plant density,

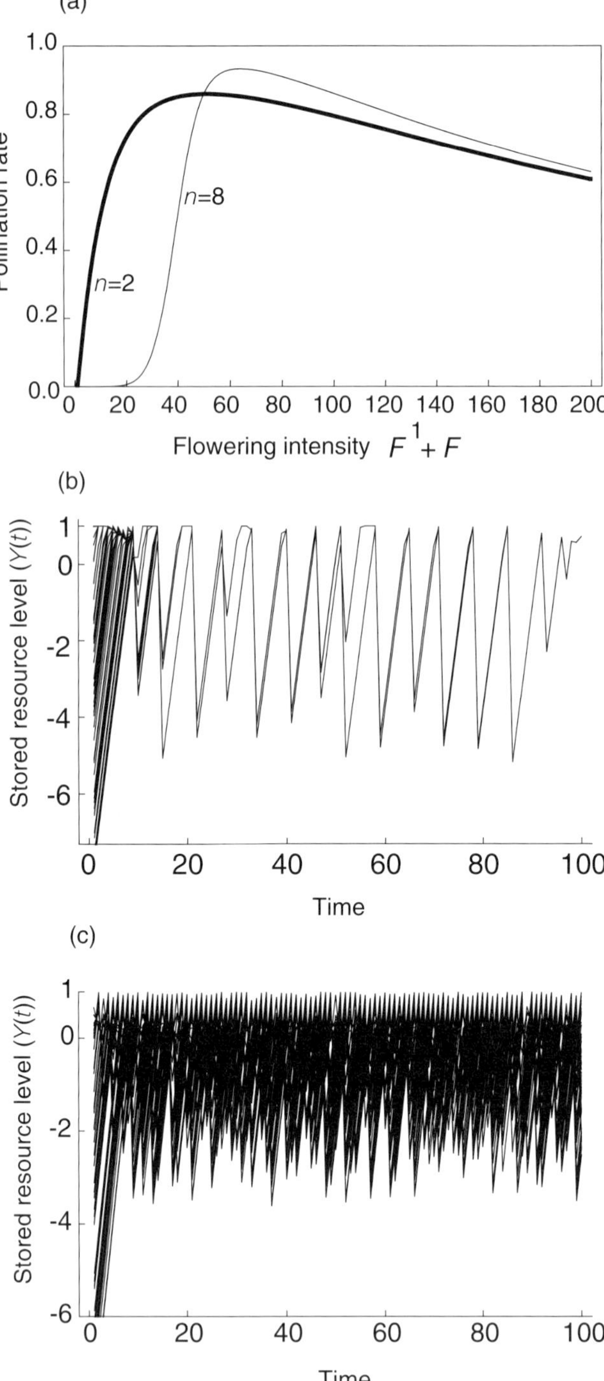

Figure 11.3 (**a**) Pollination rate for different flowering intensities. Pollination rate is defined by Equation (11.6); n is the Hill constant. (**b**) Time series of stored resource levels for 50 individuals; $n = 8.0$. (**c**) Time series of stored resource levels for 50 individuals; $n = 2.0$. Other parameters are: $N_{\max} = 200$, $\theta = 50$ and $k = 7.2$.

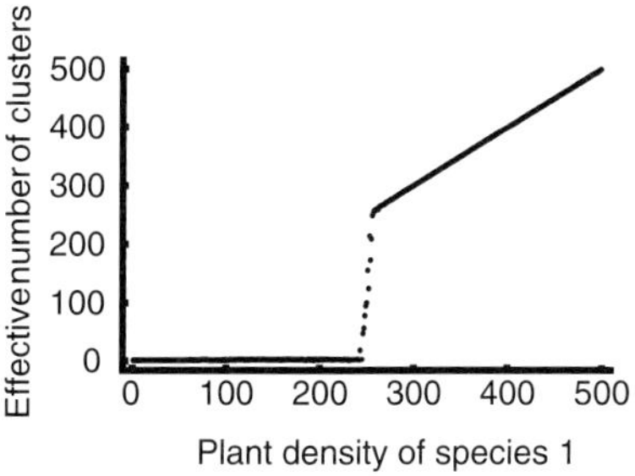

Figure 11.4 Effective number of clusters of species 1 for different numbers of individuals per hectare. Parameters are: $N_{\max} = 200$, $\theta = 50$, $n = 7.0$ and $k = 5.5$.

because even a small fraction of flowering plants is enough to provide a high pollination rate. Thus, synchronised flowering is not likely under high plant densities, and transition from synchrony, to clustering, and further to desynchrony was observed as the density of the focal species increased (Figure 11.4). When plant density is extremely high, severe competition over pollinators occurs, resulting in strong pollen limitation. In this case, synchronised flowering is induced, but plants fail to produce many fruits because of strong pollen limitation.

Synchrony between species caused by pollinator coupling

In this section, we study the situation in which synchronised reproduction among different plant species is realised. We calculated the effective number of clusters for the entire plant community, denoted as NC_{total}, that includes species 1 and 2, using the method described in the previous section. We also calculated the effective number of clusters for each species, denoted by NC_1 and NC_2 defined for species 1 and 2, respectively. We repeated the procedure 100 times, and calculated the averages $\overline{NC}_{total}$, $\overline{NC}_1$ and $\overline{NC}_2$. Based on these averaged effective number of clusters, we classified the following five phases:

1. Interspecific synchrony ($1 \leq \overline{NC}_{total} < 2$): all individuals of both species are synchronised in reproduction (dark grey region in Figure 11.5).
2. Intraspecific synchrony ($1 \leq \overline{NC}_1, \overline{NC}_2 < 2$): individuals within the same species are synchronised in reproduction, but synchrony is not realised between species (light grey region in Figure 11.5).

3. Synchrony in one species only ($1 \leq \overline{NC}_1 < 2 \cap \overline{NC}_2 > 2$ or $1 \leq \overline{NC}_2 < 2 \cap \overline{NC}_1 > 2$): synchronised reproduction is realised within one of the two species but not within the other species (black region in Figure 11.5).
4. Desynchrony: individuals of neither species show synchronised reproduction (white region in Figure 11.5).
5. Annual reproduction: all individuals reproduce every year (shaded region in Figure 11.5).

Results illustrated in Figure 11.5 show that when the Hill constant, n, is large, either interspecific synchrony or intraspecific synchrony occurs, while all of individuals are desynchronised when the Hill constant is small. The likelihood of interspecific synchrony is dependent on the magnitude of the depletion coefficient k in a complex manner (Figure 11.5b and c), but the likelihood tends to increase as the Hill constant n and θ increase (Figure 11.5a). Thus we concluded that synchronised flowering between different species is likely to emerge when pollinators show clear switching behaviour between patches (i.e. large n) and when alternative food sources are more attractive or the abundance of pollinators is low (i.e. large θ). Phase 3, synchrony in one species only, was observed only in a very limited parameter region.

Density of species 2 is another important factor that determines the likelihood of synchronised reproduction. When the density of species 2 is low, species 1 flowers together with species 2 to attract pollinators (high correlation coefficient; dark grey dots in Figure 11.6a and b). Note that this interspecific synchrony is not perfect ($1 \leq \overline{NC}_{total} < 2$ is not satisfied), but the correlation coefficient is significantly large (Figure 11.6c). As the density of species 2 increases, species 2 starts flowering asynchronously, and finally no intra- or interspecific synchrony emerges. When density of species 2 is very high, species 1 flowers synchronously, but intraspecific synchrony does not happen. Co-flowering with species 2 significantly enhanced the pollination success for species 1 (Figure 11.6a). But an improvement of pollination rate by co-flowering disappeared as the density of plant species 1 increased, because of the escalated competition for pollinator resources (data not shown).

In summary, synchronised flowering emerged only when pollen limitation prevailed. In a pollinator-coupling situation, two mechanisms induce pollen limitation. The first is a failure to attract pollinators. When plant density is low and an alternative food source is very attractive for pollinators (large θ), pollinators are always limited. In such a situation, pollen limitation becomes important, and causes synchrony

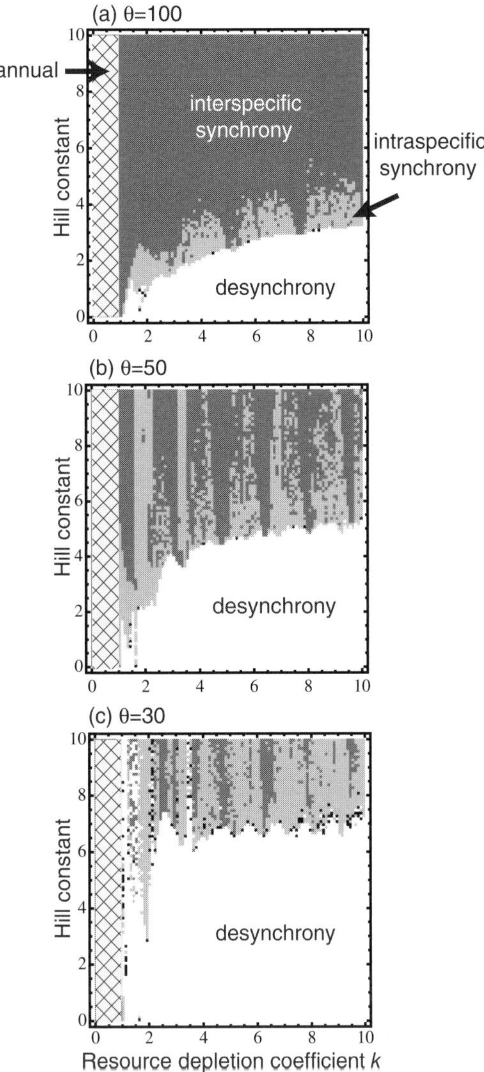

Figure 11.5 Phase diagram for two species, both having the same value of k. We classified four regions as explained in the text. Each species has 50 individuals. **(a)** $\theta = 100$, **(b)** $\theta = 50$ and **(c)** $\theta = 30$. $N_{\max} = 200$.

within or between species. In this case, the pollination rate is improved due to synchrony, and more seeds are produced. The second mechanism is the intense competition for pollinator resources. When plant density is very high, the patch attracts almost the maximum number of pollinators, but competition over pollinators is severe. This intense

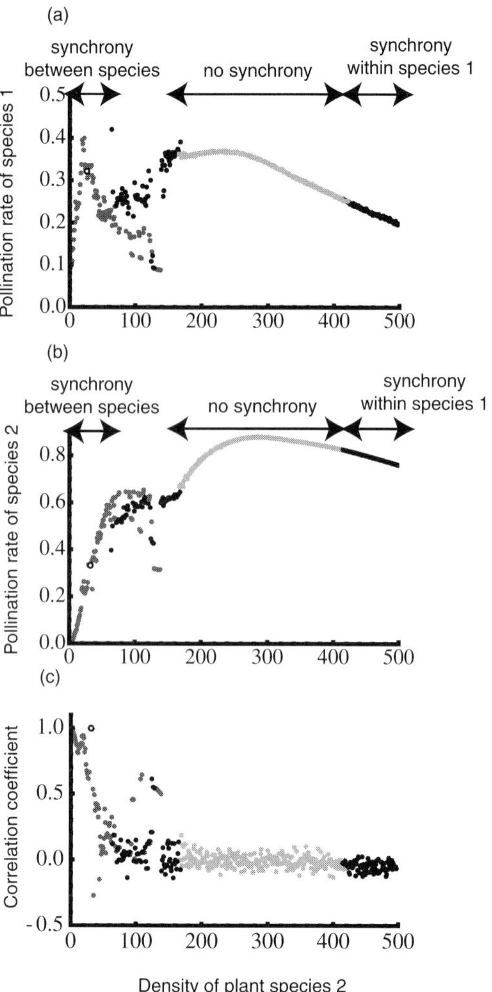

Figure 11.6 Pollination rate of species 1 **(a)** and species 2 **(b)** for different densities of species 2. **(c)** Correlation coefficient of flowering dynamics of all plants including species 1 and 2 along the density of plant species 2. Dark grey circles: intraspecific synchrony ($1 \le \overline{N}\overline{C}_1, \overline{N}\overline{C}_2 < 2$). Black circles: synchrony in a single species ($1 \le \overline{N}\overline{C}_1 < 2 \cap \overline{N}\overline{C}_2 > 2$ or $1 \le \overline{N}\overline{C}_2 < 2 \cap \overline{N}\overline{C}_1 > 2$). Light grey circles; no synchrony. The number of individuals of species 1 is 30. Parameters are: $N_{\max} = 200$, $\theta = 50$, $n = 7.0$ and $k = 4.7$. Open circles: complete synchrony between species ($1 \le \overline{N}\overline{C}_{total} < 2$).

competition leads to a reduction in pollination rate. The plants with limited pollen flower almost every year because they do not deplete resources by fruiting, which causes a synchronised flowering. But in this case, failure in fruiting always occurs.

11.3 SYNCHRONISED FLOWERING AND ENVIRONMENTAL FLUCTUATION

Synchronised flowering can be caused by environmental fluctuation that is spatially correlated. This is an example of the Moran effect (Moran 1953, Royama 1992) – population synchrony is caused by spatially correlated climatic forces. In this section, we explore the impact of environmental fluctuation in inducing synchronised flowering within a single plant species.

Correlated environmental fluctuation: Moran effect

In Satake and Iwasa (2000, 2002a), the environment is assumed perfectly constant. However, Koenig and Knops (1998, 2000) and Koenig *et al.* (1999) propose that synchronised reproduction of trees is caused by common environmental fluctuations experienced by distant trees. In this section, we examine the effect of stochastic environmental fluctuation on the degree of reproductive synchrony between trees that are coupled by pollen exchange (Satake and Iwasa 2002b).

Temperature and precipitation during the peak growing season (i.e. summer) often fluctuate across years. This among-year variation would cause yearly fluctuation of annual net production (P_s). In fact, it has been reported that unusually high temperatures in the summer enhance flowering in the following year (McKone *et al.* 1998). This situation can be modelled simply by assuming that the annual net production P_S is not constant, but fluctuates between years. Thus P_S is replaced by $P_S + u_{\mathrm{x}}(t)$ where $u_{\mathrm{x}}(t)$ is a random variable with mean value of zero. The threshold energy level for reproduction (L_T in Equation (11.1)) may also vary. For instance, in dipterocarp forests, unusually low temperatures may enhance flowering intensity in the following year by stimulating bud formation (Ashton *et al.* 1988, Appanah 1993, Yasuda *et al.* 1999). L_T is then replaced by $L_T + v_{\mathrm{x}}(t)$, in which $v_{\mathrm{x}}(t)$ is a random variable, again with mean value of zero. Because the 'noise' may differ between trees in habitats heterogeneous for nutrient level or water availability, both $u_{\mathrm{x}}(t)$ and $v_{\mathrm{x}}(t)$ are specific to individual $\mathbf{x}$. Arithmetical manipulation shows that both modes of environmental fluctuations can be represented by adding an additional term for stochastic environmental noise to the basic model in Equation (11.1).

Using a non-dimensionalised term defined as $Y_x(t) = [S_x(t) + P_S + u_x(t-1) - L_T - v_x(t-1)]/P_S$, Equation (11.1) is rewritten as

$$Y_x(t+1) = \begin{cases} Y_x(t) + 1 + \varepsilon_x(t) & \text{if } Y_x(t) \leq 0 \\ -kP_x(t)Y_x(t) + 1 + \varepsilon_x(t) & \text{if } Y_x(t) > 0 \end{cases}, \tag{11.7}$$

where the pollen availability $P_x(t)$ is given by

$$P_x(t) = \left(\frac{1}{z} \sum_{y \in U_x} [Y_y(t)]_+ \right)^{\beta}, \tag{11.8}$$

where $[Y]_+ = Y$ if $Y > 0$; $[Y]_+ = 0$ if $Y \leq 0$. U_x represents the neighbourhood of the tree at $\mathbf{x}$ (i.e. the set of neighbouring trees that are pollen donors for the focal tree), and z is the number of trees included in this neighbourhood. In this section, we consider wind-pollinated species. In Equation (11.7), the environmental fluctuation term is $\varepsilon_x(t) = [u_x(t) - v_x(t) + v_x(t-1)]/P_S$. This environmental fluctuation term $\varepsilon_x(t)$ depends both on the fluctuation of net annual production $u_x(t)$ and on the difference of the fluctuations of the reproductive threshold between two successive years: $-v_x(t) + v_x(t-1)$.

We assume here that the environmental fluctuation term of individual $\mathbf{x}$ ($\varepsilon_x(t)$) follows a normal distribution with mean 0 and standard deviation σ. This environmental noise level may be correlated among different trees. The correlation coefficient of noise is given by r, which ranges from 0 to 1. When $r = 1$, environmental noise levels of different individuals are exactly the same, while there is no correlation when $r = 0$. To generate intermediate levels of environmental correlation r, we assumed that $\varepsilon_x(t)$ is the sum of two terms, one common over all the individuals and the other independent between individuals, and controlling their relative magnitude.

The resource budget model with environmental fluctuation now has four parameters: depletion coefficient (k), coupling strength (β), the standard deviation of environmental noise (σ) and the correlation coefficient of noise between different trees (r). In the following, we study whether correlated environmental fluctuation can cause synchronised flowering.

Global and local pollen coupling and environmental noise

We consider a forest composed of 100 individuals in which the pollen availability for each tree is determined by the average flowering intensity of all the trees in the whole forest except the focal tree (z in

Equation (11.8) is $N - 1$). We call this situation global pollen coupling. We calculate the degree of synchrony as follows: the amount of seed production for each tree, denoted by $\varphi_x(t) = kP_x(t)[Y_x(t)]_+$, follows the dynamics given in Equation (11.7). We started from a random initial distribution and ran the model for 10^5 years as a transient. We then determined the seed crop for each tree over 1000 years and calculated the correlation coefficient $\rho(\varphi_x, \varphi_y)$ between the time series for each pair of trees, $\{\varphi_x(t)\}_{t=1}^{T}$ and $\{\varphi_y(t)\}_{t=1}^{T}$. We performed 100 independent trials and the average correlation coefficient over all possible pairs, $\bar{\rho} = \frac{1}{N(N-1)} \sum_{x} \sum_{y \neq x} \rho(\varphi_x, \varphi_y)$, where N is the total number of individuals in the forest, was calculated for each. The average of $\bar{\rho}$ over each independent trial was used to measure the degree of synchrony in seed production.

As discussed by Satake and Iwasa (2000), the degree of synchrony in flowering is determined by the balance between depletion coefficient and coupling strength. Even in the absence of environmental noise ($\sigma = 0$), synchronised flowering is likely when trees are strongly coupled by pollen exchange (i.e. $\beta = 1.5$; Figure 11.7a). However, as the strength of pollen coupling decreases, the correlation coefficient decreases (Figure 11.7a), meaning that synchronised reproduction becomes difficult at lower values of β (Figure 11.7a). The correlation coefficient does not show a monotonic dependence on k because the phase plane of the model has multiple windows of k for periodic fluctuation separated by intervals of chaotic fluctuation (Satake and Iwasa 2000).

In contrast, if there is no pollen coupling, correlated environmental fluctuation cannot generate synchronised flowering for parameter ranges of $k > 2$ (i.e. $\beta = 0$ and $r = 1$; Figure 11.7b). However, if there is pollen coupling, even limited correlation of environmental noise (e.g. $\beta = 1$ and $r = 0.2$; Figure 11.7b) results in some synchrony of seed production. The degree of synchrony increases as the correlation of environmental noise (r) increases.

When pollen dispersal range is limited to nearest neighbours (z in Equation (11.4) equals 8), the results are slightly changed. In the absence of environmental noise (Figure 11.8a), synchronised flowering was high when $k < 2$ (parameter range dominated by period-2 cycles) but was very low at higher values of k. This corresponds to the results of a coupled map lattice studied by Satake and Iwasa (2002), who showed that strong synchronisation over the whole forest occurs for $k < 2$. Since construction costs of flowers and fruits lead to estimates of 4.0–5.6 for k in *Fagus crenata* (Isagi *et al.* 1997), long-distance synchronisation is unlikely to be

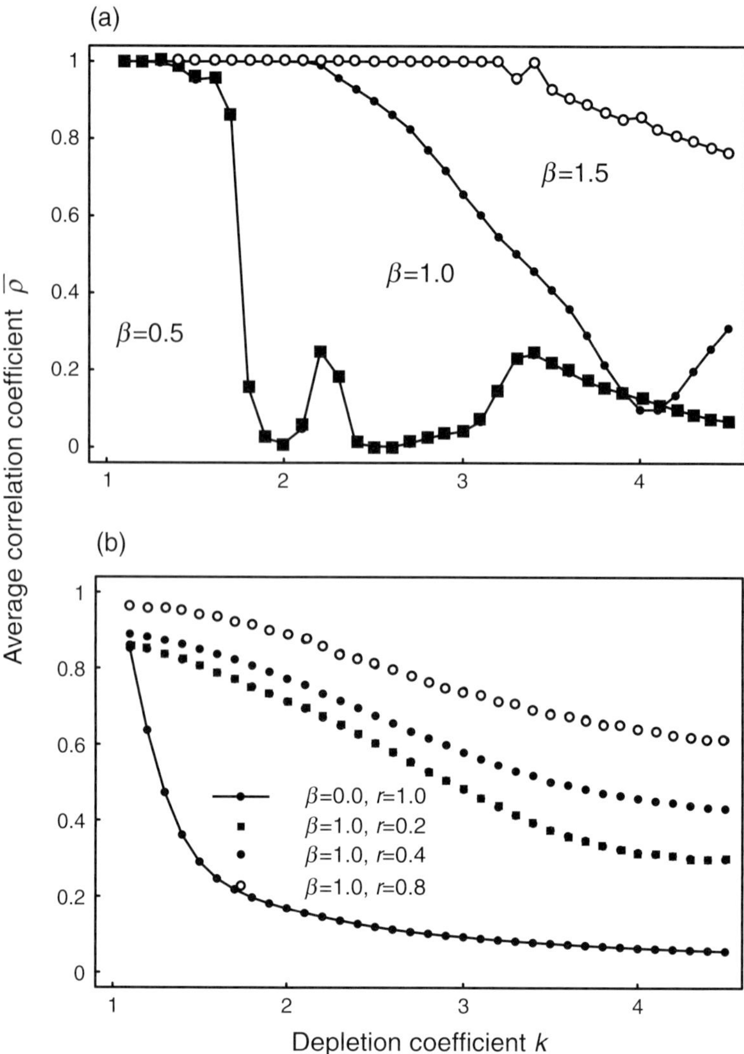

Figure 11.7 Variation of the average correlation coefficient for seed crops among trees with depletion coefficient when pollen disperses over the whole forest (global pollen coupling). Average seed crops from 100 independent trials were used. r is the correlation of environmental noise experienced by different individuals, β is the coupling strength. σ is the standard deviation of environmental noise. **(a)** $\sigma = 0.0$ (no environmental noise); **(b)** $\sigma = 0.2$. From Satake and Iwasa (2002b).

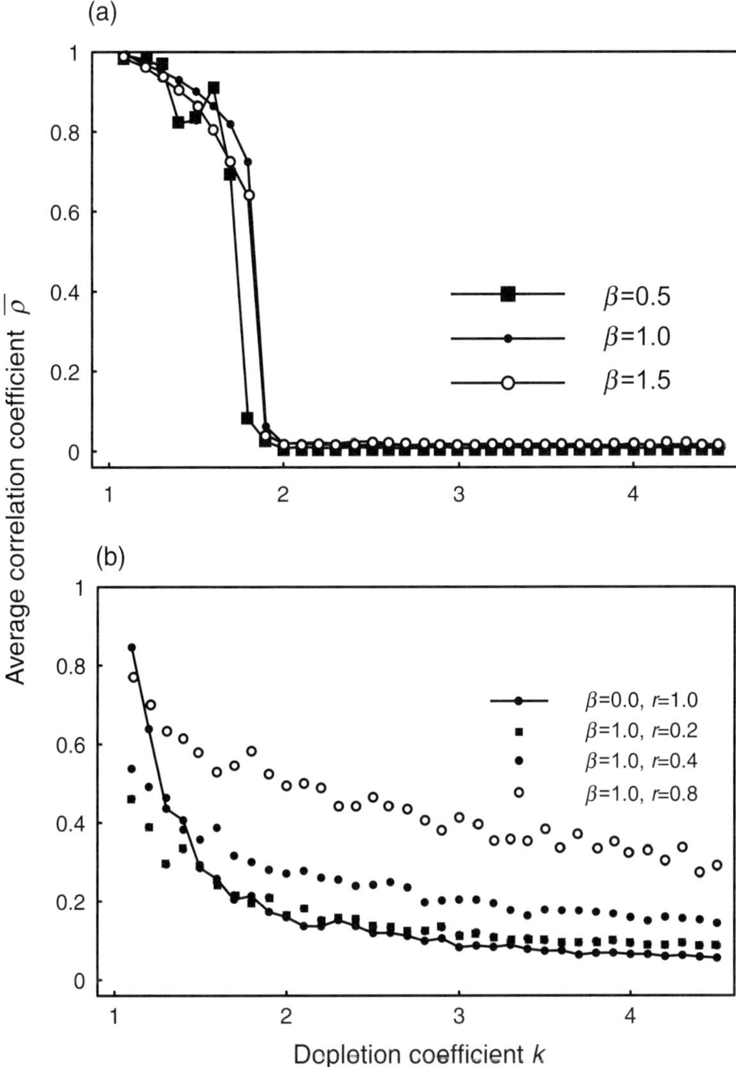

Figure 11.8 Variation of the average correlation coefficients of seed crops among trees with depletion coefficient when pollen dispersal is limited to nearest neighbours (local pollen coupling). σ is the standard deviation of environmental noise. **(a)** $\sigma = 0.0$ (no environmental noise); **(b)** $\sigma = 0.2$. From Satake and Iwasa (2002b).

explained by pollen coupling alone if pollen dispersal range is limited at a local spatial scale.

Nor did correlated environmental fluctuation lead to reproductive synchronisation in the absence of pollen coupling ($\beta = 0$, Figure 11.8b).

Even when the standard deviation of the noise was as high as 20% of the average annual net productivity ($\sigma = 0.2$), marked correlations of seed crops were only seen for $k < 2$. If pollen coupling and correlated environmental noise operate simultaneously, the model predicts much greater correlation between distant trees (Figure 11.8b).

11.4 INTRASPECIFIC VARIATION IN REPRODUCTIVE DYNAMICS

We have examined the potential for common environmental fluctuation (the Moran effect) and pollen coupling to synchronise reproduction among trees. An important finding was that when each tree shows chaotic seed production, common environmental noise alone did not induce significant reproductive synchrony even if the environmental noise was almost the same for all trees. If environmental fluctuation was correlated among individuals, and pollen coupling was at work, flowering was strongly synchronised across the entire forest.

In this section, we investigate the role of environmental conditions in creating geographic variation of reproductive dynamics within the same species. Rowan (*Sorbus aucuparia* L.), for instance, exhibits large-scale geographic variation in masting across southern Norway. Along the west coast, trees mast in alternate years, while in the east the mast crops are every three years (Kobro *et al.* 2003, Satake *et al.* 2004). Intraspecific variation in masting has also been described in snow tussocks (*Chionochloa* spp.) (e.g. Kelly *et al.* 2000), masting oaks (*Quercus* spp.) (e.g. Leibhold *et al.* 2004) and many other tree species (Koenig and Knops 1998, 2000, Schauber *et al.* 2002, Suzuki *et al.* 2005). Such variation may be induced by local adaptation to local ecological conditions or variation arising from gradients in productivity. We seek to elucidate the plausibility of these hypotheses based on the resource budget model.

Geographic variation in rowan masting

Rowan is widely distributed throughout Europe. Clusters of white flowers are produced in the early summer and berries ripen in conspicuous red clusters in the fall. It is a typical masting species – flower and berry production greatly vary between years (Sperens 1997, Kobro *et al.* 2003, Satake *et al.* 2004). From long-term data on berry production by rowan trees on commercial apple farms across southern Norway, Satake *et al.* (2004) documented geographically distinct reproductive patterns among the rowan trees: a 2-year cycle in southwest Norway and a 3-year cycle in southeast Norway. Figure 11.9 illustrates this trend based on

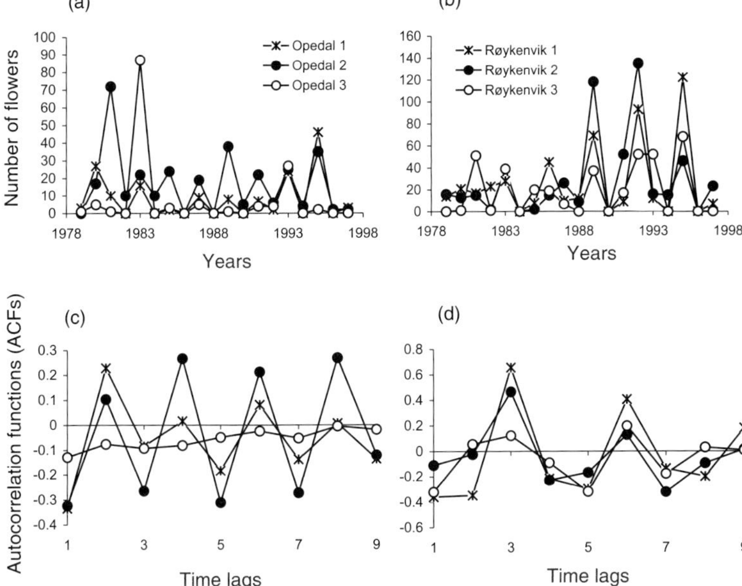

Figure 11.9 Time series data of annual flower production of rowan trees **(a)** and **(b)**, and autocorrelation functions (ACFs) **(c)** and **(d)**. Western Norway (Opedal) and eastern Norway (Røykenvik) include three individuals respectively. From Satake and Bjørnstad (2008).

data available for six individual trees between 1979 and 1997 at two different sites, Opedal (west) and Røykenvik (east). The autocorrelation functions (ACFs) of the time series (Figures 9c and 9d) show that the reproductive patterns of the Opedal trees (west) exhibit strong positive autocorrelation at lags 2, 4, 6, ... and negative autocorrelation at lags 1, 3, 5, ... (Figure 11.9c). In contrast, the trees in Røykenvik (east) show positive autocorrelations at lags 3 and 6 and negative ones at lags 1, 2, 4, 5, 7 (Figure 11.7d). Satake *et al.* (2004) provide a comprehensive description of this phenomenon across 49 geographic locations.

We reviewed the climatic information reported by the Norwegian Meteorological Institute (www.met.no) for the two regions. While the mean annual temperatures do not differ markedly between the two, the east is more continental and shows wider fluctuations. The meteorological station at Gardermoen (east; 41 km from Røykenvik), for instance, has an average of 167 days a year with minimum temperatures below freezing, as compared to 57 days for Bergen (west; 140 km from Opedal). The number of days for which the maximum temperature exceeds 20°C is 52 at Gardermoen (east) but only five at

Bergen (west). The relevant stations in the eastern region typically receive an average of 500–1000 mm precipitation a year, while the relevant stations in the west receives more than twice that; Ullensvang county (west; where Opedal is situated) reports annual precipitation of 1350 mm versus 770 mm in Gran county (east; where Røykenvik is situated). Average annual temperatures for the two counties are 6.8°C versus 3.0°C. Thus, we have good reason to assume that the environment is more suitable and productivity is higher in the west than in the east.

Building a model to explain geographic variation

We ask what are the major differences in the temporal patterns of masting dynamics when the productivity is altered in mast seeding dynamics. Here, we examine this issue by incorporating an additional level of realism into the resource budget model formalised in Equation (11.1): a minimum level of resource reserve that is achieved when a plant with a large amount of fertilised ovules produces a large amount of fruits. We assume that plants having too many fertilised ovules would show selective abortion of ovules if the amount of resource reserve is limited. This assumption may be represented by the simple biological facts that the plant's stored resource level cannot be negative (i.e. $S_i(t) \geq 0$). With the requirement that $S_i(t) \geq 0$, the non-dimensional variable $Y_i(t) = \left(S_i(t) + P_S - L_T\right)/P_S$ has the constraint

$$Y_i(t) \geq 1 - L_T/P_S. \tag{11.9}$$

This inequality implies that when the ratio of the reproductive threshold (L_T) to the rate of productivity (P_S) is very large, the minimum threshold represented by the right-hand term in Equation (11.9) can be very low after heavy reproduction. Thus, the behaviour of the modified model is similar to that of the original model because it assumes no minimum threshold. In contrast, when the ratio L_T/P_S is not very large, $Y_i(t)$ is constrained to be above the level of $1 - L_T/P_S$, even after major flowering. Such a minimum level plays an important role in affecting masting dynamics, including the possibility of a productivity-induced transition from 2- to 3-year mast intervals, as shown below.

Using the non-dimensional variable $Y_i(t)$ with the constraint of Equation (11.9), the energy allocation dynamics become:

$$Y_i(t+1) = \begin{cases} Y_i(t) + 1 & \text{if } Y_i(t) \leq 0 \\ -k[Y_i(t)]_{min} + 1 & \text{otherwise} \end{cases}, \tag{11.10}$$

where $[Y_i(t)]_{min}$ is given by:

$$[Y_i(t)]_{\min} = \begin{cases} Y_i(t) & \text{if } Y_i(t) \leq L_T/(kP_S) \\ L_T/(kP_S) & \text{otherwise} \end{cases}. \quad (11.11)$$

Equation (11.10) indicates that if the variable representing the level of stored resources, $Y_i(t)$, is negative, the plant does not reproduce; otherwise it allocates some of its resources to reproduction. However, the reproductive effort is constrained by abortion, which limits resource depletion; substitution of Equation (11.11) into (11.10) cannot yield $Y_i(t+1) < 1 - L_T/P_S$.

Local adaptation versus productivity gradient

To explain the observed geographic heterogeneity in the temporal pattern of masting trees, we have two hypotheses – local adaptation and productivity gradient. In this section, we discuss how to distinguish these two hypotheses to explain geographic variation in masting in rowan populations in southern Norway.

The local adaptation hypothesis assumes that observed intraspecific variation in masting may be caused by geographic differentiation among the genes that control the reproductive strategy. It is reasonable to assume that both the resource depletion coefficient (k) and reproductive threshold (L_T) are genetically determined, because fruit quality parameters (the size, number, nutrient quality and morphology) are important determinants of reproductive success of plants. For instance, a plant employing a larger k invests more resources on seed and flower production for a given amount of excess in resource reserve. One way to do this is to produce higher quality fruits. If L_T increases, the maximum capacity of resource investment for fruiting per year is increased, resulting in an increase in the maximum number of fruits that can be produced in any given year. The local adaptation hypothesis implies that k, L_T, or both vary locally to match the local optimum with respect to reproductive strategy given the local environmental and ecological conditions. Geographic variation in these parameters is then hypothesised to cause geographic variation in masting dynamics.

In contrast, the productivity gradient hypothesis assumes that the differences in masting dynamics result from geographic variation in productivity, P_S, due to different environmental conditions such as variation in temperature and precipitation levels. It considers that the underlying biological parameters do not vary geographically – in particular, parameters k and L_T would be the same for different geographic regions. Based on the climatic data for southern Norway, we conjecture that trees in the west have greater productivity than in the east.

Theoretical predictions

Which hypothesis is more likely to predict a shift from 2- to 3-year cycle in reproductive dynamics of the rowan populations in going from west to east in southern Norway? To answer this question, we analyse the model with minimum threshold to resource depletion, given by Equations (11.10) and (11.11).

We first note that the classic resource budget model without minimum threshold does not explain a shift from a 2- to a 3-year cycle in masting. In the classic resource budget model, plant reproductive dynamics are predicted to change from annual constant reproduction to a quasi 2-year cycle with high and low reproductive activity in alternating years, and then to irregular and strongly chaotic fluctuations as the plant invests more resources in fruiting (i.e. as the depletion coefficient, k, increases). Varying k therefore cannot produce a shift from (quasi) 2-year cycle to 3-year cycle if there is no lower bound on resource depletion. Moreover, variation in the reproductive threshold, L_T, or productivity, P_S, does not influence the reproductive dynamics qualitatively because the underlying (non-dimensionalised) resource dynamics are independent of the magnitude of L_T and P_S.

The refined model with an explicit minimum threshold on the dynamics predicts a shift from (quasi) 2-year cycle to 3-year cycle in a robust manner, as illustrated in Figure 11.10a. We calculated the power spectrum of the time series generated from the model in order to elucidate the statistical periodicities (Figure 11.8a). Recall that dominant fluctuations at a frequency of 0.5 represent a 2-year cycle (and 0.25 a 4-year cycle, and so forth). As predicted by the original model, we initially find a shift from annual, quasi-periodic 2-year cycle, to chaotic dynamics as the depletion coefficient, k, increases. However, we also observe a transition from chaotic to periodic behaviour for high values of k that is not predicted by the standard model. Importantly, there is a parameter region with a conspicuous 3-year cycle in reproduction that is absent from the model without the minimum threshold.

The results show that the transition from 2- to 3-year reproductive cycle is likely with a change of the depletion coefficient, k, (Figure 11.10a), implying a theoretically plausible mechanism for how local adaptation may give rise to geographic variation in masting dynamics. The results indicate that the transition from 2- to 3-year

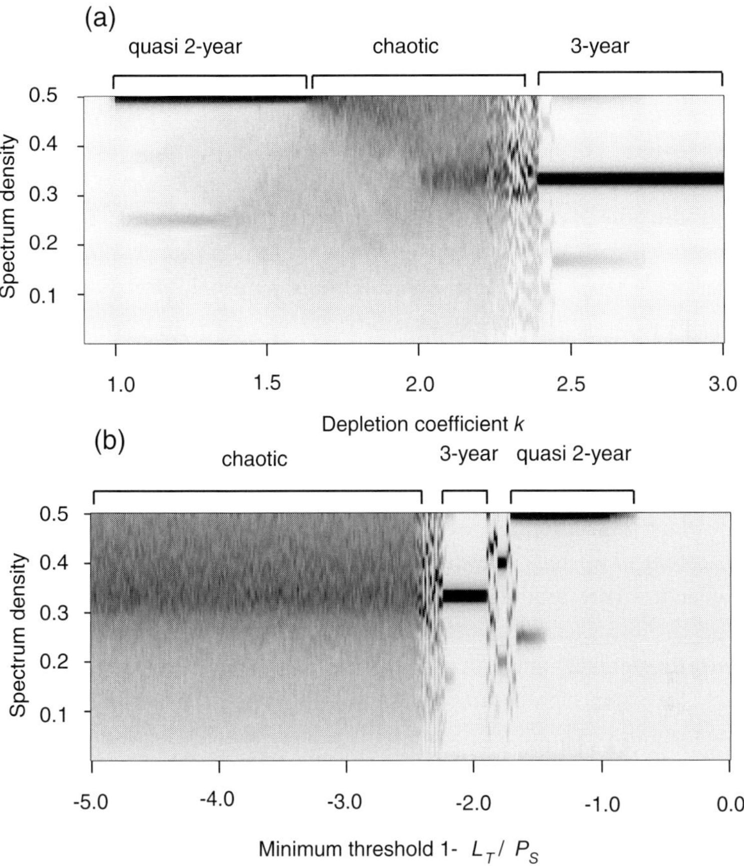

Figure 11.10 **(a)** The spectral density of reproductive dynamics of a single tree generated by the model with minimum threshold ($1 - L_T/P_S = -1.2$). Horizontal axis is depletion coefficient k. **(b)** The spectral density of reproductive dynamics of a single tree. Horizontal axis is the minimum threshold. Parameter $k = 2.5$. From Satake and Bjørnstad (2008).

reproductive cycle is possible if the rowan population in eastern Norway employs a much larger depletion coefficient (more than double) than that in western Norway (Figure 11.10a). However, such a large difference between individuals of the two areas in fruit quality seems unrealistic. Although the refined model produces a theoretically feasible transition from a 2- to a 3-year cycle because of local adaptation, the magnitude of difference required for making such a transition is too large to be realistic in explaining the geographic variation in fruit quality, and we conclude that the local adaptation hypothesis alone is unlikely to be the dominant process responsible for the observed geographic patterns of temporal fluctuation of rowan fruit production.

We now turn to explore an alternative hypothesis, a productivity gradient. We studied the spectral density as a function of the minimum threshold,$1 - L_T/P_S$. As the minimum threshold increases, the reproductive dynamics become more cyclic, showing a clear periodicity in reproduction (Figure 11.10b). The result further shows that there is a transition from 3- to 2-year cycles as the minimum threshold increases, even for a fixed value of k. The minimum threshold decreases as the reproductive threshold L_T increases. Therefore, these results imply that the observed shift in rowan populations in southern Norway is possible if the plant population in eastern Norway employs a larger reproductive threshold than that in western Norway.

Note that the minimum threshold increases with the annual productivity P_S, which means that the observed shift also is expected if the plant population in eastern Norway experiences lower productivity than that in western Norway. Based on our review of climatic data, we conjecture that productivity is in fact lower in eastern than western Norway. If this is true, the productivity gradient hypothesis is supported both theoretically and empirically.

In short, the local adaptation hypothesis considering altering the magnitude of the reproductive threshold between local populations is still a theoretical possibility, but we do not have empirical evidence to support it.

11.5 DISCUSSION

In this chapter, we have attempted to review the current theoretical studies on masting. We discussed the role of plant–pollinator interactions and environmental fluctuations in inducing interspecific or intraspecific synchrony in flowering. We have shown that synchronised and intermittent flowering among different species that share pollinators (interspecific synchrony) is likely to occur when pollinators employ a clear switching behaviour in patch choices. Social foragers such as honeybees typically show such behaviour (Momose *et al.* 1998). Colonies of social foragers have scout foragers that recruit foragers (Roubik 1989). Scout foragers monitor floral resources, ignore poor flower patches and continue seeking richer ones. When they find a patch with a larger number of flowers than a threshold, they recruit the colony members and harvest the floral resources (Roubik 1989, Roubik *et al.* 1995). Forests where such social foragers are abundant are expected to show synchronised flowering between species more frequently. Interspecific synchrony is especially likely when plant species density is low. Species diversity is rich in tropical rain forests in Southeast Asia, implying that the density of each species is low (Whitmore 1984). This low density may work to cause aggregated flowering to attract pollinators.

In Southeast Asia, plant–pollinator interactions are more complicated than the situation considered in this chapter (i.e. two plant species and one pollinator). In addition, even generalist pollinators may show significant preferences among food sources, which we neglected in the model. Having created a minimum model that incorporates a few essential factors, future work might consider how the inclusion of additional factors would alter model predictions.

Climatic variables, such as temperatures and precipitation, may influence the reproductive dynamics through various stages: floral development, pollination, seed ripening and resource acquisition. *Chionochloa pallens* (Poaceae) from the alpine zone in New Zealand is a good example to show that floral development is strongly correlated with a simple climatic variable. In this herbaceous species, it has been shown that flower induction is controlled by temperature in the summer before flower-head emergence. If January temperature is below about 11.5 °C, very few inflorescences are produced (Rees *et al.* 2002). However, the reproductive constraint should have more complex physiological processes in woody plants than the case for *C. pallens*, which has a single dominant environmental cue. Although Koenig and Knops (2000) demonstrated that the spatial autocorrelations in annual rainfall and summer temperature are similar to those exhibited by annual seed production of beech, oak, fir and pine in the northern hemisphere, this similarity cannot provide any indication to distinguish whether the spatial synchrony in reproduction was caused by climatic cueing or by simple reflection of spatially correlated productivity (resource availability) induced by climatic fluctuation. With these considerations in mind, long-term quantitative studies of reproductive behaviour of individual plants and manipulative experiments that alter nutrient level, water availability and temperature would both be useful in distinguishing different hypotheses (Kelly and Sork 2002).

A promising approach for future studies of masting is an integration of molecular, physiological and modelling approaches. Recent rapid progress in genetic and molecular analyses has provided the genetic description of well-established flowering pathways in model plants. Proposed in *Arabidopsis thaliana*, the gene regulatory model of floral transition describes complex interactions between environmental cues (such as photoperiod, low temperature) and the plant hormone gibberellin perceived by different parts or pathways. Apart from environmental cues and plant hormones, the role of sugars in plant signalling is widely accepted (Sheen *et al.* 1999). Sugars are suggested to act to signal the transition to flowering (Emert *et al.* 1995, Corbesier and Coupland 2005), and sucrose regulates flowering gene *LFY* expression (Ohto *et al.* 2001). In addition, recent analysis of transgenic plants and sugar response mutants uncovered intimate crosstalk between glucose and other plant hormone signalling pathways (Zhou *et al.* 1998). These findings provide evidence that environmental cues and resource (sugar) availability work jointly at the molecular level in determining the timing of flowering. These findings also provide opportunities to test the model assumption that plants can flower when stored resource levels exceed the threshold. Whether this assumption is reasonable or not should be tested by examining how flowering genes are regulated by resource levels.

Combining the power of genetics in unravelling the interplay between environmental cues and resource availability, and the insight from mathematical modelling in structuring and predicting underlying dynamics and evolutionary adequacy of the interplay, will provide a new means for dissecting the complicated mechanisms and evolutionary benefits of mast seeding.

ACKNOWLEDGEMENTS

This work was supported in part by a fellowship and Grants-in-Aid from the Japan Society for the Promotion of Science to AS and YI, and PRESTO of Japan Science and Technology Agency to AS. The authors thank O. N. Bjørnstad and J. Ghazoul for their helpful comments.

REFERENCES

Appanah, S. (1993). Mass flowering of Dipterocarp forest in the aseasonal tropics. *Journal of Biosciences* **18**, 457–474.

Ashton, P. S., Givnish, T. J. and Appanah, S. (1988). Staggered flowering in the Dipterocarpaceae: new insights into floral induction and the evolution of mast fruiting in the aseasonal tropics. *American Naturalist* **132**, 44–66.

Büsgen, M. and Münch, E. (1929). *The Structure and Life of Forest Trees*. London: Chapman and Hall.

Campbell, D. R. and Motten, A. F. (1985). The mechanism of competition for pollination between two forest herbs. *Ecology* **66**, 554–563.

Corbesier, L. and Coupland, G. (2005). Photoperiodic flowering of *Arabidopsis*: integrating genetic and physiological approaches to characterization of the floral stimulus. *Plant, Cell and Environment* **28**, 54–66.

Curran, L. M. and Leighton, M. (2000). Vertebrate responses to spatiotemporal variation in seed production of mast-fruiting Dipterocarpaceae. *Ecological Monographs* **70**, 101–128.

Crone, E. E., Polansky, L. and Lesica, P. (2005). Empirical models of pollen limitation, resource acquisition and mast seeding by a bee-pollinated wildflower. *American Naturalist* **166**, 396–408.

Emert, K., Wang, S.-M., Lue, W.-L. and Chen, J. (1995). Monogenic recessive mutations causing both late floral initiation and excess starch accumulation in *Arabidopsis*. *Plant Cell* **7**, 1703–1712.

Feldman, T. S., Morris, W. F. and Wilson, W. G. (2004). When can two plant species facilitate each other's pollination? *Oikos* **105**, 197–207.

Ghazoul, J. (2006). Floral diversity and the facilitation of pollination. *Journal of Ecology* **94**, 295–304.

Isagi, Y., Sugimura, K., Sumida, A. and Ito, H. (1997). How does masting happen and synchronise? *Journal of Theoretical Biology* **187**, 231–239.

Kelly, D. (1994). The evolutionary ecology of mast seeding. *Trends in Ecology and Evolution* **9**, 465–470.

Kelly, D., Harrison, A. L., Lee, W. G. *et al.* (2000). Predator satiation and extreme mast seeding in 11 species of *Chionochloa* (Poaceae). *Oikos* **90**, 472–488.

Kelly, D. and Sork, V. L. (2002). Mast seeding in perennial plants: why, how, where? *Annual Review of Ecology and Systematics* **33**, 427–447.

Kobro, S., Søreide, L., Djønne, E. *et al.* (2003). Masting of rowan, *Sorbus aucuparia* L., and consequences for the apple fruit moth, *Argyresthia conjugella* Zeller. *Population Ecology* **45**, 25–30.

Koenig, W. D. and Knops, J. M. H. (1998). Scale of mast-seeding and tree-ring growth. *Nature* **396**, 225–226.

Koenig, W. D. and Knops, J. M. H. (2000). Patterns of annual seed production by northern hemisphere trees: a global perspective. *American Naturalist* **155**, 59–69.

Koenig, W. D., Knops, J. M. H., Carmen, W. J. and Stanback, M. T. (1999). Spatial dynamics in the absence of dispersal: acorn production by oaks in central coastal California. *Ecography* **22**, 499–506.

Laverty, T. M. (1992). Plant interactions for pollinator visits: a test of the magnet species effect. *Oecologia* **89**, 502–508.

Liebhold, A., Sork, V. L., Peltonen, M. *et al.* (2004). Within-population spatial synchrony in mast seeding of North American oaks. *Oikos* **104**, 156–164.

McKone, M. J., Kelly, D. and Lee, W. G. (1998). Effect of climate change on mast-seeding species: frequency of mass flowering and escape from specialist insect seed predators. *Global Change Biology* **4**, 591–596.

Moeller, D. A. (2004). Facilitative interactions among plants via shared pollinators. *Ecology* **85**, 3289–3301.

Momose, K., Ishii, R., Sakai, S. and Inoue, T. (1998). Plant reproductive intervals and pollinators in the aseasonal tropics: a new model. *Proceedings of the Royal Society of London B* **265**, 2333–2339.

Moran, P. A. P. (1953). The statistical analysis of the Canadian lynx cycle. II. Synchronisation and meteorology. *Australian Journal of Zoology* **1**, 291–198.

Numata, S., Yasuda, M., Okuda, T., Kachi, N. and Noor, N. S. M. (2003). Temporal and spatial patterns of mass flowering on the Malay Peninsula. *American Journal of Botany* **7**, 1025–1031.

Ohto, M., Onai, K., Furukawa, Y. *et al.* (2001). Effects of sugar on vegetative development and floral transition in *Arabidopsis*. *Plant Physiology* **127**, 252–261.

Pellmyr, O. (1986). The pollination ecology of two nectarless *Cimifuga* spp. (Ranunculaceae) in North America. *American Journal of Botany* **6**, 713–723.

Rathcke, B. (1983). Competition and facilitation among plants for pollination. In L. Real (ed.), *Pollination Biology*. Orlando, FL: Academic Press, pp. 305–329.

Rathcke, B. (1988). Interactions for pollination among coflowering shrubs. *Ecology* **69**, 446–457.

Rees, M., Kelly, D. and Bjørnstad, O. N. (2002). Snow tussocks, chaos, and the evolution of mast seeding. *American Naturalist* **160**, 44–59.

Roubik, D. W. (1989). *Ecology and Natural History of Tropical Bees*. Cambridge: Cambridge University Press.

Roubik, D. W., Inoue, T. and Hamid, A. A. (1995). Canopy foraging by two tropical honeybees: bee height fidelity and tree genetic neighborhoods. *Tropics* **5**, 81–93.

Roy, B. A. (1994). The effects of pathogen-induced pseudoflowers and buttercups on each other's insect visitation. *Ecology* **75**, 352–358.

Roy, B. A. (1996). A plant pathogen influences pollinator behavior and may influence reproduction of nonhosts. *Ecology* **77**, 2445–2457.

Royama, T. (1992). *Analytical Population Dynamics*. New York: Chapman and Hall.

Sakai, S., Harrison, R. D., Momose, K. *et al.* (2006). Irregular droughts trigger mass flowering in aseasonal tropical forests in Asia. *American Journal of Botany* **93**, 1134–1139.

Sakai, S., Momose, K., Yumoto, T. *et al.* (1999). Plant reproductive phenology over four years including an episode of general flowering in a lowland Dipterocarp forest, Sarawak, Malaysia. *American Journal of Botany* **86**, 1414–1436.

Satake, A. and Bjørnstad, O. N. (2008). A resource budget model to explain intraspecific variation in mast reproductive dynamics. *Ecological Research* **23**, 3–10.

Satake, A., Bjørnstad, O. N. and Kobro, S. (2004). Masting and trophic cascades: interplay between rowan trees, apple fruit moth, and their parasitoid in southern Norway. *Oikos* **104**, 540–550.

Satake, A. and Iwasa, Y. (2000). Pollen-coupling of forest trees: forming synchronised and periodic reproduction out of chaos. *Journal of Theoretical Biology* **203**, 63–84.

Satake, A. and Iwasa, Y. (2002a). Spatially limited pollen exchange and a long-range synchronisation of trees. *Ecology* **83**, 993–1005.

Satake, A. and Iwasa, Y. (2002b). The synchronised and intermittent reproduction of forest trees is mediated by the Moran effect, only in association with pollen coupling. *Journal of Ecology* **90**, 830–838.

Schauber, E. M., Kelly, D., Turchin, P. *et al.* (2002). Masting by eighteen New Zealand plant species: the role of temperature as a synchronising cue. *Ecology* **83**, 1214–1225.

Sheen, J., Zhou, L. and Jyun-Chyun J. (1999). Sugars as signaling molecules. *Current Opinion in Plant Biology* **2**, 410–418.

Sperens, U. (1997). Fruit production in *Sorbus aucuparia* L. (Rosaceae) and pre-dispersal seed predation by the apple fruit moth (*Argyresthia conjugella* Zell.). *Oecologia* **110**, 368–373.

Suzuki, W., Osumi, K. and Masaki, T. (2005). Mast seeding and its spatial scale in *Fagus crenata* in northern Japan. *Forest Ecology and Management* **205**, 105–116.

Tachiki, Y. and Iwasa, Y. (2008). Role of gap dynamics in the evolution of masting of trees. *Evolutionary Ecology Research* **10**, 893–905.

Thomson, J. D. (1981). Spatial and temporal components of resource assessment by flower-feeding insects. *Journal of Animal Ecology* **50**, 49–59.

Thomson, J. D. (1982). Patterns of visitation by animal pollinators. *Oikos* **39**, 241–250.

Waller, D. M. (1979). Models of mast fruiting in trees. *Journal of Theoretical Biology* **80**, 223–232.

Waser, N. M. and Fugate, M. L. (1986). Pollen precedence and stigma closure: a mechanism of competition for pollination between *Delphinium nelsonii* and *Ipomopsis aggregata*. *Oecologia* **70**, 573–577.

Whitmore, T. C. (1984). *Tropical Rain Forests of the Far East*. 2nd edn. Oxford: Clarendon Press.

Yasuda, M., Matsumoto, J., Osada, N. *et al.* (1999). The mechanism of general flowering in Dipterocarpaceae in the Malay Peninsula. *Journal of Tropical Ecology* **15**, 437–449.

Zhou, L., Jang, J. C., Jones, L. and Sheen, J. (1998). Glucose and ethylene signal transduction cross-talk revealed by an *Arabidopsis* glucose-insensitive mutant. *Proceedings of the National Academy of Sciences, USA* **95**, 10294–10299.

MICHAEL G. BOWLER, FELIX BREDEN AND COLLEEN K. KELLY

12

Temporal dynamics and the spread of insect resistance transgenes

12.1 INTRODUCTION

Crops can be made resistant to pestiferous insects by genetic modification. The possibility that the modified genetic material may escape into a population of wild relatives is a threat looming over all such endeavours. Since insect infestations fluctuate from year to year, or over longer periods (locusts are an extreme example), the fitness of transgenes escaped from agriculture will be affected by this temporal dynamic and that is the subject of this chapter.

There are two ways in which genetically modified crops may pose a threat to the environment. The first is by spreading beyond arable land and simply outcompeting wild relatives by virtue of the insect resistance modification. The second is by hybridising with wild relatives so that the insect resistance transgene invades and transforms the wild population. The first is unlikely; crops are selected for yield and require tender loving care in order to thrive. The second threat is more insidious and has usually been addressed through population genetics. The role of ecological interactions has been comparatively neglected and the effect on relative fitness of temporal fluctuations in various factors, such as herbivore densities, falls into this category.

Here we develop a generic model for competing annual plant populations sustained by a seedbank and subject to levels of insect attack, primarily herbivory, which fluctuate with time. Within this general framework, we consider the cases of feral crops competing with

Temporal Dynamics and Ecological Process, ed. C. K. Kelly, M. G. Bowler, G. A. Fox.

wild relatives on the one hand and competition and interbreeding among the wild population and its modified variants on the other.

The importance of the temporal dynamics is that plants carrying an insect resistance transgene can be threateningly superior during a season of insect infestation but their spread may be halted or reversed when the plague is in abeyance. All that is required is sufficient cost associated with the production of insecticide.

The structure of this chapter is as follows. We first establish in more detail the background to the ecological problem. Then the basic seedbank model is introduced, first with competition between two varieties which do not interbreed. More complicated is the seedbank model with interbreeding – a single-locus, diploid model in which three interbreeding varieties drift across a Hardy–Weinberg landscape under ecological pressure. This evolution can commence with deposition of modified seeds, or alternatively input of pollen to the wild from the modified crop in the fields.

Examples of all three scenarios are presented in figures and the implications of our findings are then reviewed.

Background

Insect resistance (IR) transgenes offer significant advantages to agriculture and the economic pressure to use them is increasing. They have the advantage of reducing pesticide use and thus reducing the impact of agriculture on the natural environment (Dale *et al.* 2002, Stewart *et al.* 2003). However, there is concern that where close relatives of crop species occur in nearby natural communities, IR transgenes may 'escape' through gene flow from crops into native plants and disrupt normal community function (Arnaud *et al.* 2003, GM Science Review Panel 2003, Raybould and Gray 1994). Effective gene flow from a crop into a non-crop population is a composite of the compatibility of crop and non-crop types, the availability of pollen and ecological interactions of non-crop individuals with and without the 'new' genes that have been gained through crop pollen and possibly seed. Although ecological interactions may be the most important driver predicting the spread of transgenes (Rieseberg and Burke 2001), the focus thus far has been primarily from a population genetics perspective (e.g. Arias and Rieseberg 1994, Haygood *et al.* 2004, Linder *et al.* 1998, Whitton *et al.* 1997). To better weight the balance of understanding, we here turn our attention to the ecological action of transgenes in a natural community.

The details of the ecological processes involved in gene flow are generally 'black-boxed', as the relative difference in seed set of individuals with and without crop genes. However, a single such 'snapshot' may not be sufficient to predict the long-term behaviour of the interaction, through a lack of all the relevant determinants of fitness (Crawley 1999). It has been shown in a rapidly increasing number of ecological studies that the relative success of competing plant types may vary with a periodicity longer than the duration of such studies, or relative advantage may play out over an arena greater than that measured by seed set alone, or both (Chesson 2003). Seed set alone *could* be an accurate estimator of the relative role of ecology in determining gene flow, but that must be determined, not assumed.

Our model addresses the behaviour of the transgene in the natural habitat in terms of the costs and advantages incurred by possession of the transgene. With our model, it is possible to assess the potential that plants with an insect resistance Bt (IRBt) allele may have for displacing untransformed plants in the natural community, making it an important tool in evaluating the potential of IR transgenes to disrupt wider foodweb dynamics. If plants with the IRBt allele do not persist or do not form a large part of the population of host plants used by dependent herbivores, then a large impact of the transgene on the general community is unlikely. Our model describes change over time, allowing more realistic dynamic estimations of the direct and indirect impact of transgenes on higher trophic levels as well as on herbivores. Given the necessary data, the model can be used easily in risk assessment frameworks, answering recent criticism on the lack of quantitative predictions in previous assessment practices; its parameters are defined and measurable, relating to ecological endpoints and 'trigger values'.

Our model explicitly includes temporal variability in herbivore levels, a well-documented phenomenon but one rarely incorporated in analytical treatments of the impact of herbivory. In the simplest case, if the IRBt allele were to provide a consistent benefit of insect resistance to its carrier, it would exclude the wild-type allele. However, year-to-year fluctuation in numbers of insect pests is an innate property of both crop and natural systems and the advantage of resistance to herbivores will vary over time (Price 1997, NERC Centre for Population Biology 1999). The effect on plants of temporal fluctuation in a selective agent is described by storage dynamics (Chesson and Huntly 1989, Chesson 1990, 2003, Kelly and Bowler 2002, 2003, 2005). 'Storage' refers to recovery of a population from a period of low recruitment, being 'stored' in the reproductive capacity of long-lived individuals or in resting propagules. Storage theory tracks the impact of a selective factor on recruitment; this makes it especially well suited for herbivory, where the probability of mortality from herbivory is likely to be greater at the seedling than the adult stage. In the storage model of Kelly and Bowler (2002, 2005), competitors are differentially sensitive to a controlling environmental factor (Fenner *et al.* 1999, Hanley and Lamont 2001, Kelly and Hanley 2005), as would be the case for plants with and without the IRBt allele. Our model follows Kelly and Bowler (2002, 2005), incorporating a modification of the differential sensitivity model so that population persistence rests upon 'storage' of seeds in the soil seed bank *sensu* Chesson (1990) and Chesson and Huntly (1989). We have applied our model to oilseed rape (OSR) and its herbivore diamondback moth (*Plutella xylostella* syn. *P. maculipennis*). We have selected this system because the majority of IR genes that have been commercialized are Lepidopteran-specific and because year-to-year data on population levels of the pest happen to be readily available.

12.2 THE MODEL

The model assumes that seedling establishment happens in a single wave at the start of the growing season, when the initial proportion of seedlings of each type present is determined by the proportion of available seeds germinating. As the season progresses, seedlings are differentially eliminated by insect herbivores (Brown and Gange 1989, Hanley *et al.* 1995) or through competition (Ford 1975). At the end of the season all suitable sites are filled by mature plants producing seeds, which enter the seedbank to germinate in following seasons.

The IR allele may drive the wild-type allele extinct or the IR allele may merely become so common as to threaten the stability of those herbivores depending on the plant. Both possibilities are of interest. The speed and extent to which the transgene may displace the wild-type allele depends on the proportion of seasons that herbivory is low (f) or high ($1 - f$) and the relative advantage/disadvantage of the modified and unmodified types at high and low levels of herbivory. This relative relationship is largely embodied as a single quantity a^*; a^* is the time-dependent relative advantage of the transgene lineage over the unmodified lineage under target conditions and a *ratio* of several competitive factors (presented in Equation (12.6); Table 12.1). When herbivory is low, the advantage is represented as a^{*+}, and is less than unity because of the cost of resistance for the resistant form. When herbivory levels

Table 12.1 *Guide to terms used in this chapter.*

y_i	fraction of adults of variety *i* per adult site
z_i	ratio of seed loss to seed production for variety *i*
x_i	number of seeds of variety *i* per adult site in the seedbank before germination
E_i	fraction of seeds germinating
s_i	fraction of seeds in the seedbank (left after germination) which survive until next year
Y_i	seed yield per adult, measured as viable seeds in the next season
Y^*	ratio of seed yields $\frac{Y}{Y_{ww}}$
p_i	probability that a plant will be pollinated by variety *i*
η_i	amount of pollen produced by variety *i*
η^*	ratio of pollen production $\frac{\eta}{\eta_{ww}}$
α_i	probability of a germinating seed reaching the adult stage in the absence of competition
β_i	seedling competitive factor
K_i	long-term rate at which seeds are lost from the seedbank
a^*	relative advantage of the transformed over the untransformed type
f	fraction of time herbivory is low (at other times it is high)
L	a measure of the fractional increase of the genetic modification, averaged over many steps

are high and have a high impact on seedling mortality, a^* indicates the relative advantage of the protected form of the plant. Reproductive capacity is stored in a seedbank and at a given time the number of seeds of genotype *i* present in an area which can sustain one adult plant is x_i. Its value x_i^+ at the next time step depends on the loss of seeds from the seedbank and the recruitment of seeds following seed set in adult plants. Genotype *i* sets Y_i seeds per adult plant and the probability that an adult of type *i* occupies a suitable site is represented by y_i. Vegetation is dense and we impose $\sum y_i = 1$; all available places are taken by an adult of one of the genotypes. For the population under natural conditions, the number of seeds in the seedbank in year $t + 1$, x^+, is the sum of the number remaining L and the number recruited R

$$x^+ = L + R, \tag{12.1}$$

where L has the form $(1 - K)x$ and where K is the long-term rate at which seeds are lost from the seedbank; the loss of seeds from the seedbank in one time step is x_iK_i where

$$K_i = 1 - \{1 - E_i\}s_i, \tag{12.2}$$

with E_i the fraction of seeds germinating in any one year and s_i the number surviving to the next season. Generally

$$x_i^+ = (1 - K_i)x_i + R_i, \tag{12.3}$$

where the recruitment term R_i depends on the adult populations and on seed set. It is in general a function of competition and it is here that we insert the effects of fluctuating insect damage.

Without interbreeding the evolution equations for the seedbank are

$$\begin{aligned} x_W^+ &= (1 - K_W)x_W + Y_W y_W \\ x_G^+ &= (1 - K_G)x_G + Y_G y_G \end{aligned}, \tag{12.4}$$

where the second term on the right is recruitment to the seedbank in terms of the probability of an adult of genotype i and fecundity Y_i, W denoting the wild type and G the feral genetically modified crop.

With interbreeding, the recruitment is not so simply described. The recruitment term is broken into a sum of factors of form

$$R = \sum [d][m], \tag{12.5}$$

where d is the number of seeds endowed with the appropriate alleles by females and m is the frequency with which pollen containing the complementary allele is encountered. Ours is a single-locus diploid model with a genetically manipulated allele (g) and a 'gap' (w) analogous to a wild-type allele determining anti-herbivore defence levels. We suppose that at some stage a genetically modified crop has successfully crossed with a wild variety, and denote individuals without the transgene (wild type) by ww, individuals carrying one copy of the modified allele (hemizygous) by wg and plants carrying two copies of the modified allele by gg. As a result of repeated crosses between wild and wild-hybrid individuals, the transgene will be operating within the genetic background of the native species. We consider the evolution of the population composed of three different single-locus genotypes, first in the absence of a source of external pollen. The evolution equations, which describe the drift through the Hardy–Weinberg landscape under ecological pressures, become

$$\begin{aligned} x_{ww}^+ &= (1 - K_{ww})x_{ww} + [Y_{ww}y_{ww} + 0.5Y_{wg}y_{wg}]\,[p_{ww} + 0.5p_{wg}] \\ x_{wg}^+ &= (1 - K_{wg})x_{wg} + [Y_{ww}y_{ww} + 0.5Y_{wg}y_{wg}]\,[p_{gg} + 0.5p_{wg}] \\ &\quad + [Y_{gg}y_{gg} + 0.5Y_{wg}y_{wg}]\,[p_{ww} + 0.5p_{wg}] \\ x_{gg}^+ &= (1 - K_{gg})x_{gg} + [Y_{gg}y_{gg} + 0.5Y_{wg}y_{wg}][p_{gg} + 0.5p_{wg}]. \end{aligned} \tag{12.6}$$

In the equation defining x^+_{ww}, the first term in square brackets is the number of seeds a female can endow with a *w* allele and the second term in square brackets is the frequency with which that female encounters *w* pollen, and so on through the three equations. The p_i are the probabilities of a single plant being pollinated by variety *i*. It is important to note that these equations follow the number of each genotype, as opposed to standard population genetics models which predict change in gene frequencies. Following absolute numbers facilitates the incorporation of ecological factors into the model.

Comparison of Equations (12.5) and (12.6) shows that recruitment to the seedbank of any one variety is not controlled by a single fecundity parameter. The total seed yield, which is the sum of the terms in square brackets in Equation (12.6), has a simple form

$$Y_{ww}y_{ww} + Y_{wg}y_{wg} + Y_{gg}y_{gg}$$

which is approximately equal to the sum

$$K_{ww}x_{ww} + K_{wg}x_{wg} + K_{gg}x_{gg}$$

if the total number of seeds is approximately constant.

Equations (12.4) and (12.6) need to be completed by specifying the relationship between the seed densities x_i, pollination probabilities p_i and the probabilities y_i of an adult of variety *i*. The latter depends on x_i, the germination fraction E_i, the probability that a germinating seed produces an adult in the absence of competition with other plants α_i and on competitive factors following germination

$$y_i \propto \alpha_i \beta_i E_i x_i.$$

The effects of competition are included by writing

$$y_i = \frac{\alpha_i \beta_i E_i x_i}{\sum_j \alpha_j \beta_j E_j x_j}, \tag{12.7}$$

so that $\sum y_i = 1$. The quantities β_i are possible additional competitive factors, such as effects of overtopping of one variety by another or, in this application, the costs of producing insecticide. We are particularly concerned with the time-dependent effect of herbivores on the α_i parameters. This equation also illustrates that the success of one genotype over another is governed by factors other than simple fecundity.

The pollination probabilities p_i must sum to 1 and might simply be the y_i; we can be more general by setting

$$p_i = \frac{\alpha_i \beta_i E_i \eta_i x_i}{\sum_j \alpha_j \beta_j E_j \eta_j x_j},$$

where η_i measures pollen production for an adult plant of type i.

For the case where there is, at least some of the time, input of g pollen from neighbouring crops the Equations (12.6) are modified. The nature of the modification is to add to the terms in $[m]$ (Equation (12.5)) an external source of g pollen. The within-population pollination probabilities in Equations (12.6), p_i, are changed in $[m]$ as follows:

$$\begin{array}{l} p_{ww} \to p_{ww}(1-\varepsilon) \\ p_{wg} \to p_{wg}(1-\varepsilon) \\ p_{gg} \to p_{gg}(1-\varepsilon)+\varepsilon \end{array},$$

where ε is the probability of successful pollination by pollen arriving from outside the community. Equations (12.6), which describe the evolution with time of the density of seeds x_i of variety i, then become

$$\begin{aligned} x_{ww}^{+} &= (1-K_{ww})x_{ww} + (1-\varepsilon)[Y_{ww}y_{ww} + 0.5Y_{wg}y_{wg}][p_{ww} + 0.5p_{wg}] \\ x_{wg}^{+} &= (1-K_{wg})x_{wg} + (1-\varepsilon)\left\{\begin{array}{l} [Y_{ww}y_{ww} + 0.5Y_{wg}y_{wg}][p_{gg} + 0.5p_{wg}] + \\ [Y_{gg}y_{gg} + 0.5Y_{wg}y_{wg}][p_{ww} + 0.5p_{wg}] \end{array}\right\} \\ &\quad + \varepsilon[Y_{ww}y_{ww} + 0.5Y_{wg}y_{wg}] \\ x_{gg}^{+} &= (1-K_{gg})x_{gg} + (1-\varepsilon)\{[Y_{gg}y_{gg} + 0.5Y_{wg}y_{wg}][p_{gg} + 0.5p_{wg}]\} \\ &\quad + \varepsilon[Y_{gg}y_{gg} + 0.5Y_{wg}y_{wg}]. \end{aligned} \tag{12.8}$$

Given the parameters, the seed population equations can be iterated from desired starting conditions but as they stand they are rather indigestible. It is convenient to rewrite the equations in terms of variables z related to the seed densities x by the relation $z = Kx/Y$. These variables are the ratio of seed loss rate to seed production per individual of that variety. They have a simple interpretation; when things are changing slowly (in (quasi-) equilibrium) they are just the fractions of available sites occupied, the quantities y_i. This follows immediately from Equations (12.3) and (12.4). If there is no interbreeding (i.e. there are no heterozygotes) the sum of (two) z_i becomes unity after a few steps. This remains true in the single-locus diploid model if all three Y_i are equal; if any two z_i are small the third is approximately unity. This property is useful in working out analytic approximations as in Equations (12.13–12.16) below. The greatest advantage of these variables is that Equations (12.6) and (12.8) can be recast in a form which depends entirely or almost entirely on ratios of parameters characterising the two alleles, as set out in Equations (12.10) and (12.12) below.

For the case of a feral crop (G) competing with a wild relative (W) without gene flow, the evolution Equations (12.4) become in the z variables

$$z_W^+ = (1 - K_W)z_W + \frac{K_W z_W}{z_W + a^* z_G}$$
$$z_G^+ = (1 - K_G)z_G + \frac{K_G a^* z_G}{z_W + a^* z_G} \quad (12.9)$$

$$a^* = \frac{\alpha_G \beta_G E_G Y_G K_W}{\alpha_W \beta_W E_W Y_W K_G}, \quad (12.10)$$

where the single quantity a^* is a product of ratios of traits and is time dependent. It is the variations of a^* which determine whether W or G becomes dominant, or the two coexist.

If a few plants or seeds of the feral crop G exist in a background of the wild relative W, the G type will grow and progressively exclude W provided that the dimensionless measure of the fractional increase each step $L = < \ln(z_G^+/z_G) >$ is greater than zero, or $< \ln(1 - K_G + a^* K_G) >$ is greater than zero. If K_G is not too close to unity, the condition is that $<a^*>$ must be greater than unity. The average is taken over many time steps to allow for the fluctuating values of a^*.

In order to illustrate the evolution with time of the populations of three varieties in the single-locus model with hybridising, we specialise to the case where wg and gg look the same to the world (i.e. g is fully dominant) and use the variables

$$z_{ww} = \frac{K_{ww} x_{ww}}{Y_{ww}}, \; z_{wg} = \frac{K x_{wg}}{Y}, \; z_{gg} = \frac{K x_{gg}}{Y}.$$

Equations (12.6) and (12.8) now become

$$z_{ww}^+ = (1 - K_{ww})z_{ww} + \frac{K_{ww}(1-\varepsilon)}{D_2}\left[\frac{z_{ww}}{a^*} + 0.5Y^* z_{wg}\right]\left[\frac{z_{ww}}{a^*\eta^*} + 0.5z_{wg}\right]$$

$$z_{wg}^+ = (1 - K)z_{wg} + \frac{K(1-\varepsilon)}{D_2}\left[\frac{z_{ww}}{a^*}\left(\frac{1}{Y^*} + \frac{1}{\eta*}\right) + z_{wg}\right]\left[z_{gg} + 0.5z_{wg}\right]$$
$$+ \frac{K\varepsilon}{D_1}\left[\frac{z_{ww}}{a^* Y^*} + 0.5z_{wg}\right]$$

$$z_{gg}^+ = (1 - K)z_{gg} + \frac{K(1-\varepsilon)}{D_2}\left[z_{gg} + 0.5z_{wg}\right]^2 + \frac{K\varepsilon}{D_1}\left[z_{gg} + 0.5z_{wg}\right], \quad (12.11)$$

where

$$D_2 = \left(\frac{z_{ww}}{a^*} + z_{wg} + z_{gg}\right)\left(\frac{z_{ww}}{a^*\eta^*} + z_{wg} + z_{gg}\right)$$

$$D_1 = \frac{z_{ww}}{a^*} + z_{wg} + z_{gg}.$$

Y^* is the ratio of seed yields Y/Y_{ww}, a^* the relative advantage of the transformed over the untransformed type and η^* the ratio of pollen production η / η_{ww}. If $\varepsilon \rightarrow 0$, Equations (12.11) describe the situation in which there is no external input of gg pollen; the parent equations are those of (12.6) rather than (12.8).

As written, Equations (12.11) are valid provided K_{ww}, K, Y_{ww} and Y do not vary from step to step. Thus Equations (12.11) apply to foliage herbivory, where plants are damaged before flowering. The equations are easily modified for herbivory which damages the seed or directly affects seed production by infested plants (Burke and Rieseberg 2003, Snow *et al.* 2003), or for pathogens that alter seed survival in the seedbank (see below). The quantity a^* is the dominant factor in the overall advantage of modified plants, carrying g, over the natural variety *ww*. The quantity Y^* is the ratio of seed yields $\frac{Y}{Y_{ww}}$ and η^* is the ratio of pollen production $\frac{\eta}{\eta_{ww}}$. The former is also a factor in a^*, which is given by

$$a^* = \frac{\alpha\beta EY}{K}\frac{K_{ww}}{\alpha_{ww}\beta_{ww}E_{ww}Y_{ww}}, \qquad (12.12)$$

and will fluctuate with time with the ratio of any pair of parameters, such as α/α_{ww}.

If a few hybrids (*wg*) exist in a background of *ww*, the hybrid population will grow (and generate *gg*), provided that the dimensionless measure of the fractional increases each step

$$L = \left\langle \ln\frac{z_{wg}^{+}}{z_{wg}} \right\rangle_{z_{ww}\approx 1} > 0, \qquad (12.13)$$

where the average over many time steps in Equation (12.13) is to allow for environmental fluctuations in, for example, the α parameters (which might be caused by varying herbivory). This is easily obtained from the second of Equations (12.11), yielding the result that the transgene will spread against the wild variety provided

$$L = \left\langle \ln\left[1 - K + 0.5K\left\{\frac{\eta*}{Y^*} + 1\right\}a^*\right]\right\rangle \qquad (12.14)$$

is greater than 0. If $\frac{\eta^*}{Y^*} \cong 1$ and K not close to 1, the condition is $\langle a^* \rangle > 1$. The parameters in Equation (12.14) apply to the heterozygotic variety; Equation (12.14) is not affected by the possibility that the *gg* homozygote might look different.

(At the other end, suppose the *gg* variety has become prevalent, $z_{gg} \cong 1$. If the wild variety is not to be driven extinct, both z_{ww} and z_{wg} must be able to grow when almost gone. The condition for this is approximately

$$\left\langle \frac{1}{a^*} \left[\frac{1 + \frac{Y^*}{\eta^*}}{2} \right] \right\rangle > 1, \quad (12.15)$$

or if $\frac{Y^*}{\eta^*} \cong 1$, Equation (12.9) is $\approx \left\langle \frac{1}{a^*} \right\rangle > 1$.

The wild variety, the hybrid and the *gg* variety will coexist between these limits, which can be made possible by fluctuations in herbivory. The limit Equation (12.15) is of little real interest because the timescale for recovery is very long.)

Doubling time – a useful number for assessing the rate of spread of *g* – is the time taken for some initially small population of *wg* to double in a background of *ww*. This is given by

$$\frac{\ln 2}{L}, \quad (12.16)$$

where L is the average defined in Equation (12.14).

The doubling time Equation (12.16) calculates the amount of time it takes for the proportion of hemizygotes to double (while they comprise approximately less than 20% of the population). The same information is needed for the doubling time equation and for Equations (12.11); if the doubling time is long, then the patterns generated by iterating Equations (12.11) give no more information than does doubling time alone.

12.3 EXAMPLES

We have iterated the model using various values of herbivore frequencies, transgene advantage and potential costs for the transgene in a non-agricultural background. Field estimates indicate diamondback moth levels are high in approximately 35% of years, ranging between 20% and 40% (Harcourt 1963, Markkula 1965, Vanholder 1997). We have used high herbivore frequencies of 35% and 65% to illustrate the effect of frequency in the selective factor on progress of the transgene in the host plant population. Crop studies suggest that possessing the transgene may allow as much as 20 times the productivity of that of untransformed plants in an agricultural context (Stewart *et al.* 1997). In contrast, a study on BC_1 plants revealed advantages no greater, on average, than 1.55 and in some circumstances not significantly different from unity (Snow *et al.* 2003). We have therefore also examined a range of advantage levels. We have illustrated the potential impact of the transgene carrying a cost under natural conditions by applying reasonable levels for the costs of the defences that transgenes might bear (for example, through protein manufacture) in natural systems (Herms and Mattson 1992, Bergelson and Purrington 1995). For all iterations with the diploid model we have set η^* and Y^* at unity. (It is however the case that these quantities do modify the effect of the dominant factor a^* – for example, male sterility in g carrying plants would correspond to very low values of η^*.)

We present the results of our iterations of the model in Figures 12.1–12.3 and Table 12.2. Figure 12.1 illustrates the case of competition between a wild variety and a feral crop, Figure 12.2 the time evolution of three varieties characterised by *ww*, *wg* and *gg* without external pollen input. The case of agricultural *gg* pollen being supplied over the first 10 years and then cut off is shown in Figure 12.3. It will be noted that in Figures 2 and 3 the population of the hemizygote has maxima at 0.5, a well-known feature of the Hardy–Weinberg world. In all three figures, the first two rows show cases where there is no cost to possessing the transgene and in the third through fifth rows the cost decreases from a 20% fitness decrement to a 1% fitness decrement. In Figures 12.1 and 12.2 the frequency at which herbivore levels are high increases from left to right. With the diploid model, although we performed iterations with both 1% and 10% hemizygote starting proportions and give the output for both in Table 12.2, we show, for clarity, figures for only the higher initial proportion.

The figures all tell the same story. When there is no cost to possessing the transgene, the effect of increased frequency of high herbivore levels is primarily quantitative, speeding the rate at which transformed plants come to dominate the population. At 35% high herbivore levels, however, transgene costs at a level that have been observed in natural systems of defence (Bergelson and Purrington 1995) are sufficient, even when quite low (1%), to greatly slow the rate at which transformation of the population takes place. If the IR advantage at high levels of herbivory is 1.2, then a disadvantage of 0.8 at low levels means no cause for concern, 0.9 is marginal and 0.99 should be avoided. In at least one instance at 35% high herbivore frequency, the transformed type is driven out (panel (c) in Figures 12.1–12.3). In (d),

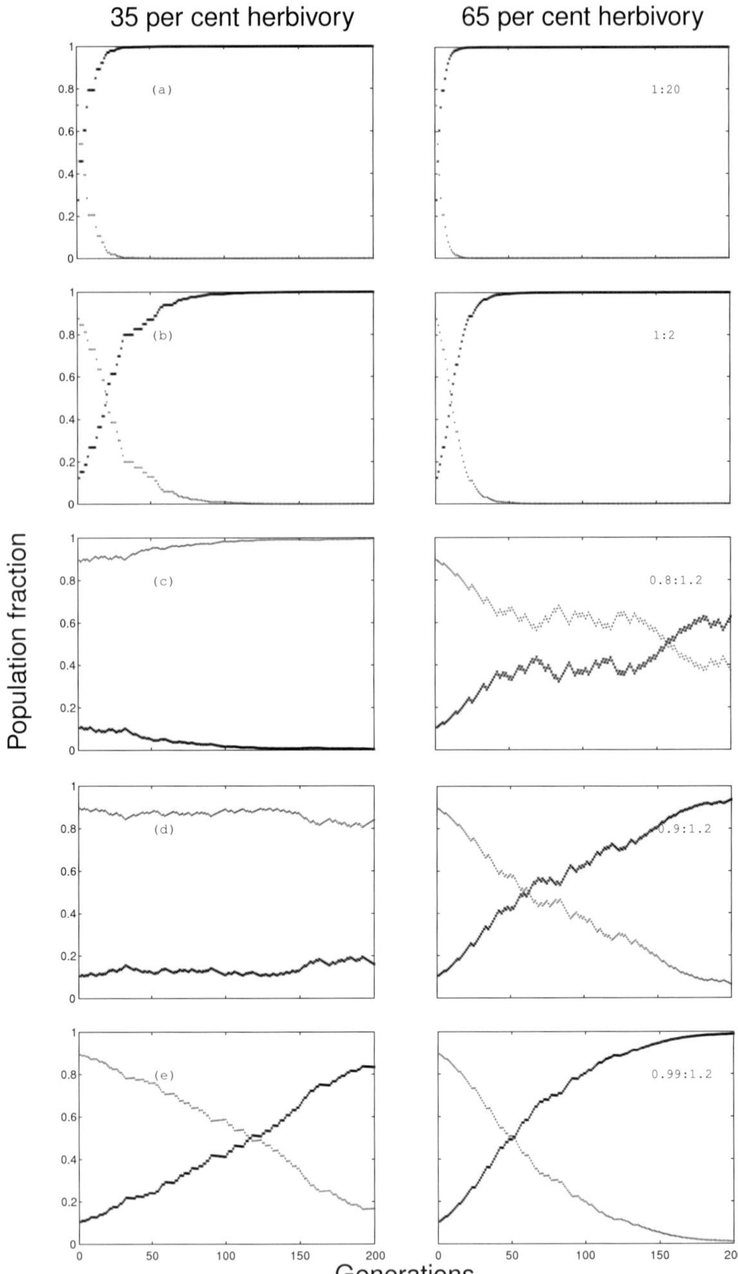

Figure 12.1 The change with time of the populations of wild type (+) and insect resistant feral crop (*) when herbivory levels are high for 35% (as observed for *Plutella*) and for contrast 65% of years (left- and right-hand

where the transgene carries a 10% fitness cost, the doubling rate of transformed plants under 35% high herbivore levels is so slow as to be indistinguishable from stable coexistence within the 200-year span of the iteration. At a greater frequency of high herbivore levels, however, all three of the iterations where transgenes carry a cost ($a^{*+} < 1$) show a steady decrease over time in the proportion of wild-type individuals; the costs are not high enough to keep the transgene in check.

As far as a period of sustained injection of agricultural pollen is concerned, iterations of the model over 200 generations show that the progress of the transgene in the wild population does not differ hugely if *g* pollen is input for 10 years or for only 1 year (Figure 12.3). For example, if the long-term rate at which seeds are lost from the seedbank K is equal to 0.3 (30% per year), the seed yield per adult Y_i is equal for wild and transgene types and the probability of successful pollination from an external *g* pollen source is 3%, then $z_{wg} \approx 10^{-2}t$ (only about one third of seeds in the bank are the latest crop). If the external *g* pollen source were removed after 10 years, about 10% of plants in the wild would be the *wg* variety. Once the *g* pollen source is removed, $\varepsilon \rightarrow 0$ and the progress of the transgene would follow the trajectory mapped in Figure 12.2 and Table 12.2.

However, the rate of *g* pollen input – whether low *or* high – can become irrelevant very rapidly. If the modified allele has an overwhelming advantage or suffers no disadvantage, then it will take over the wild-type population (Figure 12.3a) rather quickly from a very low level.

Caption for figure 12.1 (cont.) columns, respectively) under different levels of IR advantage for the modified feral crop. The relative advantage when herbivory is high is represented by a^{*+} and when herbivory is low by a^{*-}. In **(a)** and **(b)** a^{*-} has been assigned advantages of protection from *Plutella* that have been recorded for crops. In both cases it has been assumed that the IR allele confers no disadvantage to the feral crop when herbivory is low, $a^{*+} = 1$. In **(c–e)** the advantage the feral crop possesses at high herbivory has been set to $a^{*-} = 1.2$, meaning for example that the feral crop produces 20% more seed than the wild relative. The effect of a competitive disadvantage to the crop at low herbivory is shown for values of $a^{*+} = 0.8, 0.9, 0.99$. The values of these parameters are given in the right-hand panels. In all cases $K_W = K_G = 0.3$ (typical values of the long-term rate at which seeds are lost from the seedbank). The initial value of the feral population, w_G, was taken to be 10%. These figures can be directly compared with Figure 12.2, which shows the development over time of the populations in the diploid model, in which it is assumed that invasion by an IR allele commences with a hemizygote crossbreed.

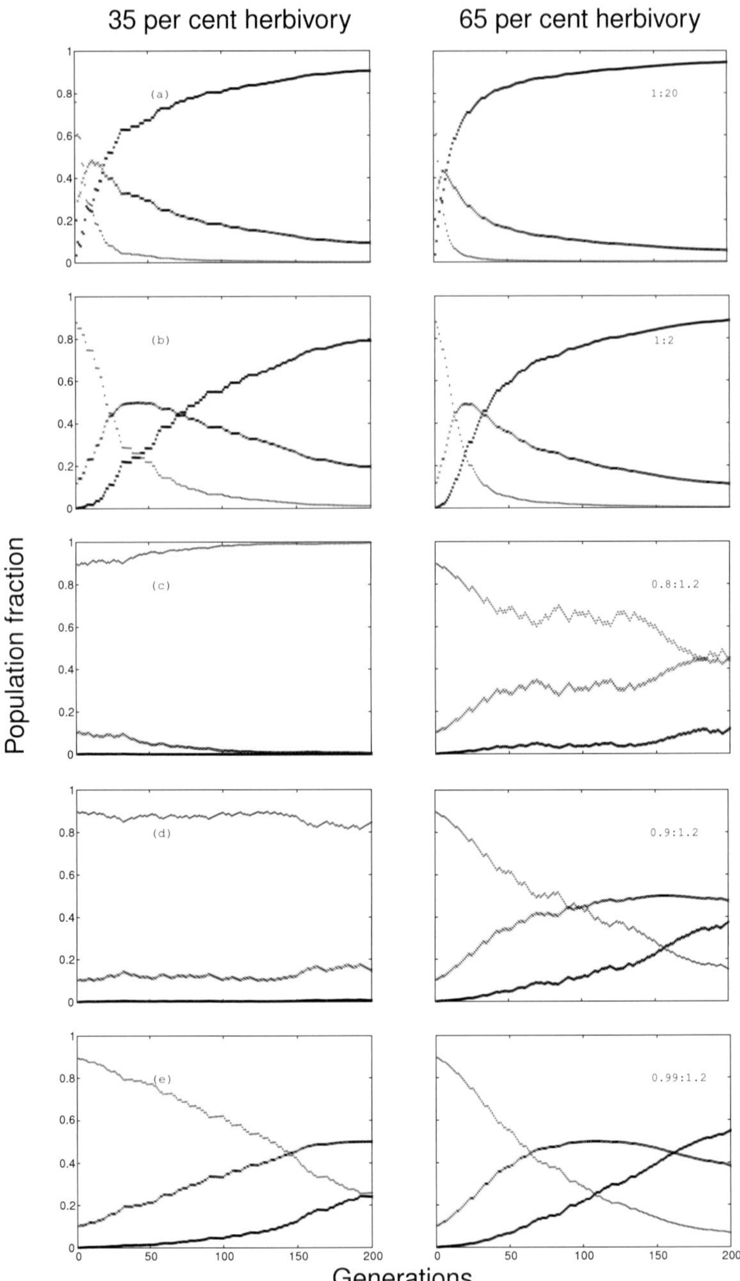

Figure 12.2 Rate of change of wild-type (+), homozygous GM (*) and hemizygote (x) genotypes when herbivory levels are high for 35% (as observed for *Plutella*) and 65% of years (left and right columns, respectively)

Where neither of these conditions hold, the results from various levels of continuous *g* pollen input do not differ greatly from those inferred starting from a single injection (reading across Figure 12.3). In general, it is not the pattern of pollen injection that determines whether untransformed plants can be maintained at a high level (>75%; reading from right to left in Figures 12.3c and d), but rather the differences in costs incurred by the transgene in the background of the natural genome (compare Figure 12.3d with Figure 12.3e, with costs equal to 10% and 1%, respectively).

In the calculations presented in this chapter, we have assumed that there are two levels of herbivory and correspondingly two values of the convenient parameter a^*. Herbivory is low during a proportion f of years and we selected the appropriate value of a^* at random, year by year, from this distribution. While this is not unrealistic, the equations in our model are potentially much more general. Correlations between years of high herbivory are trivial to include – if it is known what correlations apply – and it is not necessary for a^* to be two valued. The time dependence of a^* could be represented by a constant to which is added a sinusoidal variation or indeed a full Fourier decomposition.

Using the model to assess a target transgene

In selecting a transgene for commercial use, the a^* value might be chosen to decrease the possibility of transgene takeover under natural conditions. In our model, the advantage a^* depends linearly on a number

Caption for figure 12.2 (cont.) under differing levels of IR advantage. The relative advantage of resistance when herbivory is high is represented by a^{*-}; the relative advantage of the IR allele when herbivory is low, i.e. the cost of the IR allele, is a^{*+}. **(a)** and **(b)** The effect of levels of advantage from protection from *Plutella* herbivory recorded for crop systems. a^* values (given in the right-hand panels) are calculated as the ratio of seed set of oilseed rape with and without protection from herbivory. In both cases it is assumed that there is no disadvantage to the IR allele when herbivory is low ($a^{*+} = 1$; (Brown *et al.* 1999; Stewart *et al.* 1997). **(c–e)** Effects of differing levels of cost to carrying IR alleles when herbivory is low (a^{*+}) projected for non-crop species. Here, the IR advantage at high levels of herbivory (a^{*-}) is assumed to be 1.2 (i.e. at high herbivory, the wild-type plant produces 80% of the number of seed per seed sown of individuals carrying the IR allele). In all cases, $K_{ww} = K = 0.3$ (K = long-term rate at which seeds are lost from the seedbank; Roberts and Boddrell 1983). These figures can be directly compared with Figure 12.1 for a feral crop competing with a wild relative and with Figure 12.3.

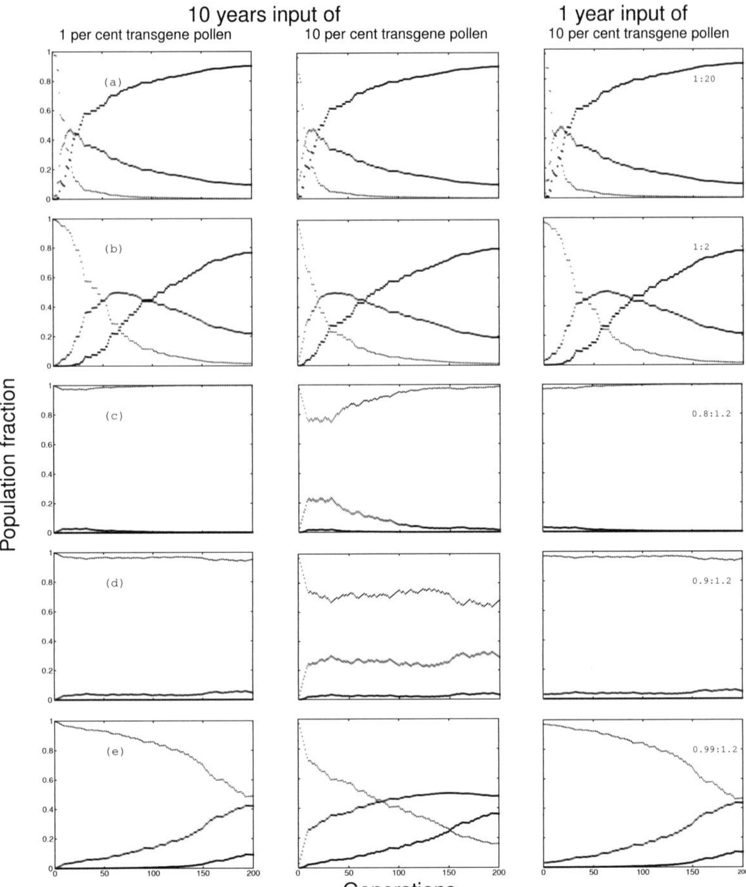

Figure 12.3 Continuous pollen input. Change over 200 generations of wild-type (+), homozygous GM (*) and hemizygote (x) genotypes under differing levels of IR (insect resistance) advantage when herbivory levels are high 35% of the time. The column on the left shows proportions of genotype in the population when there is 1% continuous input of transgene pollen from external sources over 10 years; the centre column shows genotype proportions when external pollen flow is 10% for 10 years; the column on the right shows external pollen flow at 10% for 1 year with no external g pollen received thereafter. The advantage a^* and seed loss parameters K are as in Figure 12.2; this figure is to be compared with the left-hand column of Figure 12.2.

of factors (seedling resistance to herbivory α, seedling competitive ability β, viable seed production Y), but its dependence on the germination fraction E and the dormancy parameter s is nonlinear (see Figure 12.4,

Table 12.2 *Prevalence of genotypes in the population. The values shown here for 10% initial* wg *correspond to Figure 12.1.*

			Initial %*wg* at 35% high herbivory		Initial %*wg* at 65% high herbivory		Doubling time[1] (generations)	
a^{*-}	a^{*+}		10	1	10	1	35% high herbivory	65% high herbivory
20	1	Generation at which *ww* falls below 50%	3	5	3	4	≈1[2]	0.56[2]
		%*ww* and *wg* at 200 generations	0.2/9.4	0.2/9.4	0.08/5.3	0.07/5.3		
2	1	Generation at which *ww* falls below 50%	20	45	11	21	7.5	4.1
		%*ww* and *wg* at 200 generations	1.3/20.0	1.7/22.6	0.4/11.5	0.4/12.4		
1.2	0.8	Generation at which *ww* falls below 50%	–	–	160	>200	*wg* goes to extinction	42.5
		%*wg* and *wg* at 200 generations %*wg* and *wg* %*ww* and *wg*	99.6/0.4	99.96/ 0.04	39.3/46.7	78.6/20.1		
1.2	0.9	Generation at which *ww* falls below 50%	>200	>>200	65	171	1150	23.7
		%*ww* and *wg* at 200 generations	83.5/15.7	98.1/1.9	14.1/46.8	36.7/47.2		
1.2	0.99	Generation at which *ww* falls below 50%	125	>200	54	130	31	17.8
		%*ww* and *wg* at 200 generations	23.7/50.0	64.7/31.4	6.5/38.0	15.3/47.6		

[1] When heterozygote proportion is below approximately 20%.

[2] When doubling is a year or less it is only an average. What actually happens is highly dependent on herbivory levels in year 1.

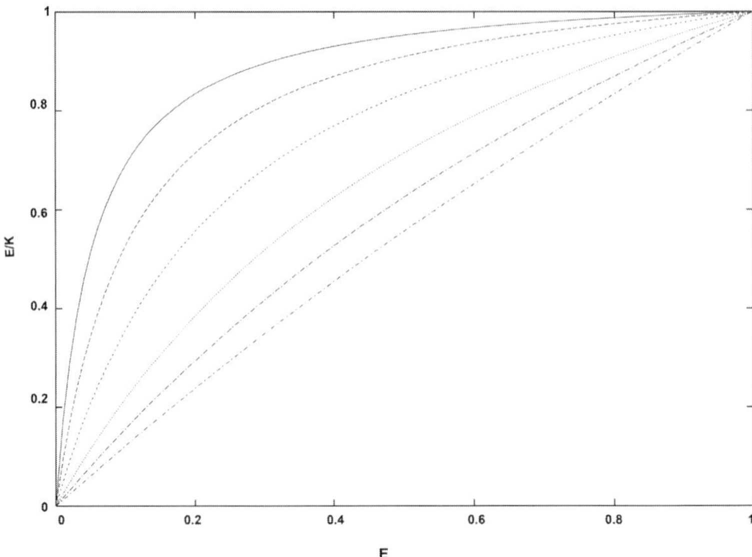

Figure 12.4 Trait manipulability. The curves show the relationship between E, the germination fraction, and the ratio E/K appearing in Equation (12.12) for a^*, where K is the function (12.2) of E and of s, the fraction of seeds in the seedbank that survive until the following season (excluding the period between year 0 and year 1). The values of s for which curves are shown are, in ascending order, 0.2, 0.4, 0.6, 0.8, 0.9 and 0.95. Note the increasingly nonlinear relationship with increasing values of s. If s is close to 1, and E is greater than 0.5, E becomes relatively unimportant in determining a^* and so an unrewarding target for manipulation. If E is small, it is very important in determining a^*. If E is close to 1, then s has little effect because nearly all seeds germinate. For non-crop Brassicaceae, s appears to be approximately 0.9 and E in the range of 0.2–0.3 (Roberts and Boddrell 1983).

Equation (12.2)). Because of this nonlinearity and given the values of s that occur in wild species (Roberts and Boddrell 1983), manipulation of E and s may not have the desired effect. On the other hand, the linearity of the relationships between a^* and α, β and Y means that the degree of change in any of these factors straightforwardly predicts the degree of change in the relative competitive abilities of the two lineages. For IR transgenes, the crop will have been selected for a large value of α under agricultural conditions and this parameter may not be available for further manipulation.

Of β and Y, β represents physical traits such as overtopping which confer a competitive advantage, or the costs of producing insecticide. The existence or otherwise of such costs in the *wg* variety certainly merits investigation. Y, the yield of seeds viable in year 1 per *wg* or *gg* adult, offers a trait that reliably determines the proportion of seedlings available to mature into seed-producing adults. Y appears in both a^* and Y^* in Equation (12.14). It is also clear from Equation (12.14) that male sterility associated with the *wg* variety substantially offsets a considerable advantage in a^*. We conclude that the necessary competitive disadvantage is accomplished most effectively through Y and/or η. The proportion of viable seeds may be manipulated through, for example, seed viability per se, i.e. male sterility in modified crop plants (genetic usage restriction technologies; GURT) is another possibility. However, if fertility is to be manipulated, it must be tightly linked to the IR gene complex and even then will eventually become dissociated from the insecticide allele through recombination. The optimal plan to 'control' spread through natural populations of genes of ecological consequence may thus best combine use of GURTs with the sorts of negative tradeoff addressed in our model.

There are two variants of the model that allow it to be applied to transgenes developed for other sorts of natural enemies. In the first variant, seed rather than seedling survivorship in the sensitive species is more affected by herbivory, e.g. predation of seeds on parent plant. This means quantifying the relative differences in production of undamaged (viable) seed for the modified and unmodified lineages under high and low seed predation levels. The second variant is closer to Chesson (1990) in that the fraction of seeds germinating varies due to the effect of seed predation in the soil or of a soil pathogen affecting the seed. To apply the model to this variant, relative differences in germination fraction of transgene-containing and wild-type seed cohorts under high and low levels of the environmental factor must be assessed.

In the first case the Y parameters are functions of environmental fluctuations, in the second case fluctuations in the E or s parameters affect the K in Equation (12.2). Equations (12.6) and (12.8) remain valid. Equations (12.11) require the generic replacement

$$z^+ \to \frac{KY^+z^+}{YK^+}$$

on the left-hand side (where the $+$ indicates the value at the next step). Nonetheless, Equation (12.14) and the expression for the doubling time Equation (12.16) remain valid.

The model is ecologically more general than this application to herbivory; the same dynamics and solutions should also apply to anti-pathogen transgenes.

12.4 DISCUSSION

Continuous input of the transgene into the wild population over a realistic period of time does not produce a hugely different result to single-season input of the transgene. We conclude that, in general, IR transgenes need not pose an immediate threat to natural communities. At lower frequencies of high herbivore levels, it takes very little disadvantage to slow or halt the progress of the transgene in the population, and it seems unlikely that there will be no such disadvantage under natural conditions (Herms and Mattson 1992). Population profiles for Lepidoptera available in the NERC Global Population Dynamics Database (NERC Centre for Population Biology 1999) indicate the probability of high herbivory levels to be mostly less than 0.5. The only relevant data we have found show fitness advantage of an IRBt-containing plant under natural conditions to be similar to that used in our simulations (Snow *et al.* 2003). Thus transgene carrying can be restricted to a low proportion of the natural population of plants provided that there is some suitable level of 'cost' associated with the transgene – which can be assessed. Persistently low levels of transformed plants will limit or eliminate potential 'knock-on effects' to other parts of the natural community such as herbivores and their predators and parasites.

The possibility that as much as 10% of pollination will be derived from sources outside the natural population (Figure 12.1, central column) is likely to be extreme. Studies of pollen movement in OSR indicate such movement levels to be unlikely (Cresswell *et al.* 2002, 2004, Cresswell and Osborne 2004). Moreover, such pollen may be less effective than wild-type pollen and even when effective, may produce offspring capable of producing fewer seed than can the pure wild type (Hauser *et al.* 1998, Pertl *et al.* 2002). We have chosen 10 years as the possible commercial lifetime of an OSR transgene although it is not unlikely that it will be less (J. Durkin, personal communication).

Regardless of any general expectation, particular transgenes should continue to be assessed individually. However, our model offers the significant advantage of clarifying what needs knowing and how well it needs to be known for such assessment. It has been suggested that temporal variability in the selective factor may affect the influx of transgenes into natural populations (Burke and Rieseberg 2003); our iterations demonstrate that to be the case for herbivore resistance transgenes. Indeed, our model indicates that a failure to account for temporal variability in herbivore levels would produce at best an incomplete, and more probably a misleading, projection of transgene behaviour.

Transgenes may be best manipulated through viable seed production, yet viable seed production is far from a sufficient measure of advantage. Differential seedling survival also contributes to fitness and may be expected to have an especially large role in the impact of IR transgenes. Damage from herbivory that may or may not decrease seed set in a mature plant will kill a seedling. That being said, an IR transgene in a wild-type background is expected to impart less advantage than the same transgene in a crop genome. Non-agricultural species possess a number of mechanisms to defend against herbivory (Belsky *et al.* 1993) and the transgene will be only one more; the productivity of a crop species with an IR transgene has been shown to be as much as 20 times that of the untransformed crop under the same conditions (Stewart *et al.* 1997) while the greatest relative advantage thus far shown for a transgene in a 75% wild-type background (BC_1) has been 1.55:1 (Snow *et al.* 2003). The two studies used different genera (*Brassica* versus *Helianthus*) and the first study encompassed seedling mortality while the second did not, but this is the best information currently available, not a test of the idea.

A small cost to carrying the transgene can greatly influence its progress through the natural population, so it is essential that the existence and degree of any costs to possession of a transgene under natural conditions be determined with certainty. Transgene costs are more likely in the context of the complex demands of a wild-type genome operating in the low resource availability of a natural habitat (Herms and Mattson 1992, Stamp 2003), and the limited data that exist do not refute cost to the transgene under natural conditions (Snow *et al.* 2003). Unfortunately, appropriate experiments have not, to our knowledge, been performed. In the best study to date, sample sizes are too small to reveal costs to transgene possession as great as 10% decrement in productivity (relative to untransformed plant). Furthermore, estimates of cost have not accounted for potential differential seedling survivorship (Snow *et al.* 2003). In the absence of herbivory, IRBt-containing seedlings may be less competitive than untransformed seedlings, affecting β, and in consequence suffer higher mortality in mixed communities.

Earlier models have dealt with environmental variation by asking what might be the relative benefits of producing a cohort of highly variable offspring versus a cohort of offspring with the same geometric mean fitness but not variable, as strategies for dealing with a stochastically variable environment (e.g. Gillespie 1972). Our storage model is asking what are the relative benefits of insect resistance with a physiological cost versus non-IR where herbivory, the off-setting benefit of the

IR trait, varies over time. Thus two different things are variable: in Gillespie (1972), the number of offspring varies stochastically, while in our model, the environmental factor, herbivory, varies. Also, we model the variable parameter differently, in that it does not depend on a normal distribution of high–low fluctuations; it thus possesses an advantage over mean–variance tradeoff models in that the distribution of fluctuations can be chosen by the researcher.

Our model, as it is presented, is not age-structured, although it can be made so. We have run our iterations using a constant probability of emergence from the seedbank, a reasonable approximation of the available data, but it could easily be made a non-uniform function to include age-structuring in the seedbank *sensu* Kalisz and McPeek (1992).

Our model indicates that natural populations may maintain a significant proportion of untransformed individuals for more than the lifetime of a transgene cultivar (approx. 12 years). These individuals could constitute a significant 'refugium' for a target herbivore *sensu* Carriére and Tabashnik (2001) and thereby delay herbivore adaptation to the IR transgene. The possibility of this will depend on the nutritional value of the non-crop plant to the herbivore and the size of the natural population versus that of the crop. Although these are factors outside the aegis of our model, our results suggest that they may be worth investigating in the interest of IR management strategies.

We have explicitly examined the case only for simple dominance, but our model is not restricted to this assumption. The general case is contained in Equations (12.6) and (12.8). By inserting values appropriate to the target relationship, it is possible to produce predictions for other specific cases such as codominance or particular levels of incomplete dominance, but even qualitative results can be valuable. For example, we expect full lethality to the target herbivore from a single transformed allele (hemizygote) for commercialized IR genes; this would ensure the highest level of effectiveness of the transgene under crop conditions. A 'dosage effect' with the plant producing twice the amount of defence in the homozygote would be irrelevant to the relative advantage of the homozygote over the hemizygote: a dead herbivore is dead. On the other hand, although twice as much defensive product may have little negative effect on the plant in a highly nutrient-enriched agricultural field, it could double the expected cost for the homozygous IR individual in the more constrained budget of a plant under nutrient-poor natural conditions. Under codominance of this sort, the IR gene would increase even more slowly in the natural community than for simple dominance (but the doubling time from a very low level, Equation (12.16), would not be affected).

When applying our model as a tool to investigate the spread of a given transgene, all the important differences controlling interactions between lineages with and without the transgene are relative (ratios), and the influence of particular trait values in potential transgene spread could therefore be evaluated under controlled conditions (Linder 1998). The similarity of controlled and real conditions may be calibrated through wild-type lineages, thus allowing appraisal of the danger of transgene runaway with minimum exposure of the transgene to the environment and the risk of unplanned release.

Our goal has been to clarify the components of fitness in the interaction between individuals with and without IRBt alleles in nature, with and without the action of recurrent gene flow, explicitly in relation to the ecology of temporal variability. We have also developed an effective tool for efficiently identifying transgene types that have the ecological capability to invade wild-type communities. While an advance in risk assessment and regulation of transgenes, our model is also of significance to the basic ecology of crop–native plant communities. Its predictive power should serve to increase greatly the precision of regulation and mitigation of GM and non-GM crops with regard to risks associated with gene flow. From a practical point of view the greatest difficulties in implementing this work are likely to be assembling the necessary trait parameters. However, we have demonstrated that the population dynamics of transgenes can be strongly affected by the circumstances when herbivory is low and at the very least this may be relevant for the interpretation of population dynamics yet to be observed.

ACKNOWLEDGEMENTS

This article was adapted by the authors from Kelly *et al.* (2005), Kelly *et al.* (2006) and hitherto unpublished work by Kelly and Bowler.

REFERENCES

Arias, D. M. and Rieseberg, L. H. (1994). Gene flow between cultivated and wild sunflowers. *Theoretical and Applied Genetics* **89**, 655–660.

Arnaud, J.-F., Viard, F., Delescluse, M. and Cugen, J. (2003). Evidence for gene flow via seed dispersal from crop to wild relatives in *Beta vulgaris* (Chenopodiaceae): consequences for the release of genetically modified crop species with weedy lineages. *Proceedings of the Royal Society of London B* **270**, 1565–1571.

Belsky, A. J., Carson, W. P., Jensen, C. L. and Fox, G. A. (1993). Overcompensation by plants: herbivore optimization or red herring? *Evolutionary Ecology* **7**, 109–121.

Bergelson, J. and Purrington, C. B. (1995). Surveying patterns in the cost of resistance in plants. *American Naturalist* **148**, 536–558.

Brown, J., McCaffrey, J. P., Harmon, B. L. *et al.* (1999). Effect of late season insect infestation in yield, yield components and oil quality of *Brassica napus*, *B. rapa*, *B. juncea* and *Sinapis alba* in the Pacific Northwest region of the United States. *Journal of Agricultural Science, Cambridge* **132**, 281–288.

Brown, V. K. and Gange, A. C. (1989). Differential effects of above and below-ground insect herbivory during early plant succession. *Oikos* **54**, 67–76.

Burke, J. M. and Rieseberg, L. H. (2003). Fitness effects of transgenic disease resistance in sunflowers. *Science* **300**, 1250.

Carriére, Y. and Tabashnik, B. E. (2001). Reversing insect adaptation to transgenic insecticidal plants. *Proceedings of the Royal Society of London B* **268**, 1475–1480.

Chesson, P. (2003). Quantifying and testing coexistence mechanisms arising from recruitment fluctuations. *Theoretical Population Biology* **64**, 345–357.

Chesson, P. L. (1990). Geometry, heterogeneity and competition in variable environments. *Philosophical Transactions of the Royal Society of London B* **330**, 165–173.

Chesson, P. L. and Huntly, N. (1989). Short-term instabilities and long-term community dynamics. *Trends in Ecology and Evolution* **4**, 293–298.

Crawley, M. J. (1999). Bollworms, genes and ecologists. *Nature* **400**, 501–502.

Cresswell, J. E., Davies, T. W., Patrick, M. A. *et al.* (2004) Aerodynamics of wind pollination in a zoophilous flower, *Brassica napus*. *Functional Ecology* **18**, 861–866.

Cresswell, J. E. and Osborne, J. L. (2004). The effect of patch size and separation on bumblebee foraging in oilseed rape: implications for gene flow. *Journal of Applied Ecology* **41**, 539–546.

Cresswell, J. E., Osborne, J. L. and Bell, S. A. (2002). A model of pollinator-mediated gene flow between plant populations with numerical solutions for bumblebees pollinating oilseed rape. *Oikos* **98**, 375–384.

Dale, P. J., Clarke, B. and Fontes, E. M. G. (2002). Potential for the environmental impact of transgenic crops. *Nature Biotechnology* **20**, 567–574.

Fenner, M., Hanley, M. E. and Lawrence, R. (1999). Comparison of seedling and adult palatability in annual and perennial plants. *Functional Ecology* **13**, 546–551.

Ford, E. D. (1975). Competition and stand structure in some even-aged plant monocultures. *Journal of Ecology* **63**, 311–333.

Gillespie, J. H. (1972). The effects of stochastic environments on allele frequencies in natural populations. *Theoretical Population Biology* **3**, 241–248.

GM Science Review Panel (2003). *GM Science Review (First Report): an Open Review of the Science Relevant to GM Crops and Food Based on Interests and Concerns of the Public*. London: Department of Trade and Industry, pp. 109–114.

Hanley, M. E., Fenner, M. and Edwards, P. J. (1995). An experimental field study of the effects of mollusc grazing on seedling recruitment and survival in grassland. *Journal of Ecology* **83**, 621–627.

Hanley, M. E. and Lamont, B. B. (2001). Herbivory, serotiny and seedling defence in Western Australian Proteaceae. *Oecologia* **126**, 409–417.

Harcourt, D. G. (1963). Major mortality factors in the population dynamics of the diamondback moth, *Plutella maculipennis* (Cust.) (Lepidoptera: Plutellidae). *Memoirs of the Entomological Society of Canada* **32**, 55–66.

Hauser, T. P., Jørgensen, R. B. and Østergård, H. (1998). Fitness of backcross and F_2 hybrids between weedy *Brassica napa* and oilseed rape (*B. napus*). *Heredity* **81**, 436–443.
Haygood, R., Ives, A. R. and Andow, D. A. (2004). Population genetics of transgene containment. *Ecology Letters* **7**, 213–220.
Herms, D. A. and Mattson, W. J. (1992). The dilemma of plants: to grow or defend. *Quarterly Review of Biology* **67**, 283–335.
Kalisz, S. and McPeek, M. A. (1992). Demography of an age-structured annual: resampled projection matrices, elasticity analyses, and seed bank effects. *Ecology* **73**, 1082–1093.
Kelly, C. K. and Bowler, M. G. (2002). Coexistence and relative abundance in forest tree species. *Nature* **417**, 437–440.
Kelly, C. K. and Bowler, M. G. (2003). Communications arising (reply): competition and coexistence in forest trees. *Nature* **422**, 587.
Kelly, C. K. and Bowler, M. G. (2005). A new application of storage dynamics: differential sensitivity, diffuse competition and temporal niches. *Ecology* **86** 1012–1022.
Kelly, C. K., Bowler, M. and Breden, F. (2006). An analytical model assessing the potential threat to natural habitats from insect resistance transgenes: continuous transgene input. *Biology Letters* **2**, 293–297.
Kelly, C. K., Bowler, M. G., Breden, F., Fenner, M. and Poppy, G. M. (2005). An analytical model assessing the potential threat to natural habitats from insect resistance transgenes. *Proceedings of the Royal Society of London B* **272**, 1759–1767.
Kelly, C. K. and Hanley, M. E. (2005). Juvenile growth and palatability in congeneric British herbs. *American Journal of Botany* **92**, 1586–1589.
Linder, C. R. (1998). Potential persistence of transgenes: seed performance of transgenic canola and wild × canola hybrids. *Ecological Applications* **8**, 1180–1195.
Linder, C. R., Taha, I., Seiler, G. J., Snow, A. A. and Rieseberg, L. H. (1998). Long-term introgression of crop genes into wild sunflower populations. *Theoretical and Applied Genetics* **96**, 339–347.
Markkula, M. (1965). The pests of cultivated plants in Finland 1965. *Maatal Ja Koetoim* **20**, 185–195.
NERC Centre for Population Biology, Imperial College. (1999). *The Global Population Dynamics Database*. Available at: http://www3.imperial.ac.uk/cpb/databases/gpdd.
Pertl, M., Hauser, T. P., Damgaard, C. and Jørgensen, R. B. (2002). Male fitness of oilseed rape (*Brassica napus*), weedy *B. rapa* and their F_1 hybrids when pollinating *B. rapa* seeds. *Heredity* **89**, 212–218.
Price, P. (1997). *Insect Ecology*. 3rd edn. New York: John Wiley and Sons.
Raybould, A. F. and Gray, A. J. (1994). Will hybrids of genetically modified crops invade natural communities? *Trends in Ecology and Evolution* **9**, 85–89.
Rieseberg, L. H. and Burke, J. M. (2001). The biological reality of species: gene flow, selection, and collective evolution. *Taxon* **50**, 47–67.
Roberts, H. A. and Boddrell, J. E. (1983). Seed survival and periodicity of seedling emergence in eight species of Cruciferae. *Annals of Applied Biology* **103**, 301–304.
Snow, A. A., Pilson, D., Rieseberg, L. H. *et al.* (2003). A Bt transgene reduces herbivory and enhances fecundity in wild sunflowers. *Ecological Applications* **13**, 279–286.
Stamp, N. (2003) Out of the quagmire of plant defense hypotheses. *Quarterly Review of Biology* **78**, 23–55.

Stewart, C. N., All, J. N., Raymer, P. L. and Ramachandran, S. (1997). Increased fitness of transgenic insecticidal rapeseed under insect selection pressure. *Molecular Ecology* **6**, 773–779.

Stewart, C. N., Halfhill, M. D. and Warwick, S. I. (2003). Transgene introgression from genetically modified crops to their wild relatives. *Nature Reviews: Genetics* **4**, 806–817.

Vanholder, B. (1997). Belgian migrating Lepidoptera. Available at: http://www3.imperial.ac.uk/cpb/databases/gpdd.

Whitton, J., Wolf, D. E., Arias, D. M., Snow, A. A. and Rieseberg, L. H. (1997). The persistence of cultivar alleles in wild populations of sunflowers five generations after hybridization. *Theoretical and Applied Genetics* **95**, 33–40.

GORDON A. FOX, MICHAEL G. BOWLER AND
COLLEEN K. KELLY

13

Concluding remarks

13.1 INFERENCES ABOUT TEMPORAL PROCESSES

Niche dynamics

The chapters in this volume convey several important messages. Perhaps the most obvious of these is that there is mounting – and compelling – empirical evidence for the importance of temporal niche differentiation in natural communities. The contributions by Adler, Chesson *et al.*, Fowler and Pease, Hanley and Sykes, Kelly *et al.* and Venable and Kimball provide strong evidence that temporal niche differentiation occurs, that it can be documented and that in at least some cases we can understand many of its mechanistic underpinnings. While there is still much to be learned, these contributions – together with earlier published work (see citations in chapters) – have established the temporal niche as a vital area of research and built a foundation for the study of temporal processes in nature.

It is essential to the better understanding of temporal process that empirical evidence continues to be developed, and with a variety of approaches. For many ecologists, temporal niches (and differentiation between them among competing species) represent something new to the way we have been accustomed to think and to conduct research. Persuading ecologists that this is an important area of research (and that temporal niche differentiation is an important part of species coexistence) is likely to require more than a single book or critical experiment, but here we present a considerable body of

Temporal Dynamics and Ecological Process, ed. C. K. Kelly, M. G. Bowler, G. A. Fox.
Published by Cambridge University Press.

evidence developed in a variety of communities, using multiple approaches to make inferences.

Several strategies – which are not mutually exclusive – are on display here. Adler (this volume) sought evidence that a system satisfies the theoretical assumptions of the storage effect. Using two extant data sets, he fitted models to the data, and then used model behaviour to infer that storage was occurring in the Kansas prairie but not in the Idaho sage-steppe community. The use of long-term data sets helps to make this research persuasive. His approach to inference – showing that these systems do (or don't) meet the criteria for storage to occur – is powerful, but sceptics may remain unconvinced because models are not the same things as the systems they mimic (see Wimsatt 1980, Morris 1990) and because one can never really check whether all assumptions are met. We believe that such objections do not invalidate the contribution made by this research – given the data used, Adler's is a productive and informative approach. However, such objections do underline the value of multiple lines of evidence and approaches to inference.

A different inferential strategy was used by Venable and Kimball (this volume). Previous work with their long-term study (on the demography of a number of species of annual plants in the Sonoran Desert) provided convincing evidence for increased long-term fitness via bet hedging in the form of seed dormancy (Venable 2007). Subsequent work revealed the physiological basis of this variation between species: those with high fitness variance grow rapidly and have low WUE, while these traits are reversed (slow growth and high WUE) in species with low fitness variance. This array of traits strongly suggested that the storage effect might mediate coexistence in this community, and Angert *et al.* (2009) demonstrated that indeed this tradeoff is sufficient to assure coexistence. We suggest that this demonstration of the storage effect is so persuasive because it links demographic performance to underlying physiological mechanisms, and as a result, it is possible in this system to move from the general construct of 'good' and 'bad' years, to a specific solution that ties the real performances of these populations to variation in moisture availability.

Kelly *et al.* employ still another approach to inference. Studying tree populations in a dry tropical forest in Mexico, they could not readily estimate lifetime fitness. Instead, they examined the age structures of populations, finding evidence for episodic recruitment (Kelly and Bowler 2002). Critically for their argument, pairs of coexisting congeners invariably differed in their evident sensitivity to environmental fluctuations, with the more sensitive species always being the better competitor. Since this is precisely the condition for coexistence predicted by their model – and not otherwise expected – they provide strong evidence for temporal niche differentiation. Here, in order to address questions of community-level structure of temporal niche dynamics, they synthesise more recent work from both their original site in Mexico (Kelly and Bowler 2009b, Kelly *et al.* 2008) and from Panama's Barro Colorado Island (Kelly *et al.* 2010), the site that has played such a strong part in the argument for neutral communities raised by Hubbell (2001) and others, to explore the larger question of community-level structure of temporal dynamics. Both of these analyses are persuasive because they confirm predictions made by a coexistence model that are not otherwise expected, and in so doing, demonstrate a strategy for studying coexistence in communities of long-lived organisms. Aspects of these results are addressed by Schwinning *et al.* (this volume), who discuss implications of the possible structure of temporal niche dynamics for ecosystem function and response to environmental change.

Several other papers in this volume demonstrate the 'real life' ecological permutations of a temporal storage dynamic, enlarging our understanding of the expression of temporal processes. Chesson *et al.* examine evidence from two systems that are rather different. Both involve plants that are not closely related to one another. In their Chihuahuan Desert system, however, there are some important ecological similarities between the winter-germinating annuals they studied. In their Australian wet sclerophyll forest understorey system, the niche partners are more dissimilar: a perennial tussock grass and a stoloniferous herbaceous perennial herb that tend to be active at different times of year. To make inferences about the operation of storage effects, the authors develop a novel prediction of differences in the total variance in growth between plots with both species and plots with only one species, if the storage effect is occurring; estimation of variance components from their data provides support to the hypothesis.

Fowler and Pease provide an intriguing – and perhaps surprising – result in their study of a Texas grassland community: annual fluctuations in densities of seven (out of eight) species of perennial grasses and forbs are described better by a model with fixed growth rate but stochastic fluctuation in density dependence, rather than by a model of stochastic variation in growth rate. They point to their results as evidence that their study species have substantial differences in their temporal niches (and indeed, the density fluctuations are at most very weakly correlated between species). These analyses identify an underlying requirement of temporal dynamics, pinpointing a system ripe for further, directed research able to determine the extent to which the coexistence of these species is due to temporal niche processes.

Hanley and Sykes introduce a novel set of issues in their study on the effect of fluctuations in the intensity of herbivory as a mediator of coexistence of closely related plant species (see also Hanley 1998, Fenner *et al.* 1999). These results – tied to the demonstration of a tradeoff between growth rate and palatability between the close relatives (Kelly and Hanley 2005) – raise several interesting sets of issues for further research. First, how much feedback is there between the dynamics of the plant and herbivore populations? If herbivore dynamics depend strongly on plant population dynamics, it may prove useful to study models of their joint dynamics rather than simply thinking of herbivory as an instance of environmental stochasticity. A second issue of potential importance is the possible role of responses to herbivory as an isolating mechanism between these closely related species. Since herbivory can contribute to differences in flowering time (Lyons and Mully 1992, Elzinga *et al.* 2007), differential responses to herbivory may be subject to assortative mating, thus contributing to the reproductive isolation of plants with different responses to herbivores (Devaux and Lande 2009). Finally, the interaction between herbivory and competition, and the ways they affect diversity, can be complex (Chesson and Kuang 2008). Because some of the mechanisms underlying the responses to herbivory are known here, and because the organisms are relatively easy to manipulate, this may provide a useful model system for further studies of these interactions.

We observed in the Introduction that one difficulty in studying temporal niche differentiation is that it occurs over time, and sometimes involves brief or uncommon episodes. But these are characteristics that it shares with evolution, suggesting some important lessons. In both cases, our ability to study the process of interest often faces a problem: our lack of a time machine. In studying either evolution or temporal dynamics, we are able to observe directly the processes of interest only with short-lived organisms such as bacteria or annual plants. These are valuable cases for study. Unfortunately, though, these study systems may not inform us about some of the important issues in either evolution or temporal dynamics of long-lived organisms. In both lines of research, we are interested in the underlying process; however, with long-lived organisms, although it is difficult to observe these processes, it may be relatively easy to observe resulting patterns, which are not generally unique. Inferences may be even more elusive (in both disciplines) because patterns may result from relatively infrequent events, or from the interplay between those infrequent events and more common ones. For example, the relative frequency (and magnitude) of uncommon events is crucial in understanding selection for bet-hedging life-history strategies (Seger and Brockmann 1987, Philippi and Seger 1989), and it provides the key to coexistence mediated by the storage effect as well.

Our understanding of evolutionary processes is much stronger as a result of having several different lines of evidence – not only because they sometimes provide different types of insights, but also because differences among lines of evidence have repeatedly led to new scientific questions. There are important lessons for community ecology in this. It seems clear that understanding the mechanisms underlying community structure will likewise require multiple approaches; there is no single best way to study the storage effect or any other coexistence mechanism. Numerous kinds of studies will be informative, and it seems likely that, as with evolutionary biology, our understanding will be much richer if we have multiple types of studies.

To date, inferences regarding temporal processes have come from (1) theoretical studies considering the conditions for coexistence (Bowler *et al.*); (2) parameter estimates for models asking whether conditions for the storage effect have been met (Adler, Chesson); (3) observation of patterns best explained by the storage effect (Kelly *et al.*); (4) observation of mechanisms expected to cause temporal niche differentiation (Hanley and Sykes, Venable and Kimble); and (5) observation of fluctuations consistent with temporal niche differentiation (Fowler and Pease, Venable and Kimble). We expect that all of these approaches – as well as others not yet tried – will prove valuable in the future.

A research agenda for temporal niche dynamics

Temporal niche dynamics (TND) promise to be a significant factor in community assembly and as such, a potentially important component in predictive modelling of community response to perturbation and climate change. The complementary nature of recruitment in a temporal dynamic suggests the possibility of replacement of one species into the slot filled by its niche partner(s) in response to changing conditions. The following three questions flow from the chapters included in this volume and in the current literature to point out the groundwork needed for this next very vital step in determining the role of TND in community structure and function.

What are the predominant physical regulators of temporal storage dynamics?

Identification of the primary determinant as well as the required frequency of interannual variation in that determinant would be an aid to predictive modelling of community response to perturbation. As with the 'limiting resources' of classic competition models, 'interannual variation' encompasses a wide range of possible factors. Under some conditions, an estimate of the frequency of 'good' versus 'bad' conditions as perceived by the competing species may be extracted from census (relative abundance) or long-term growth data without explicitly identifying the specific operative factor (Adler *et al.* 2006, 2009, Huxman *et al.* 2008, Kelly and Bowler 2009b, Kelly *et al.* 2010, Williams and Kelly 2013, Adler this volume, Venable and Kimball this volume). For estimates of process at this scale, any such calculation is subject to a number of assumptions, raising caveats for inference at any one site, but may be useful for comparison among sites or communities, or as an initial investigative tool for pinpointing actual determinants.

Nonetheless, exactly pinpointing regulators may be desirable in specific instances. Average or total resource delivery, schedule of resource availability over the physiologically active season or schedule of initiating event all have the potential to regulate a temporal niche, and it is widely expected that pattern change in precipitation will be at least as important to communities as shifts in average temperature or precipitation (Knapp *et al.* 2008, McCluney *et al.* 2012). The distribution of seasonal schedules for events such as early seasonal rainfall or late frosts affect the timing of seed germination is seen in desert annuals (Kimball *et al.* 2010) and is suggested in the population structure of birch (*Betula*) (Kelly *et al.* 2003). Furthermore, gross interannual variation and event scheduling are not necessarily independent of one another: for example, late initiation of the growing season could also reliably signify a short season. In all, there are multiple complex permutations on climate – and the relevant one may well differ for different niche partners within any one community (Ellis *et al.* unpublished manuscript). A better understanding of the regulators of temporal dynamics is likely to be a necessary step in effectively generalising the response of particular ecosystems to projected climate change.

What is the structure of temporal dynamics in natural communities?

Community function, stability and productivity are all factors affected by the identity, form and frequency of assembly mechanisms, and the coexistence of functionally similar species – as temporal niche partners inherently must be – is likely to have fundamental consequences for community response to environmental change (Gonzalez and Loreau 2009, Schwinning and Kelly 2013, Schwinning *et al.* this volume). In terms of temporal dynamics, it is theoretically possible to have an arbitrary number of species coexisting on a single resource provided that there are (appropriate) fluctuations in time (Koch 1974a, b, Armstrong and McGehee 1976, 1980). Thus, although the initial formulation of temporal storage dynamics was presented as a simple two-species lottery model, it was simplified for the sake of clarity rather than as an assumption that storage dynamics necessarily operated in nature as a two-species dynamic (Sale 1977, 1978, Chesson and Warner 1981); subsequent theoretical work has been directed at opening out this two-species environment (Chesson and Huntly 1997, Chesson 2000).

However, it is to be expected that in natural communities, the number of species that can divide up a temporal niche will have some maximum value. For a temporal dynamic to be stable, the advantage a species has during its own best temporal segment must be greater than the average advantage of all the other species with which it interacts. Furthermore, the fraction of time in which each species occurs must be sufficient for stable population maintenance. In consequence, a dynamic encompassing many species would require a greater degree of fine-tuning of any one temporal niche than seems likely in natural systems.

Quite a lot is already known about the structure of temporal dynamics in closely related species (Kelly *et al.* this volume), and it would be tempting to exploit the inherent similarity of congeners to infer something general about how similarity works in a temporal dynamic. However, the similarity that allows coexistence is differently derived in close versus distant relatives. In the pairwise, focused dynamic of congeneric species, similarity is a function of close relatedness, with divergence principally along the temporal axis. In contrast, unrelated species in a temporal dynamic have converged on the ecological conditions that regulate their coexistence, and their overall similarity is only so much as to allow them to persist as adults under the same general regime. These observations have two implications. In the first instance, distantly related niche partners may be very different from each other (e.g. Chesson *et al.* this volume). Second, for close relatives, the dynamic itself may be quite predictable, being constrained at some point by the requirements of sympatric speciation (Kelly *et al.* this volume), but no such single constraint would necessarily operate for distantly related niche partners. In consequence, temporal dynamics for distantly related niche partners may not be easily generalised or predicted.

How do temporal niche dynamics interact with other assembly mechanisms?

Processes determining community characteristics do not occur in isolation from one another. Saying that TND are operating does not exclude other mechanisms (e.g. classical niche partitioning, spatial processes, pathogens or predators) from also operating, or from interacting with one another in conditional ways (Chesson and Kuang 2008). For example, specialist predators and pathogens can create or augment the fluctuating dynamics of a stable temporal process in competing plant species (Kelly and Bowler 2009a); conversely, generalist predators can destabilise a temporal dynamic (Kuang and Chesson 2009), However, add in a growth-rate–defence tradeoff between competing plant species, and the preferences of a fairly generalist predator are capable of supporting a temporal dynamic (Hanley 1998, Fenner *et al.* 1999, Kelly and Hanley 2005, Hanley and Sykes this volume). Additionally, fluctuating behaviour at one level of the trophic web, whether caused by biotic or abiotic factors, can influence processes further up the trophic structure (Kuang and Chesson 2009). Also, in terms of process interactions, Snyder (this volume) works with theory to show that temporal storage dynamics are much more stable when a spatial storage component is also present.

There are also more pragmatic reasons to study multiple mechanisms and their interaction. Failure to account for multiple mechanisms may seriously bias estimates of the importance of processes affecting coexistence. By focusing on single coexistence mechanisms, one risks inflating the error estimate and thus falsely concluding that coexistence mechanisms are less important than they may be. For example, a study considering only TND might ascribe variation due to spatial niche partitioning to 'error', reducing the apparent importance of TND. If interactions between coexistence mechanisms are important, then clearly one could never understand them if considered one at a time. By the same token, evidence against the importance of TND or any other coexistence mechanism can never be satisfactory without studies that allow for their simultaneous – and possibly interactive – effects.

13.2 INTERNALLY INDUCED FLUCTUATION

The chapters by Iwasa *et al.* and Satake *et al.* on masting, the phenomenon of intermittent and synchronised reproduction in trees, offer a fascinating review of complicated time-dependent phenomena emerging from a model that is very simple, both biologically and mathematically. These chapters are concerned with intermittent and synchronised reproduction of forest trees – masting. They depend upon elaboration of a very simple biological model for resource reserves, the idea being that if a tree depletes its resources severely in an episode of fruiting it will have no resources to spare for reproduction for several years thereafter.

The chapter by Iwasa *et al.* introduces the simplest version of the model. The environment is constant and if there is no coupling between trees then they either reproduce every year or intermittently, but with no pattern of synchrony. The trees in a forest are likely to benefit from outcrossing, and this form of coupling indeed can result in synchronised reproduction. There are regions in the parameter space of the equations where synchronised reproduction takes place over the whole forest, with either chaotic or periodic time sequences, and even in clusters of trees. If masting occurs in particular regions of parameter space, those regions must be evolutionary attractors if the mathematical model is to have biological relevance. The treatment is extended beyond the dynamics of resource depletion and it is found that masting is evolutionarily favoured provided that an active seedling bank exists. Specialist seed predators have a minor role at best.

The chapter by Satake *et al.* starts by examining the coupling of reproduction of trees in more than one species, through the action of generalist pollinators. Generalists will be concentrated by lots of flowers and again there are regions of parameter space in which reproduction is synchronised not just within a single species but with different species reproducing together. The possible role of environmental fluctuations (of an abiotic kind) in inducing synchronised reproduction over a landscape is also examined in the context of the general resource reserves model. The fluctuations are stochastic and cannot induce synchrony even within a single species unless trees are coupled. Pollen coupling can induce synchrony without environmental fluctuation, but its effect can be enhanced in the presence of environmental fluctuations. Finally, there is the phenomenon of a species exhibiting periodic synchronised reproduction, but with different periods in different climates. Within resource depletion dynamics this is traced to different levels of productivity.

These temporal dynamics are not directly related to one of the themes of this book, the role of environmental fluctuation in promoting coexistence, but there may well be an indirect link. The phenomenon of masting generates a fluctuating environment for seed predators and fluctuating levels of infestation constitute a fluctuating environment for the species of insects which prey upon the larvae, and on up the trophic web.

ACKNOWLEDGEMENTS

GAF was supported by grant DEB-0615024 from the US National Science Foundation, and CKK by grants from NSF (SGER) and from the National Geographic Society. Chamela Biological Station has provided CKK with many years of hospitality and general helpfulness. CKK thanks Merton College for continuing support and hospitality.

REFERENCES

Adler, P. B., HilleRisLambers, J., Kyriakidis, P. C., Guan, Q. F. and Levine, J. M. (2006). Climate variability has a stabilizing effect on the coexistence of prairie grasses. *Proceedings of the National Academy of Sciences, USA* **103**, 12793–12798.

Adler, P. B., HilleRisLambers, J. and Levine, J. M. (2009). Weak effect of climate variability on coexistence in a sagebrush steppe community. *Ecology* **90**, 3303–3312.

Angert, A. L., Huxman, T. E., Chesson, P. and Venable, D. L. (2009). Functional tradeoffs determine species coexistence via the storage effect. *Proceedings of the National Academy of Sciences, USA* **106**, 11641–11645.

Armstrong, R. A. and McGehee, R. (1976). Coexistence of species competing for shared resources. *Theoretical Population Biology* **9**, 317–328.

Armstrong, R. A. and McGehee, R. (1980). Competitive exclusion. *American Naturalist* **115**, 151–170.

Chesson, P. (2000). Mechanisms of maintenance of species diversity. *Annual Review of Ecology and Systematics* **31**, 343–366.

Chesson, P. and Huntly, N. (1997). The roles of harsh and fluctuating conditions in the dynamics of ecological communities. *American Naturalist* **150**, 519–553.

Chesson, P. and Kuang, J. J. (2008). The interaction between predation and competition. *Nature* **456**, 235–238.

Chesson, P. and Warner, R. R. (1981). Environmental variability promotes coexistence in lottery competitive systems. *American Naturalist* **117**, 923–943.

Devaux, C. and Lande, R. (2009). Selection on variance in flowering time within and among individuals. *Evolution* **64**, 1311–1320.

Elzinga, J. A., Atlan, A., Biere, A. *et al.* (2007). Time after time: flowering phenology and biotic interactions. *Trends in Ecology and Evolution* **22**, 432–439.

Fenner, M., Hanley, M. E. and Lawrence, R. (1999). Comparison of seedling and adult palatability in annual and perennial plants. *Functional Ecology* **13**, 546–551.

Gonzalez, A. and Loreau, M. (2009). The causes and consequences of compensatory dynamics in ecological communities. *Annual Review of Ecology, Evolution and Systematics* **40**, 393–414.

Hanley, M. E. (1998). Seedling herbivory, community composition and plant life-history traits. *Perspectives in Plant Ecology, Evolution and Systematics* **1**, 191–205.

Hubbell, S.P. (2001). *The Unified Neutral Theory of Biodiversity and Biogeography*. Princeton, NJ: Princeton University Press.

Huxman, T. E., Barron-Gafford, G., Gerst, K. L. *et al.* (2008). Photosynthetic resource-use efficiency and demographic variability in desert winter annual plants. *Ecology* **89**, 1554–1563.

Kelly, C. K. and Bowler, M. G. (2002). Coexistence and relative abundance in forest trees. *Nature* **417**, 437–440.

Kelly, C. K. and Bowler, M. G. (2009a). Investigating the role of enemies in temporal niche dynamics: differential sensitivity, competition, and stable coexistence. *Theoretical Population Biology* **76**, 278–284.

Kelly, C. K. and Bowler, M. G. (2009b). Temporal niche dynamics, relative abundance and phylogenetic signal in coexisting species. *Theoretical Ecology* **2**, 161–169.

Kelly, C. K., Bowler, M. G., Joy, J. B. and Williams, J. N. (2010). Temporal niche dynamics of closely related tree species on Barro Colorado Island, Panama. **ArXiv:1008.2527v1**

Kelly, C. K., Bowler, M. G., Pybus, O. G. and Harvey, P. H. (2008). Phylogeny, niches and relative abundance in natural communities. *Ecology* **89**, 962–970.

Kelly, C. K., Chase, M., Fay, M. F., de Bruijn, A. and Woodward, F. I. (2003). Temperature-based population segregation in birch. *Ecology Letters* **6**, 87–89.

Kelly, C. K. and Hanley, M. E. (2005). Juvenile growth and palatability in congeneric British herbs. *American Journal of Botany* **92**, 1586–1589.

Kimball, S., Angert, A. L., Huxman, T. E. and Venable, D. L. (2010). Contemporary climate change in the Sonoran Desert favors cold-adapted species. *Global Change Biology* **16**, 1555–1565.

Knapp, A. K., Beier, C., Briske, D. D. *et al.* (2008). Consequences of more extreme precipitation regimes for terrestrial ecosystems. *Bioscience* **58**, 811–821.

Koch, A. L. (1974a). Coexistence resulting from an alternation of density dependent and density independent growth. *Journal of Theoretical Biology* **44**, 373–386.

Koch, A. L. (1974b). Competitive coexistence of two predators utilizing the same prey under constant environmental conditions. *Journal of Theoretical Biology* **44**, 387–395.

Kuang, J. J. and Chesson, P. (2009). Coexistence of annual plants: generalist seed predation weakens the storage effect. *Ecology* **90**, 170–182.

Lyons, E. E. and Mully, T. W. (1992). Density effects on flowering phenology and mating potential in *Nicotiana alata*. *Oecologia* **91**, 93–100

McCluney, K. E., Belnap, J., Collins, S. L. *et al.* (2012). Shifting species interactions in terrestrial dryland ecosystems under altered water availability and climate change. *Biological Reviews* **87**, 563–582.

Morris, W. F. (1990). Problems in detecting chaotic behavior in natural populations by fitting simple discrete models. *Ecology* **71**, 1849–1862.

Phillipi, T. and Seger, J. (1989). Hedging one's evolutionary bets, revisited. *Trends in Ecology and Evolution* **4**, 41–44.

Sale, P. F. (1977). Maintenance of high diversity in coral reef fish communities. *American Naturalist* **111**, 337–359.

Sale, P. F. (1978). Coexistence of coral reef fishes: a lottery for living space. *Environmental Biology of Fishes* **3**, 85–102.

Schwinning, S. and Kelly, C. K. (2013). Plant competition in water-limited environments and implications for ecosystem function and adaptability to climate change. *Functional Ecology* **27**, 886–897.

Seger, J. and Brockmann, H.J. (1987). What is bet-hedging? *Oxford Surveys in Evolutionary Biology* **4**, 182–211.

Venable, D. L. (2007). Bet hedging in a guild of desert annuals. *Ecology* **88**, 1086–1090.

Williams, J. N. and Kelly, C. K. (2013). Deconstructing the signal: phylogenetic structure, elevation change and the implications for species coexistence. *Evolutionary Ecology Research* Published online 14 August 2013.

Wimsatt, W. C. (1980). Randomness and perceived-randomness in evolutionary biology. *Synthese* **43**, 287–329.

Index